The Electronic Laboratory

The Electronic Laboratory

Tutorials and Case Histories in Laboratory Automation

Raymond E. Dessy

AMERICAN CHEMICAL SOCIETY
WASHINGTON, D.C. • 1985

Library of Congress Cataloging-in-Publication Data

Dessy, Raymond E., 1931–
The electronic laboratory.

Bibliography: p.
Includes index.

1. Chemical laboratories—Automation. 2. Chemical laboratories—Data processing.

I. Title.

QD51.D47 1985 542'.8 85-15111
ISBN 0-8412-0928-6
ISBN 0-8412-0929-4 (pbk.)

PRINTED IN THE UNITED STATES OF AMERICA

About the Author

Raymond E. Dessy is a professor of chemistry at the Virginia Polytechnic Institute and State University (VPI&SU). His research group is currently interested in computer automation of instruments, the design of new types of detectors, new methods for data acquisition and handling, and more effective ways to teach chemists about computers and laboratory automation.

He began teaching at the University of Cincinnati in 1957 as an organic chemist studying the acidities of weak acids. This research led him into the study of organometallic reactions. In 1966 he came to VPI&SU and continued research in organometallic electrochemistry and fast reaction mechanisms. The computer was a logical step in the search for new and faster ways to detect, acquire, and process data.

He received his B.S. in pharmacy in 1953 and his Ph.D. in chemistry in 1956 from the University of Pittsburgh. He was a postdoctoral fellow with Mel Newman at Ohio State University and a Sloan Fellow from 1962–64. He is a past chairman of the ACS Computer Division, and has been an ACS tour speaker since 1969. Since 1971 he has taught more than 3000 people in the United States, Sweden, Australia, the People's Republic of China, and Pakistan about computers and laboratory automation, primarily through ACS-sponsored short courses. He presented the third Breakthrough Lecture at the 1981 ACS New York Meeting on chemistry and the microcomputer revolution. He is editor of the "A/C Interface" column which has appeared in *Analytical Chemistry* since 1982. He is chairman elect of the Laboratory Automation Subsection of the ACS Computer Division.

Contents

Preface

The material presented in this book is equally divided between tutorial presentations and case history reports. The tutorials are useful to readers desiring an introduction to the concepts and vocabulary of modern laboratory computerization and automation and its many facets. The case histories are important to beginners as well as to mature practitioners of the art.

Current popular issues such as local area networks (LAN) and laboratory information management systems (LIMS) are examined. The laboratory scientists are also urged to explore their needs in more basic issues such as operating systems (OS) and languages appropriate to the laboratory.

A computer can extend a scientist's capabilities in many dimensions. A series of tutorials addresses robots in the laboratory that can extend the human frame; a laboratory *son et lumière* that explores voice input and output, graphics, and image analysis; and expert systems that emulate a portion of our mental skills.

Developments in hardware and software that have an impact on the workplace are examined in tutorials dealing with laboratory workstations and magnetic and optical disks. The key issue of standards in OS, buses, disk formats, and data structures is raised. The vexing problems of security, protection, archiving, and validation are analyzed, and potential solutions are suggested.

Finally, a tutorial on managing the electronic laboratory focuses on the problems that impede success in half of the attempts to establish a coherent, synergistic, electronic laboratory.

Each tutorial is accompanied by a series of four or five capsule reports written by industrial scientists who have identified and solved problems in that specific area of laboratory automation. They give the reader a variety of solutions to a related problem area and allow the synthesis of a solution appropriate to each reader's environment. These capsule reports provide a unique insight into practices that have proved valuable in some of the most successful electronic laboratories. Much of this material is unavailable in any other form.

The series was developed with the encouragement of Professor George Morrison, editor of *Analytical Chemistry*. Recognizing the role computerization and automation were playing in changing the practice of analytical chemistry, he championed the establishment of a series called "A/C Interface" in the journal. Rather than adopt the traditional route of presenting long articles dealing with singular solutions appropriate to a specific environment, *Analytical Chemistry*'s editorial advisory board approved the tutorial–case history approach that led to the articles in this volume.

ment for his help in the final editing of the manuscripts and the conversion of the rough drawings to the forms that appear in this volume. His critical analysis of topics and approaches was essential to the creation of the continuum that exists between articles.

Many of the philosophies presented resulted from discussions with members of the laboratory automation and instrument design group in the chemistry department of the Virginia Polytechnic Institute and State University.

Finally, gratitude is extended to the *Analytical Chemistry* art and production staff that works, unseen and unsung, to make such publications possible.

Raymond E. Dessy
Virginia Polytechnic Institute and State University
May 13, 1985

Originally published in
ANALYTICAL CHEMISTRY **1982** *54*(11) 1167A
and **1982** *54*(12) 1295A.

Local Area Networks

Part 1

It all starts with a wire—a wire connecting a terminal to a small microcomputer-based instrument that allows the two to communicate. Another wire connects a second terminal to a modem and then through telephone lines to a time-shared public data base. Yet another wire connects a third terminal to the corporate mainframe computer.

It is immediately obvious why local area networking is essential. Terminal costs, the expense of purchasing and installing long cable runs, and the human price paid to coordinate access to the various systems force the chemist to consider a solution in which a single terminal can access all computing facilities. This is possible when all systems are connected in a fully intercommunicating local area network (Figure 1).

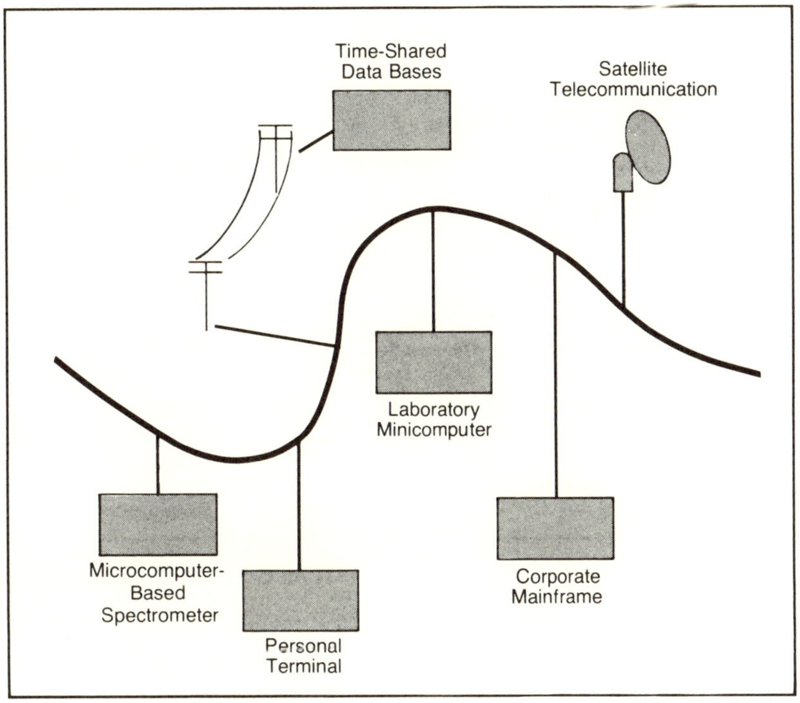

Figure 1. Local area network

The following tutorial material provides the technological base and vocabulary of such networks. It is presented to help the interested chemist better understand the concept of networks in the laboratory. In the next article (October issue) in this new series, fellow chemists from the laboratories of Union Carbide, Upjohn, Mary Kay Cosmetics, and Burroughs Wellcome will describe their laboratory computing systems in capsule form.

To understand the pressures exerted toward local area networks, examine the contacts an analytical chemist might have with computer facilities as he or she begins to probe the cause of a product failure. The failure has been generated by an organic surface-active agent plus small inorganic particulate matter somehow being incorporated into the product.

The first step would be a bibliographic literature search for relevant data via a terminal/modem accessing abstract services like DIALOG or ORBIT. Electronic retrieval becomes necessary as paper retrieval methods become unjustifiably costly. With some literature information in hand, a spectroscopic examination of the organic material putatively responsible is conducted by a computer-based mass spectrometer or NMR. This leads to the need to search a local library of compounds stored on disk in a search/match scan to identify the organic compound involved.

Telecommunications could connect the spectroscopist to on-line second-generation data bases, such as the Chemical Information Systems (CIS), having more extensive libraries of spectroscopic information. If the compound has never been examined before, or is a new composition of matter, one may request the service of third-generation data bases such as the STIRS MS program from the Cornell Computing Center. Using probability-based matching, this program suggests substructures in a true unknown for use by the chemist in proposing a possible structure.

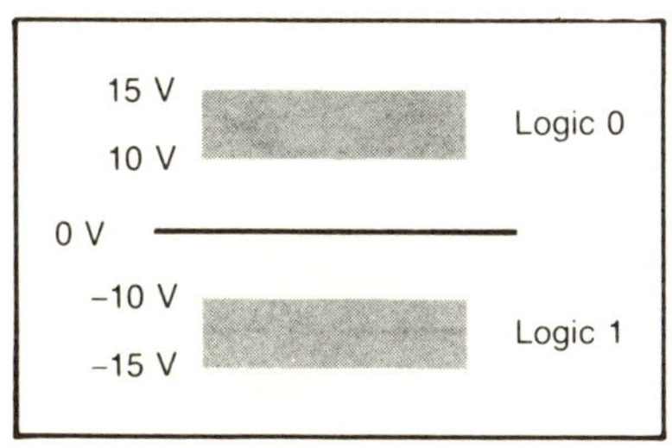

Figure 2. RS-232C electrical protocol

The inorganic components may re-

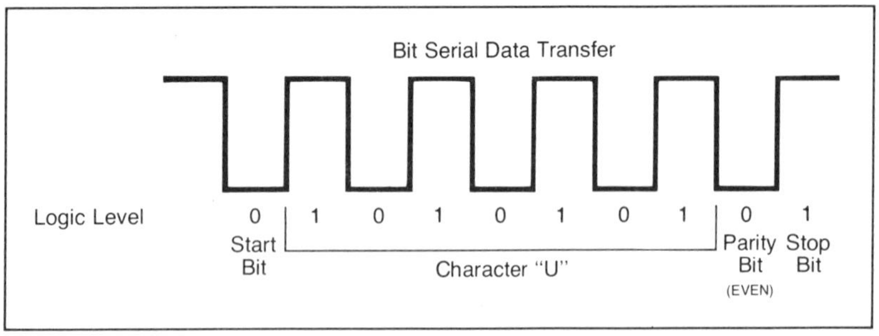

Figure 3. ASCII (American Standard Code for Information Interchange) communications transmission protocol

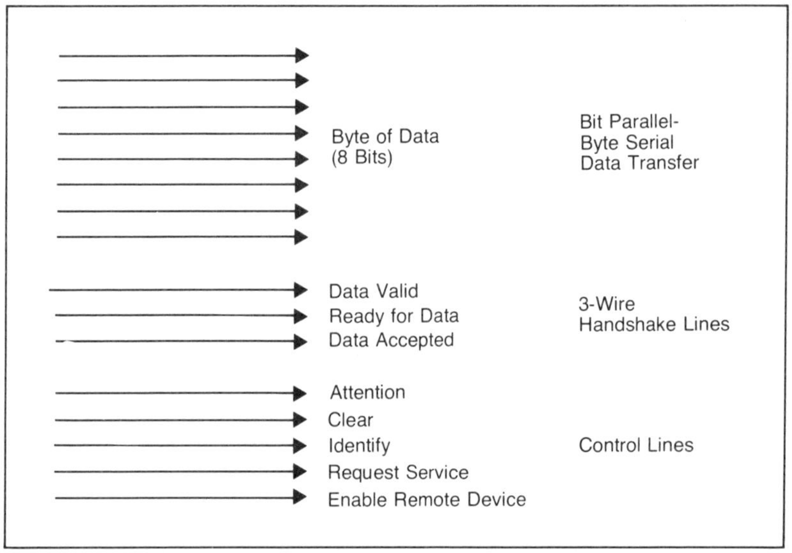

Figure 4. IEEE-488 communication and flow control protocol

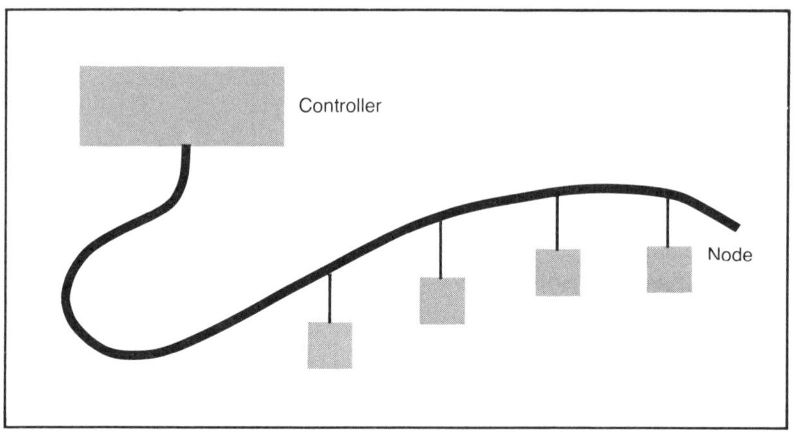

Figure 5. Multidrop bus (polling)

quire the use of intelligent (computer-controlled) SEM or X-ray fluorescence spectrometers, with the application of both data reduction programs and library searches. Imaging systems containing interactive computer-generated graphics produce particle size distribution information.

When the identity of the contaminants and their source are known, a self-optimizing reactor can examine the response surface for the manufacturing/separations sequence. Experimental design methods can be performed automatically to report the best conditions under pilot or lab bench conditions. Once full production is implemented by extending these findings to the plant, the analytical chemist is faced with an increasing number of quality-control samples.

He or she must have access to statistical packages that help abstract more meaningful data from analytical results. Sample log-in, sample flow, sample tracking, work lists, backlog reports, interim summary sheets, and final reports are all capable of being computer generated if programs and access to utilities are user friendly. Finally for long-term purposes, massive archival storage and sophisticated data management programs are necessary.

Multiple levels of computer resources are required. The span is from the small microprocessors embedded in the analytical equipment to large corporate facilities. A communication channel must connect them all.

Local Area Network

A local area network (LAN) may be inexactly defined as a common communication conduit that allows various terminals, computers, disks, printers, copy machines, intelligent instruments, and many other electronic devices to intercommunicate over short distances. Transmission length may range from 500 m to tens of kilometers at data rates between 10 K bits/s and 10 M bits/s. The wide range of values is due to the varying technologies used in implementing the LAN.

The idea was conceived for the office and business environment. It was driven by the need to interconnect word processors, copy machines, and personal computers with larger corporate computer facilities, and fired by the marketing potential involved. Now the concept has spread to laboratory environments as they attempt to respond to the technological revolution created by microprocessors and inexpensive computers. The technological part merely refers to the VLSI (very large scale integrated) circuit technology that has dramatically cut the cost of computing power. The revolution part is just that—a flight of the masses from the tyranny imposed by systems analysts and from the arcane languages of big systems and their lack of user cordiality. Less emotionally, large systems cannot provide the real-time capability required by the laboratory except under carefully controlled conditions.

Small computers have become excellent servants in the laboratory when dedicated to a single or, at most, a few tasks. They produce prodigious amounts of high-quality data—data that is beginning to challenge our ability to collect, store, analyze, manipulate, and retrieve it. The quantity of data plus fiscal and legal imperatives has stimulated the drive toward LANs that will allow resource and data sharing. Data conduits are essential from point of data capture, through intermediate nodes providing record management and report generation, to nodes where archiving and extensive data management can occur.

Many chemical companies are already installing LAN facilities. Others are in the planning phase. The concepts are very simple. The vocabulary is perhaps a bit new and foreign but understandable by simple analogy.

RS-232C Point-to-Point

Early attempts at intercommunication revolved around point-to-point nets that work like an intercom. Individual wires link each node to others with which it must communicate. Many computer networks use what is called an RS-232C protocol to establish similar computer intercommunication. This is a voltage convention in which characters and letters are sent down a twisted pair of wires as a series of ones (−10 to −15 V) and zeros (+10 to +15 V) (Figures 2, 3). This is called serial transmission since the bits are sent one at a time. Seven bits of information allow 128 possible combinations. This is enough to encode the letters, numerals, and grammatical symbols we use. The packet of seven bits is accompanied by a bit of information used to determine if an error has occurred at the receiving end. This eight-bit string (or byte) is encapsulated by a preceding start bit and a following stop bit (Figure 3). These are electrical and transmission protocols. The RS-232C protocol also includes the shape of the computer/peripheral mating plugs required. A typical speed of transmission is 9600 bits/s, or about 1 ms transmission time per alphanumeric character. The Laocoön nest of wires produced in complex systems has prompted the use of other protocols.

IEEE-488 Multidrop

One of these protocols transmits the eight bits representing alphanumeric information or chemical data in a parallel manner, i.e., all at once rather than in the serial progression of one bit at a time. This transmission is called bit parallel-byte serial (Figure 4). The method requires some control elements, just as the bit serial method requires start and stop bits. The IEEE-488 Committee standard protocol uses three lines to implement the handshake between transmitter and receiver and five lines to coordinate data transfer (Figure 4). Thus the conduit for information transfer, called a bus, is a 16-wire cable. In short-haul configuration it can transmit 250 K bytes/s (K = 1024). This approach permits devices, each with a unique address, to share the bus. Each of them hangs off of the common 16-wire conduit to a limit of 15 devices on a bus 20 m long. (Active repeaters are needed for extensions.) This is referred to as a multidrop system (Figure 5). An analogy is the old country-lane telephone line where each subscriber was identified by a different ring pattern, address, or number. Each device is polled when control or information is requested. The polling is done by a centralized controller.

Error Detection

As the amount of data increases, error detection/recovery is necessary. The single bit used in serial communication to detect errors is called a parity bit. Parity convention for a device

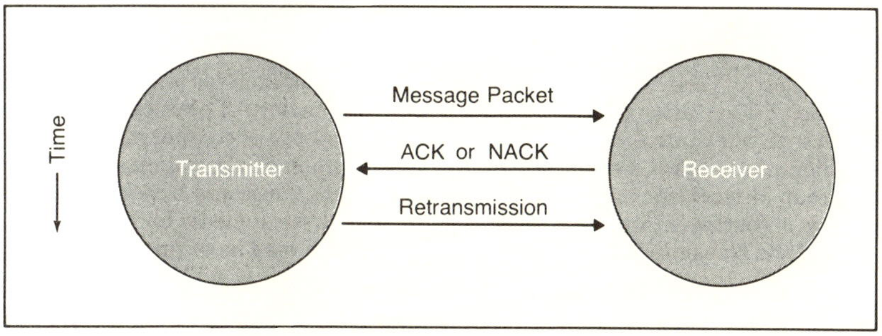

Figure 6. Error detection/positive acknowledgment/retransmission

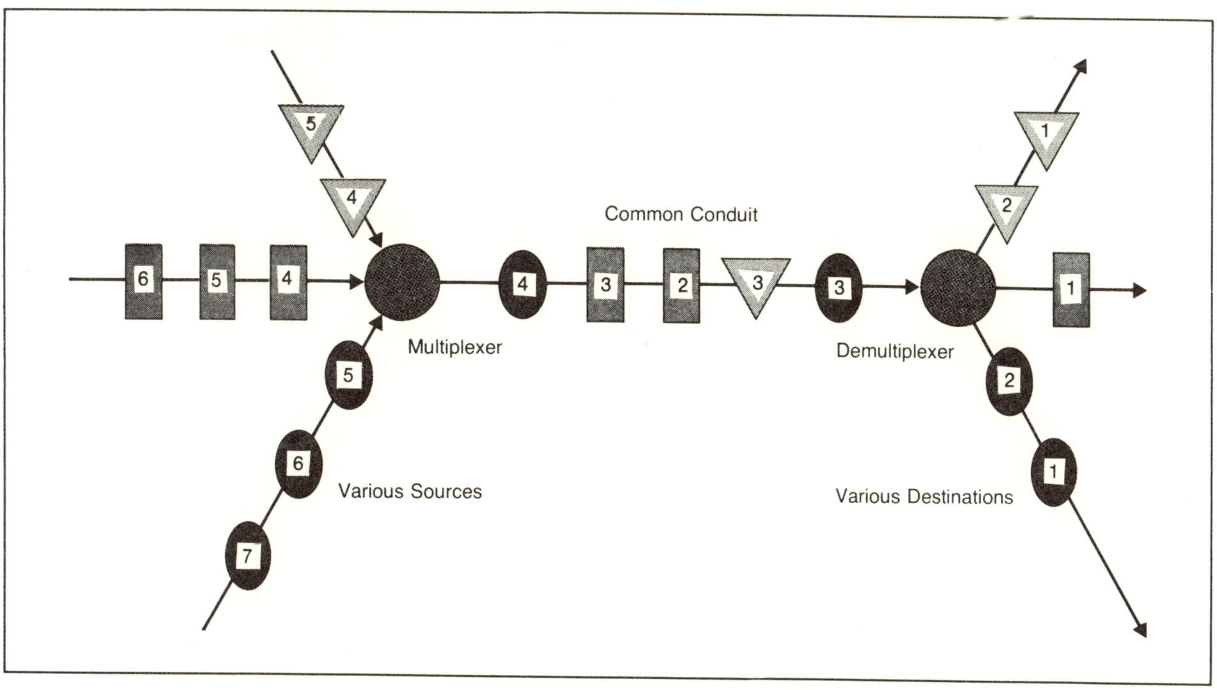

Figure 7. Packet switched communication

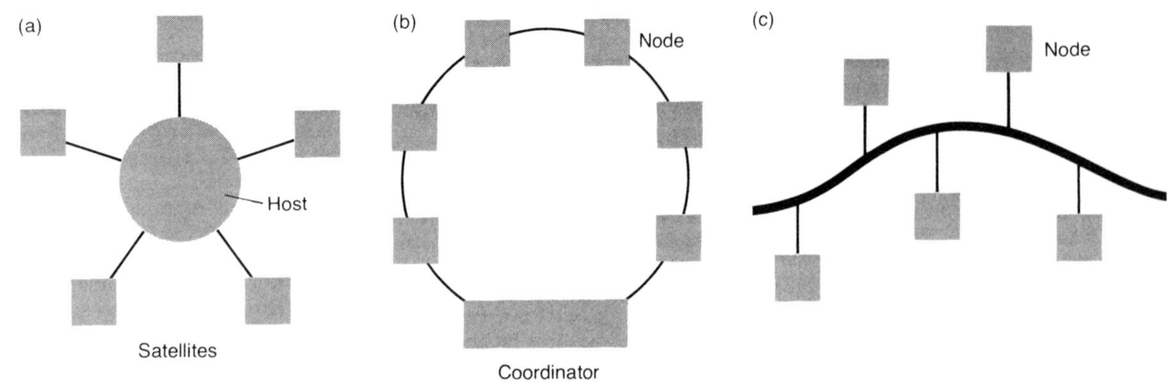

Figure 8. Networking topologies, arbitration schemes, and multiplexing approaches may be intermixed in any combination. The most common associations are shown here. (a) Star network (centralized control, space division multiplexing); (b) ring network (token passing/reservations, time division multiplexing); (c) bus network [contention, time division multiplexing (baseband), frequency division multiplexing (broadband)]

may be ODD or EVEN. If it is ODD, the transmitting device will set the parity bit to a one or zero so that the sum of ones in a given eight-bit byte (seven data bits plus parity) is ODD. If the parity convention for a device is EVEN, the transmitting device will set the parity bit to a one or zero so that the sum of ones in a given eight-bit byte is EVEN. (Figure 3 uses EVEN parity.)

Using EVEN parity, if the receiving end sees an ODD number of bits set to a one, then an error has occurred during transmission. One of the bits has been altered by noise. This detection of an error causes a "not acknowledge" (NAK) to be sent from the receiver to the transmitter that repeats the transmission (Figure 6). An even number of multiple errors cannot be detected by this method.

Another method for error detection involves summing vertically all of the numbers transmitted. The transmitter sends a copy of the least significant 8 or 16 bits of the sum to the receiver, which also calculates a "check" sum. If the difference between these two numbers is zero, the transmission is correct and an "acknowledgment" (ACK) signal is sent back to the transmitter. Otherwise, a NAK is sent. Again multiple even errors cannot be detected.

Better error detection mechanisms involve expressing a string of n bits as a polynomial with its coefficients the ones and zeros of the string. This creates a polynomial M(x) of degree n-1. A second smaller polynomial G(x) (degree y) is defined, again with one and zero coefficients. M(x) is multiplied by x^y and then divided by G(x). The polynomial remainder R(x) has a series of one and zero coefficients that defines a cyclic redundancy check (CRC). This bit pattern can be used like the check sum described above. However, it has a unity chance of detecting multiple errors.

Message Packets

With multiple nodes in a system the inclusion of destination address information with our control or data message allows us to select a single receiver, a group of receivers, or all receivers. This is routing information. The source of the transmission should also be included in the message. We have built a message "packet." It includes identity, routing, and error checking information, as well as a number of characters representing data or instructions. There are three categories of message packets. Supervisory message packets tell receivers what to do. Manager packets tell receivers how to perform their function. Data or program packets also need to be transported about the system.

To avoid monopolization of the network, the length of a single packet may be set so that all are equal, or have an upper limit. Since each complex transmission may involve large amounts of data it may be broken into a number of packets. These packets can then be numbered to ensure the sequence is transmitted and received in the proper order.

If multiple users could now share a communication channel each transmitter could switch its current packet onto the single channel in an orderly way. We now have packet switched communication (Figure 7). Intelligent hardware assembles (multiplexes) the packet train and disperses the packets (demultiplexes).

Over the past few years intense interest has arisen in developing protocols that would allow a large number of users to be connected together to achieve the multidrop, positive/negative acknowledgment, request-to-retransmit-on-error, packet protocols described above at moderate communication speeds.

In general, such networks can be built in a variety of physical shapes using a variety of conduit materials. The common configuration of nets involve stars, rings, and buses (Figure 8). The physical media for intercommunication may be ordinary coaxial cable, cable TV (CATV) coaxial cable, twisted pair wire, multiconductor cable, or microwave and optical fiber technology.

Star Net

The centralized control of the star and its point-to-point cabling requirements are disadvantages except in the most limited extensions using RS-232 protocols and transmission speeds of 9600 to 19.2 K bits/s. However, the star net is the least expensive to install over moderate distances. The star can achieve very high throughput if multiconductor cable, parallel transmission, high-speed line drivers, and special interfaces are used. For very short distances the transmission speed can be >1 M byte/s.

The goal of rings and buses is to allow as many users as possible to send data packets as rapidly as possible along a *single* shared conduit. In local area networks, users are typically interconnected with coaxial cable. Sharing of this conduit by the many possible transmitters requires some form of multiplexing.

Multiplexing

A star is a space division multiplexed arrangement. Different messages travel along different wires.

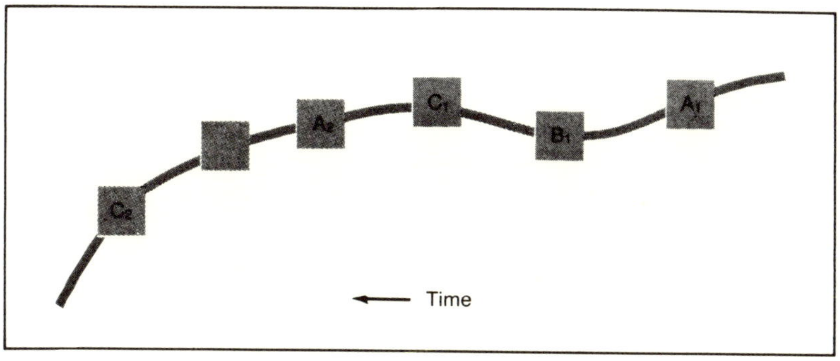

Figure 9. Time division multiplexing

Using only one wire, time and frequency division multiplexing can be implemented easily. Time division multiplexing (TDM) allocates slots in time to individual transmissions (Figure 9). As an analogy, a rotary switch moving at 1 rps could connect 10 different input signals or channels to a common output at a 1-Hz rate per channel. This technology becomes inefficient when many of the transmission sources are silent most of the time and do not use their assigned slot (cf. empty packet area in Figure 9).

Ring Net

An optimizing solution involves a constant circulation of message packets around a ring network (Figure 10). If a full packet containing data for another destination passes a node, that packet is not used but passed on. If an empty packet passes, the node can write its information into the passing stream. Routing information in the packet ensures that it is delivered to a specific node in the loop as the packets cycle around past all nodes. After one pass around the loop the original transmitter can remove its message packet from the stream since it should have been received by all concerned parties.

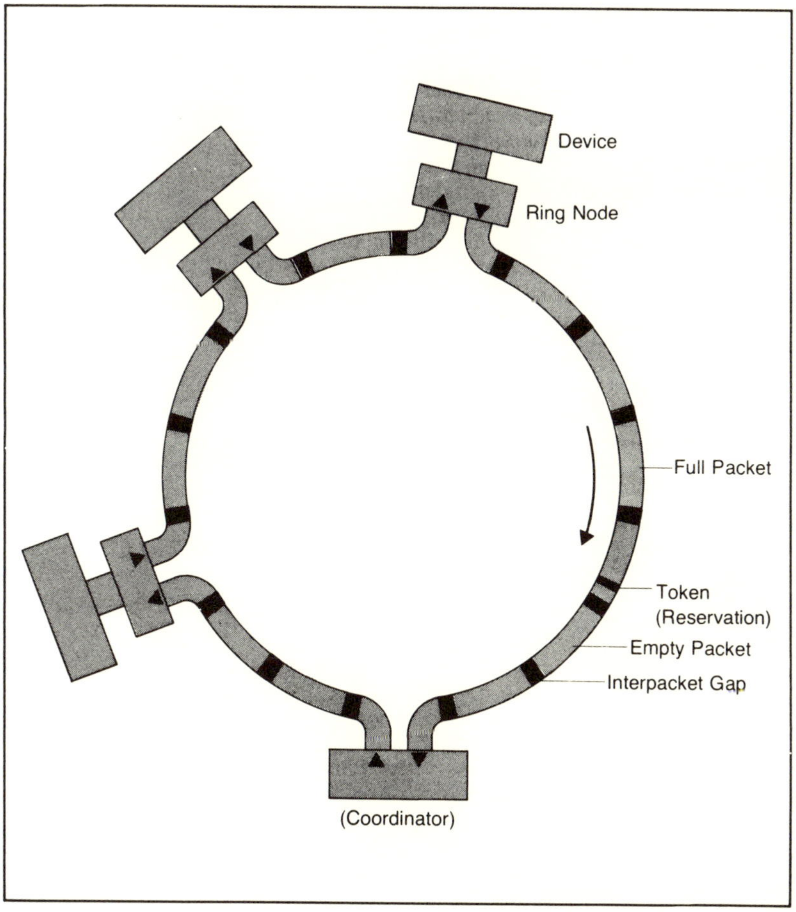

Figure 10. Ring network, packet flow

Tokens

Tokens can be placed in message packets to ensure that certain nodes can have preferential access to the net. A token is a message granting a polled node temporary but exclusive use of the medium. This node can then pass the right of exclusive use (a reservation) on to the next designated station, allowing the user to calculate what the worst-case response would be for transit of data. This is vital in real-time data acquisition and process control. Note that the central control has been replaced by a distributed polling. Integrated circuits are commercially available that implement such token access control.

Bus Net

The bus architecture (Figure 8) lends itself to another type of arbitration. Imagine that you approach a group of three people standing together at a social gathering. If someone is already talking it is best not to speak. If no one is talking you may. Detecting if someone is talking can be called "carrier sense." Nodes along a bus can perform the same operation. If all nodes along the bus contend for use of the bus there is "multiple access." If two people start talking simultaneously both should stop talking, wait a random time delay, and start talking again. The random wait prevents another verbal collision. People do it naturally. Computer networks also do it and call the process "collision detection." Carrier sense (CS), multiple access (MA), collision detection (CD) or CSMA/CD is the most commonly used arbitration method in a bus net. As with token passing, there is no centralized control.

Ethernet

What has been described is the basis of Ethernet, an outgrowth of work by the Xerox Corporation. Many think it will develop into an industry standard for contention-arbitrated bus nets. It is one of the protocols being evaluated by the IEEE-802 Committee. Commercial hardware and software support exists from a variety of vendors. Some are computer-specific boards that plug into CPU (central processing unit) back planes. Others are equipment independent, consisting of an RS-232 plug on one side and an Ethernet coaxial connector on the other.

Baseband Net

Data is transmitted on an Ethernet line by Manchester encoding (Figure

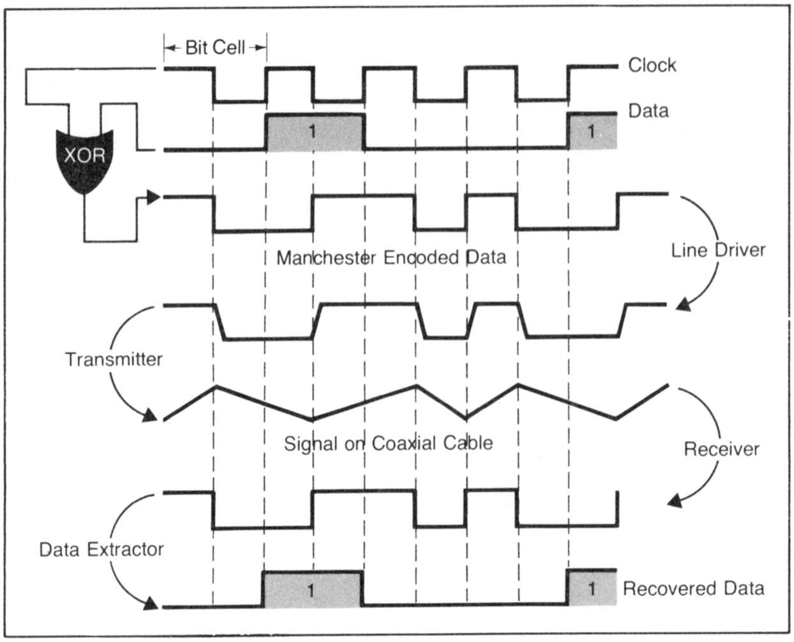

Figure 11. Baseband (Ethernet) transmission protocol

11). The clock line is exclusive ORed with the data line. This self-clocking signal is transmitted with almost no filtering. Data frequencies from 1 to 10 M bits/s are typical. An interpacket gap is followed by a 64-bit synchronizing burst, a 48-bit address field for source and destination, and a variable length data field. A 32-bit CRC error word along with the beginning of the interpacket gap terminates the transmission of the packet (Figure 12). Only one node at a time can communicate on the channel. However, the contention scheme easily supports several hundred nodes. The term *baseband* has been applied to this type of approach. Ethernet-like systems using twisted pair (RS-422) or a 16-conductor bit parallel-byte serial convention are also available. (The RS-422 protocol extends transmission quality beyond the RS-232 protocol.)

Broadband Net

Frequency division multiplexing (FDM) would be an obvious way to increase the number of channels on the coaxial line (Figure 13). Cable TV technology is well developed. The bandwidth is over 300 MHz. It is possible to assign different carrier frequencies to many individual channels and then frequency or amplitude modulate digital data onto each of these (Figure 14). The CSMA/CD techniques described above can be used on each channel. Sufficient guard regions between channels exist so all channels operate independently. Such technology, with 100 channels operating CSMA/CD at low transmission speeds, can accommodate 10 000 different nodes. Fewer channels with a larger bandwidth allow modulation at higher data rates. Two to 10 M bit/s transmission speeds on a channel are possible with commercial equipment. The term *broadband* has been applied to this approach. Several companies offer broadband system building blocks.

Considerable emotion and misinformation has arisen from a baseband/broadband controversy. Marketing strategy, self-interest, and the momentum of the technology have led to the idea that the two methods are in competition. Others have concluded that the two will coexist synergistically. The IEEE is working on establishing standards for a baseband bus and a broadband bus as well as a token

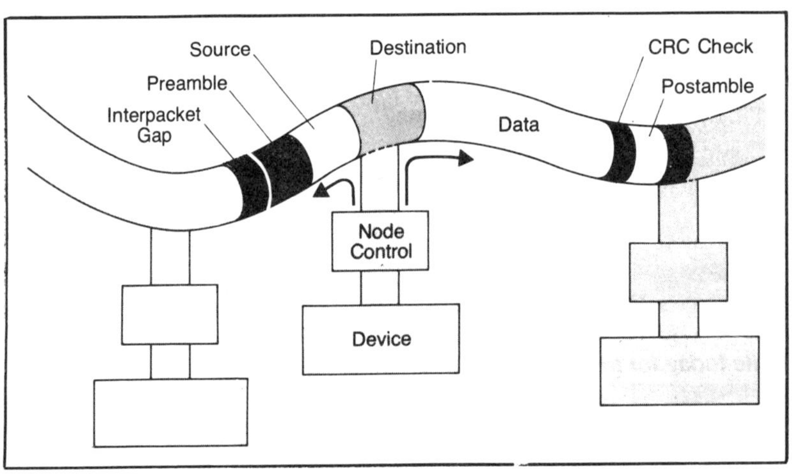

Figure 12. Baseband bus contention network packet construction

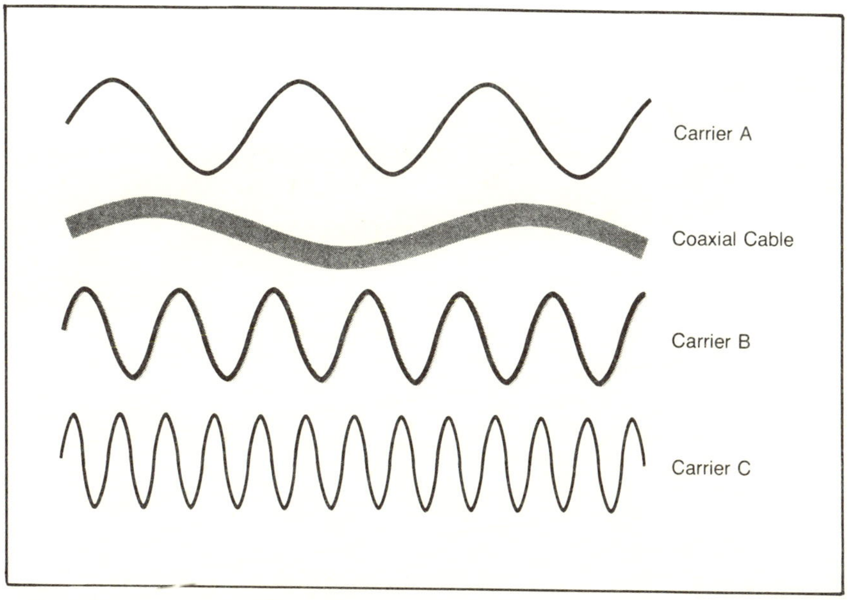

Figure 13. Frequency division multiplexing

ring. To see why both are useful we will examine their details.

• Packet delivery is a best-effort matter. There is no assurance that packets will not be lost, duplicated, or delivered out of sequence. The only assurance is that the packets accepted are correct.

• Baseband technology is bidirectional, whereas the CATV conduit (broadband) is unidirectional. In broadbands, remote nodes communicate toward a central facility called the head end. Transmission from remote end to head end requires one channel. Communication from head end to remote end requires another channel. Typically remote-end to head-end channels cluster in a low-frequency area. Head-end to remote-end channels are clustered at a higher frequency. Digital data are modulated onto the frequency specific to a channel (Figure 14). Frequency translation, or turnaround, must be performed at the head end (Figure 15) to establish a "virtual" connection.

When a limited number of large bandwidth channels are involved (e.g., five channels in each direction capable of 5 M bits/s transmission) proper mating of in-bound and out-bound channels is also done at the head end.

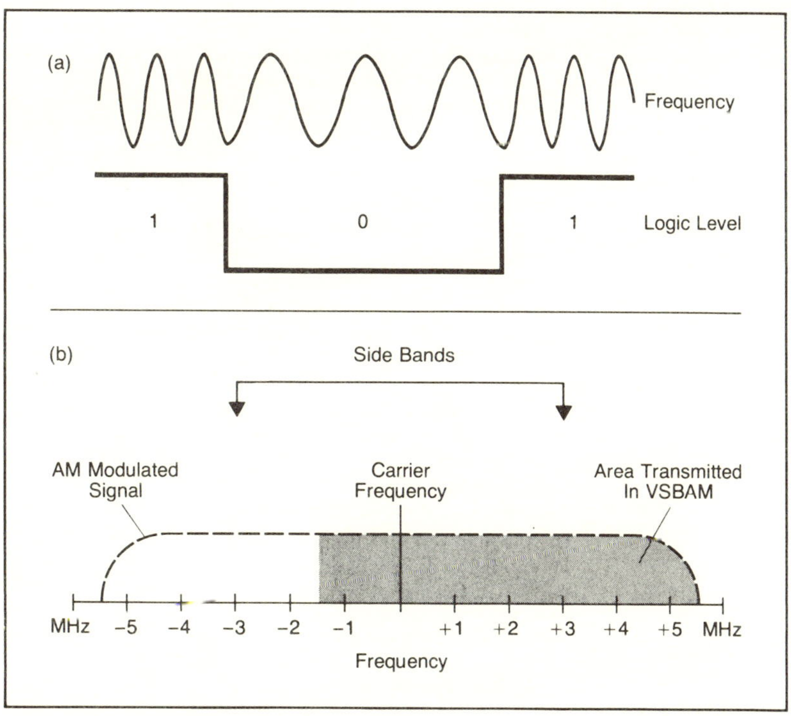

Figure 14. (a) Broadband frequency modulation transmission protocol (frequency shift keying); (b) broadband amplitude modulation transmission protocol (VSBAM—vestigial sideband amplitude modulation)

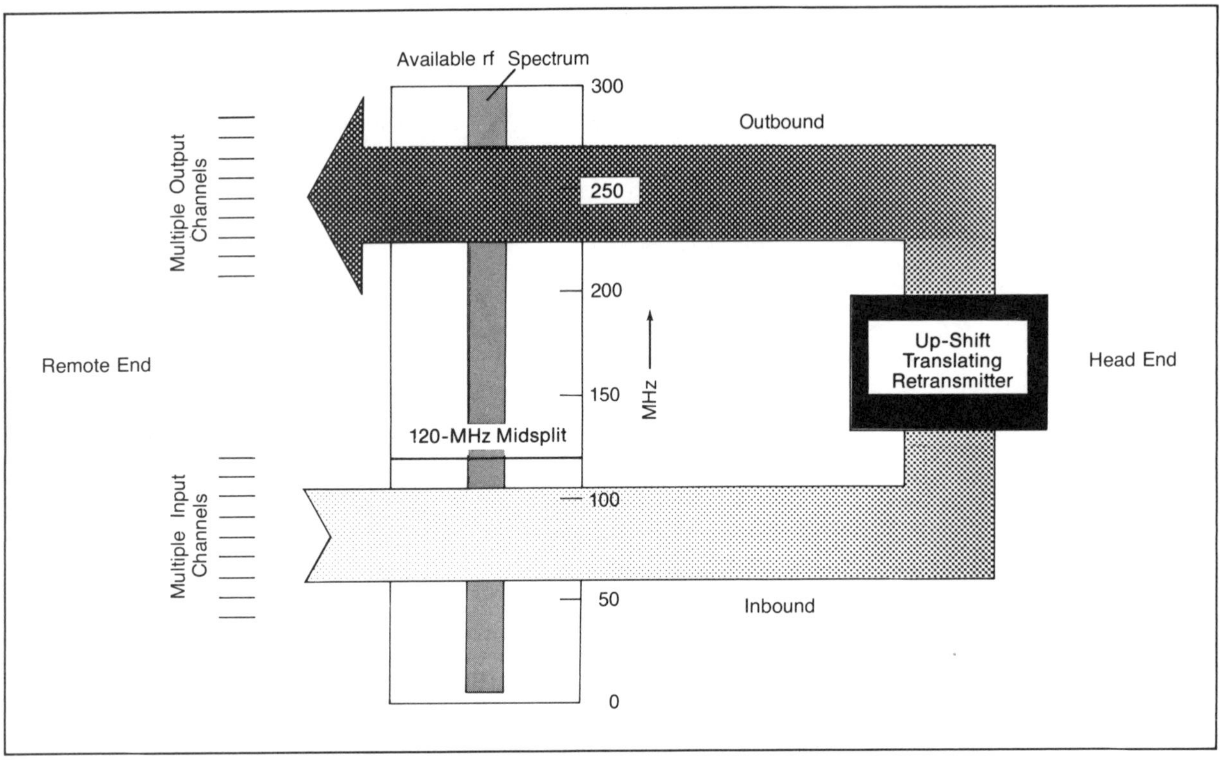

Figure 15. Typical broadband channel allotment

When large numbers of low bandwidth channels are used (typically hundreds at 128 K bits/s transmission), frequency agile modems (modems whose carrier frequency can be altered by digital commands from the head end) are used. The modems cannot be agile over hundreds of separate frequencies so vendors provide a limited group of frequencies in each modem. Virtual connection between modems belonging to the same group is a simple matter. Intergroup communication requires a second translation at the head end. These translation facilities are called bridges, and they are commercially available. (Some broadband installations use separate cables for each direction.)

- Broadband hardware is therefore more complicated than baseband hardware. Frequency modulated broadband equipment is less expensive to build than amplitude modulated (VSBAM) technology. The latter conserves considerable bandwidth and is the transmission mode used in the television industry. VLSI circuit chips implementing baseband CSMA/CD have been announced by several reputable sources. If these are delivered, the cost of baseband will drop precipitously. With VLSI technology it will eventually be market volume that determines price. Currently it costs about $1000–1500/node to establish either broadband or baseband connections.
- Installation of broadband technology involves CATV technology and connections. The latter are standardized. Most installations to date have required in-house CATV technicians to properly balance the system. Baseband connector technology has not yet matured, but the technology can be installed by almost anyone.
- Broadband communications will support longer local area networks (in the range of tens of kilometers). Similar extension of baseband leads to dispersion of the signal and loss of speed.
- Baseband requires all nodes to communicate on a common channel. To achieve reasonable per node average transmission speed the single channel must communicate at high speeds (10 M bits/s). This leads to short message times. The performance of CSMA/CD systems is proportional to the ratio of message times to propagation delay. The higher the ratio the better the performance or efficiency of the system. Long baseband cable runs increase propagation delay, and therefore the system is limited to distances of 1–3 km.
- In broadband applications two FDM channels do not carry double the capacity of a single channel. Translation or rebroadcasting exacts a price and an improvement of 1.3 is seen. Ten channels improve throughput by a factor of five.
- Both of the technologies are subject to system overhead costs. When the lab computer installed at a node is responsible for monitoring each message to see if it is targeted to its node, 1% (typical) to 20% (busiest minute) of the CPU time may vanish in such activities. Fully buffered intelligent node controllers prevent this.

The systems may be viewed as complementary. Geographically localized LANs could use baseband interconnections. Through gateways these could access broadband bus networks. Vendor technology to accomplish this interconnection is beginning to appear. Instrument vendors and computer vendors are beginning to realize that all of the equipment in a laboratory will not be made by a single source, and the lab chemists should expect instruments, terminals, networks, and computers to intercommunicate.

Hardware and software have been announced that will allow vendor-specific networks like DECNET to use Ethernet as a communication channel, and/or connect to other vendors' network protocols, like IBM's system network protocol called SNA. Beyond the local area network, gateways also allow access to long-haul communication channels such as X.25. This is a public packet switched network protocol extensively developed and used in Europe for low-speed communication of data over commercial telephone lines. Support of this protocol in the U.S. by the mini- and maxicomputer

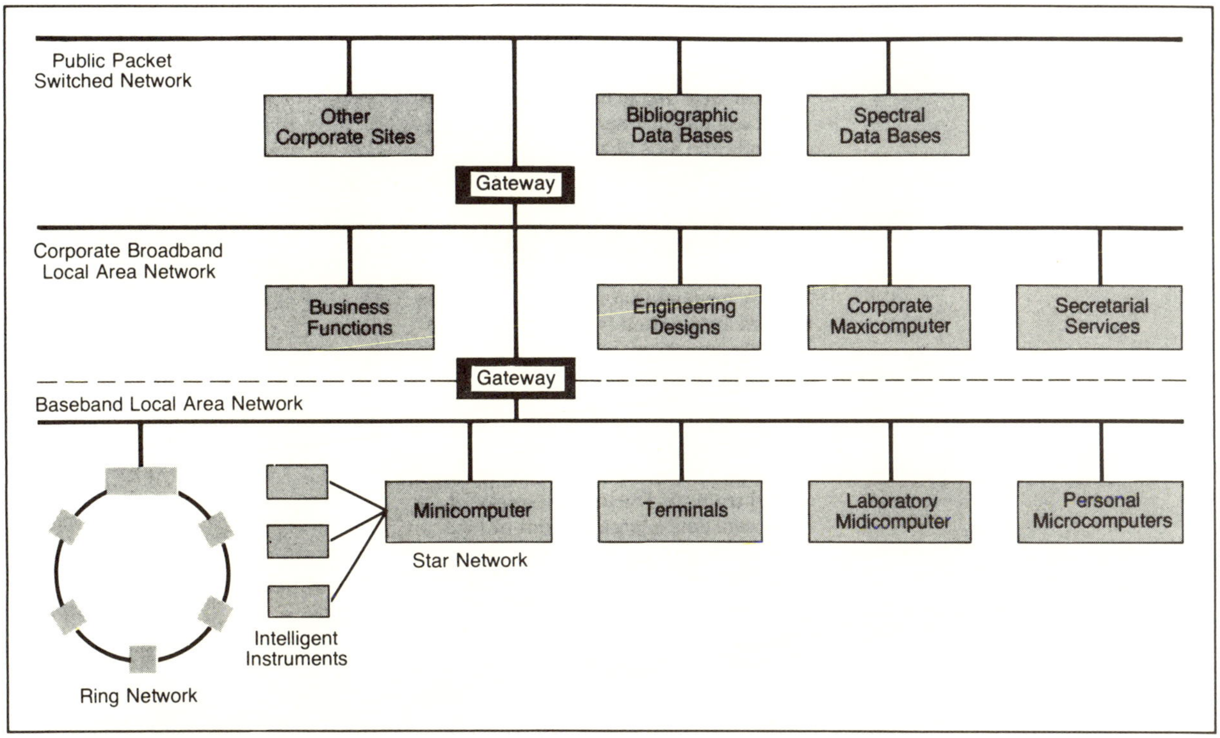

Figure 16. Hierarchical network interconnection

vendors provides the last link required for intercontinental and transcontinental communication.

Unfortunately, there are some caveats to be aware of in establishing local area networks today. No universally accepted standard exists for network protocol. A DEC-Intel-Xerox consortium is pressing for the acceptance of its baseband protocol with inexpensive licensing available from Xerox to prospective users or vendors. Other baseband protocols are now commercial. Broadband vendors are attempting to establish de facto standards by aggressive marketing strategies. The IEEE Committee is still in the working phase. The International Standards Organization (ISO) has attempted to lay down a layered set of protocols for an open system interconnection (OSI) that vendors may use in discussing their product line. It has no de jure status. The boundaries between these layers are artificial, but the nomenclature is useful in understanding various vendors' promotional material and equipment specifications.

Seven OSI layers exist between a node and the net:

Layer 1—*Physical layer* defines the mechanical and electrical interface to the net (baseband vs. broadband, coaxial vs. multiconductor cable).

Layer 2—*Data link layer* defines individual packet format and access control (address and error-checking protocols, encapsulation of the packet).

Layer 3—*Network layer* defines how to get a message from one node to another (routing control, sequencing, and flow control).

Layer 4—*Transport layer* implements a higher level of connection between network entities.

Layer 5—*Session layer* defines how to make and break a communication path, how to segment long transmissions, and how to buffer them properly.

Layer 6—*Presentation layer* implements protocol conversion, data unpacking, translation, and encryption.

Layer 7—*Application layer* is what users see as they implement transfer transactions. (Ideally, this is all they should ever see.)

Most of the available equipment addresses layers 1 and 2. Some vendor hardware supports the OSI layers to about level 4, using node controllers with built-in microcomputer and ROM-based software to perform the indicated functions. The National Bureau of Standards is attempting to establish standards above the transport layer. Thus, attempts are being made to establish de facto and de jure standards starting at both ends, reaching toward the middle. Networking still requires in-house capability in software, hardware, and/or communications technology depending on the choice of vendor. It is important that vendor strengths complement those in your laboratory.

For analytical chemists the bottom line in local area networking schemes is access, convenient simple access to all of the computing facilities attached to the network (Figure 16). With one desk terminal the analytical chemist can achieve virtual connection to the microcomputers in the laboratory using the user friendliness that prompted their acquisition. Data from them can flow upstream to larger minicomputers by operations that are completely transparent. The chemist can access the minicomputer facilities without unplugging/plugging operations by simple connect requests made at his or her terminal. Mainframe facilities are also accessible. Time-shared facilities can be reached by gateways to long-haul facilities.

It is most appropriate to use local area networking to buffer the analytical laboratory from the political and technical interest of corporate computing centers. In industrial and governmental laboratories throughout the country one sees a common scenario. Power struggles develop and class lines are drawn in the contention over who is to automate what in the labora-.ory and how it will be accomplished. One is reminded of the question of states rights vs. federal control.

Perhaps an analogy less fraught

with emotion represents the central corporate or laboratory computing center as a "made horse," well trained and capable of responding to the commands and aids of an experienced rider or trainer. Its repertoire is not easily changed nor can it tolerate the casual rider. A horse bred and trained for the track will not perform well in a three-day event or on a trail ride. Yet many propose that the proper solution to lab automation is a single large corporate computer. In the face of such attitudes the laboratory worker often goes into the foal yard and purchases a young animal. It is manageable, or at least small enough to give the sense of mastery to its new owner. But it does need training. During that period it will consume a large amount of feed, just as personal computers, although touchable and apparently controllable, consume software and time voraciously. Both the foal and microcomputer may eventually prove unsuited to the task as it evolves in complexity. Microcomputer owners may find generation of software for graphics and data base management involves skills they do not possess and capabilities their hardware does not have. At this point a good chemist may easily become a second-class systems programmer. This is not to decry the use of small personal or dedicated computers. It merely urges that they be capable of integration into the network where previously developed resources and data can be shared.

Examine Figure 16. Below the imaginary line indicated, we, as chemists, own, control, and manage the computer systems so we can perform our experiments in the way the chemistry dictates. Above that imaginary line we have access to large computing facilities that we do not wish to manage or control, merely use. A properly designed local area network leaves the chemist in charge of his or her environment.

Acknowledgment

The philosophies developed in this article are the result of a series of discussions that have involved the Laboratory Automation and Instrument Design Group of the Chemistry Department at Virginia Polytechnic Institute and State University and, in particular, Ian Chapple, Jim Currie, Steve Dueball, and Mark Thompson. The alumni of the group who have also contributed to these discussions are David Binkley, David Hooley, Wayne Nunn, Michael Starling, and Henry Wohltjen.

Part 2

The capsule reports that follow represent approaches to the problem of providing local area network (LAN) capabilities in the analytical laboratory. The contributors were invited to give the reader an exposure to alternate implementations and different site characteristics. The capsule reports build from the philosophies and vocabulary introduced in the preceding tutorial article (September issue, p. 1167 A). The purpose of these companion articles is to offer interested analytical chemists sufficient background material and to provide a spectrum of examples so they can synthesize solutions appropriate to their environments. Each laboratory has its own unique skills, needs, history, and internal structuring and requires its own individual solution. It is hoped this bipartite tutorial/capsule approach will help us better understand and control the power that computers can bring to our laboratories. The next series will address the problem of laboratory information management systems: how we store, manipulate, and retrieve the data that flow along the conduit of a local area network.

The following capsules dealing with local area networks present five quite different approaches to the same problem: providing access to computing power easily and efficiently. In sequence, the reader will be exposed to solutions involving:

- a mixed baseband bus network coupled to a star net;
- a broadband bus network;
- a ring network;
- a star network coupled to a broadband net; and
- a single multiprocessor center.

Mary Kay Cosmetics, Inc.
Manufacturing Group
1330 Regal Row
Dallas, Tex. 75247

Contributor: Ray Miller

The Quality Assurance Laboratory at Mary Kay Cosmetics provides raw material testing, bulk and finished product testing, and microbiological testing for cosmetic products and components. The two-floor laboratory is divided into analytical and microbiological sections and houses 12 scientists. Most of the analytical instrumentation is located in the lower floor lab. The upper floor contains some analytical instrumentation and all microbiological equipment. The maximum spatial separation for data transmission at this time is 100 ft.

The majority of laboratory instrumentation operates either under dedicated computer control (e.g., FTIR, pulse NMR, GC data systems) or has digital communication capabilities (balances, pH meters). Few instruments require analog data acquisition.

Automation of the laboratory is being handled in four stages:

- elimination of paper information flow through the use of a data base management system (DBMS);
- acquisition, manipulation, and transmission of data to the DBMS through a lab data system;
- automatic control of instruments and autosamplers; and
- automation of sample preparation.

A laboratory data system needs to perform a wide variety of tasks. Communications must be established with the dedicated laboratory computer systems, so that processed data can be passed to the Manufacturing Group DBMS. This information is compact and sporadic. It can be queued in the dedicated unit until transfer is permitted.

Communicating instruments that do not have raw data manipulation capabilities need more interaction with a laboratory computer system. Control information can be passed back and forth as data are received and processed for transfer to our DBMS.

Noncommunicating instruments require that a computer monitor status lines and provide control signals inside the instrument. Acquisition of digital or analog data may also be required of this computer.

The system must be redundant so that in the event of a malfunction there are backup capabilities to minimize downtime and lost data.

The DBMS is the recipient of all laboratory results from the lab data system. A 32-bit mini that can support up to 100 users over RS-232 lines is used for management information through an integrated manufacturing data base. The system has large disk storage and backup capabilities. Five user terminals and a printer are available in the lab for direct data base inspection and manipulation.

One RS-232 line is used to communicate with the lab system (Figure 1). The lab system consists of three interconnected units: a star network topology, an IEEE-488 multidrop, and a baseband network. The star network is based on a large 16-bit microcomputer with multitasking capabilities. The

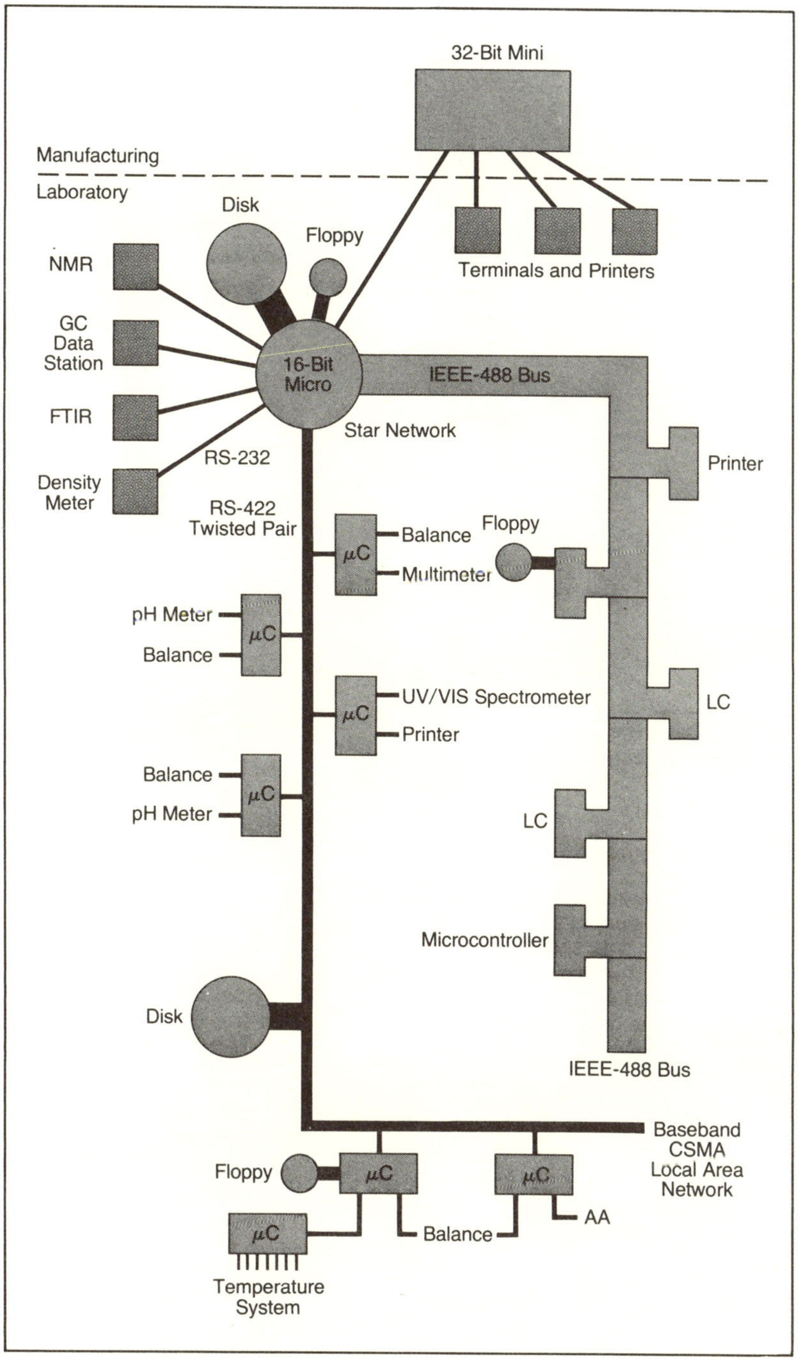

Figure 1. A mixed baseband/bus network coupled to a star net; Mary Kay Cosmetics Manufacturing Group. μC = microcomputer

system has a 10-megabyte Winchester disk, dual 1-megabyte floppies, a color printer, and a 700 × 500 pixel graphic display. Five RS-232 lines are available for controlling communications with dedicated computers on the FTIR, NMR, and GC data systems.

This microcomputer has an IEEE-488 interface controlling a multidrop bus to which liquid chromatographic equipment is attached.

The baseband carrier sense multiple access (CSMA) LAN allows for data acquisition and control of instrumentation over customized microcomputer interfaces. Each microcomputer is an 8-bit system with keyboard and 192 × 256 pixel graphics. The micro can be a user-interactive instrument controller or a terminal to the network. This baseband net implementation allows communication over a twisted pair of wires at 1 M bit/s through distances of greater than a kilometer. A 20-megabyte disk drive with backup capability will be accessible from all of the microcomputers in a transparent fashion. A total of 64 microcomputers can be supported by the network.

If the LAN connection fails, the experiment is still under control of its microcomputer; data storage can occur locally. If communications are down between the LAN and star network, data can be stored on the LAN disk. If the LAN disk is down, data can be sent to the star network for storage. The star network is not absolutely crucial since each dedicated system can hold data until communications are reestablished.

Staff members directly involved with implementation and maintenance of the system are: one full-time chemist with instrumentation background and one full-time analytical chemist with formal instrumentation training and a microcomputer hardware and software background.

The development of this system is based on two premises:

- More distributed processing results in faster response to user demands and less critical downtime for any unit.
- Software cannot circumvent hardware design limitations—highly flexible hardware allows software to be optimized for the application.

Union Carbide Technical Center
South Charleston, W. Va. 25303

Contributors: N. Bruce Angelo, Richard T. Sloan, and Wayne G. Nunn

At Union Carbide's Technical Center some 900 professionals, technicians, and support staff form the Research and Development Department for two of Union Carbide's major business divisions. Fundamental and business-directed research as well as process development are distributed among four major laboratory/office buildings and numerous satellite buildings throughout the 670-acre site.

Numerous computers are already used within R&D for process management and control, automated instrumental analysis, interactive/time-shared/batch data processing, information storage/management/query/retrieval, word processing, and various other computer-aided activities. Until 1981, communication with and among the several systems and services was by individual twisted-pair cable or telephone company circuits.

In mid-1981, following a year of study and design, an LAN employing rf broadband technology was placed in service to provide communications facilities to every laboratory module, office, and work station in all 16 buildings of R&D.

During the study phase, the choice

of the technology to be employed was governed by a variety of requirements and constraints. Communication paths were required to support: a distributed network for process management and control; a distributed network supporting intelligent instruments and research laboratory computers; a large, dedicated data collection and processing system supporting a chromatography skill center; terminal access to any of the above networks; interactive gateways to other corporate computing services; and office automation services.

The design was constrained by requirements for minimum single points of failure, high noise immunity, and ease of installation and maintenance.

The communications technologies considered were direct-connect twisted pairs, telephone circuits, microwave, fiber optics, baseband, and broadband.

The direct-connect twisted-pair option was eliminated quickly. The Technical Center's underground utility conduits were almost filled to capacity. The rate of growth in the number of computers and intelligent devices requiring communication services indicated that any reasonable addition to conduit capacity would not be adequate for very long. Direct-connect twisted pairs would offer little immunity to electromagnetic interference.

The reasons for eliminating phone circuits were essentially the same as for direct-connect twisted pairs.

Microwave transmission, at least between buildings, was considered, but appeared to be an expensive alternative. In addition, a licensed engineer would have been needed to service and maintain the equipment, and a station license would have been required.

Fiber optics appeared very attractive because of its immunity to electrical noise and consequent low bit-error rate. However, frequent repeaters would have been necessary. In addition, multidropping of fiber optics cables was not possible.

Baseband and broadband appeared to be the only viable alternatives. One drawback of baseband was its distance limitation. Another was that it allowed only one user at a time the use of the cable. Simultaneous paths were required for multiple real-time data acquisition systems and many concurrent interactive users. Radiofrequency broadband was chosen as the communication medium for the R&D LAN. Broadband permits both frequency division and contention access multiplexing. Therefore, many channels can coexist simultaneously on one cable. Broadband uses hardware common to the cable television industry, and cable, amplifiers, couplers, splitters,

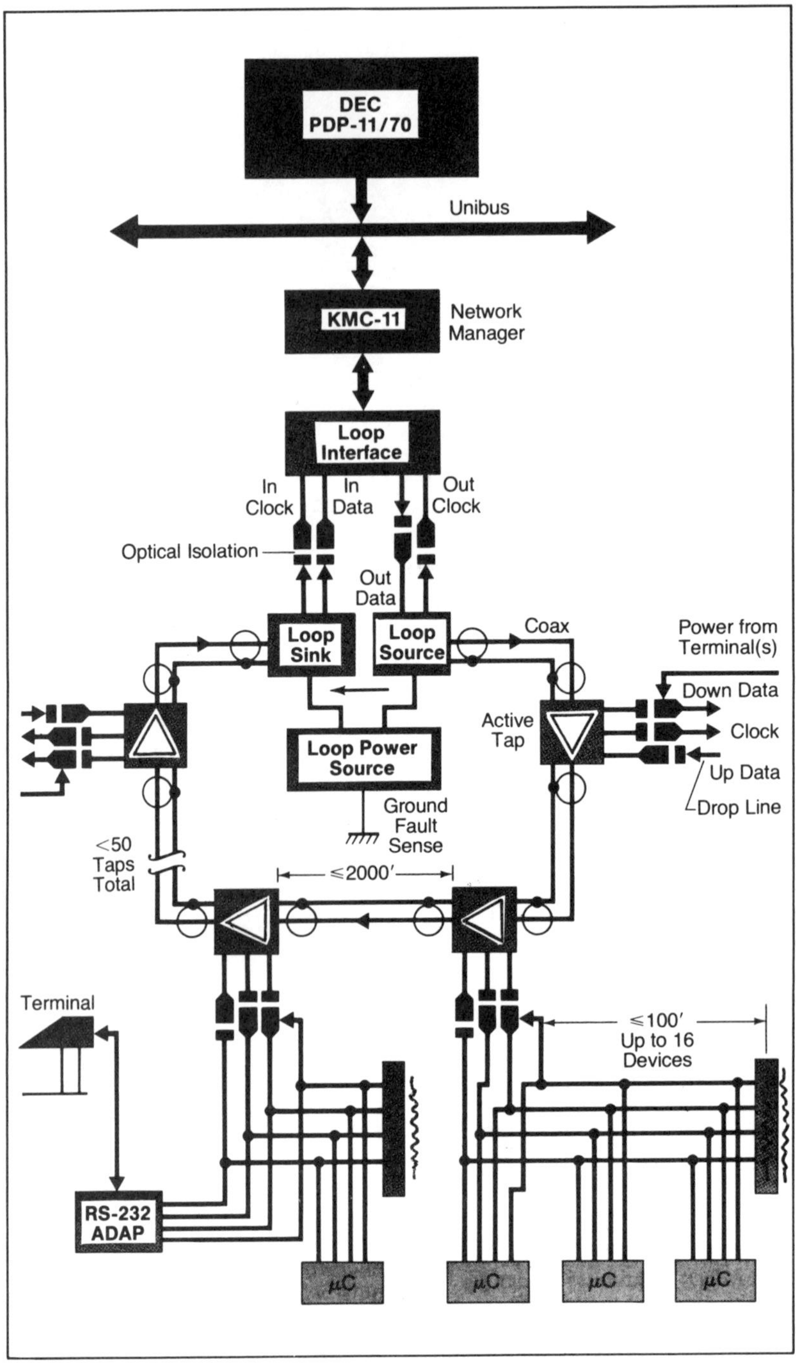

Figure 2. A ring network; Burroughs Wellcome Company

and taps were readily available and low in cost.

A mid-split, single-cable implementation was chosen since it offered more than adequate channel capacity for any reasonable growth projections and could be installed in existing utility conduits. Included in the initial installation were the Technical Control Center (head end), all interbuilding and intrabuilding trunk and feeder cables (7000 ft exterior cable, 5000 ft interior cable), certain individual distribution cables, all distribution amplifiers and associated splitters, couplers, and taps. Critical elements in the Technical Control Center, including power supplies and translators, were duplicated and provided with automatic switchover units.

Design and hardware specifications were provided by an independent contractor. Installation was done by on-site craft persons after minimal instruction in handling and splicing techniques. Fabrication of the head end, tuning of the installed network, and initial debugging were performed by a communications engineer on the R&D staff. This engineer and a sup-

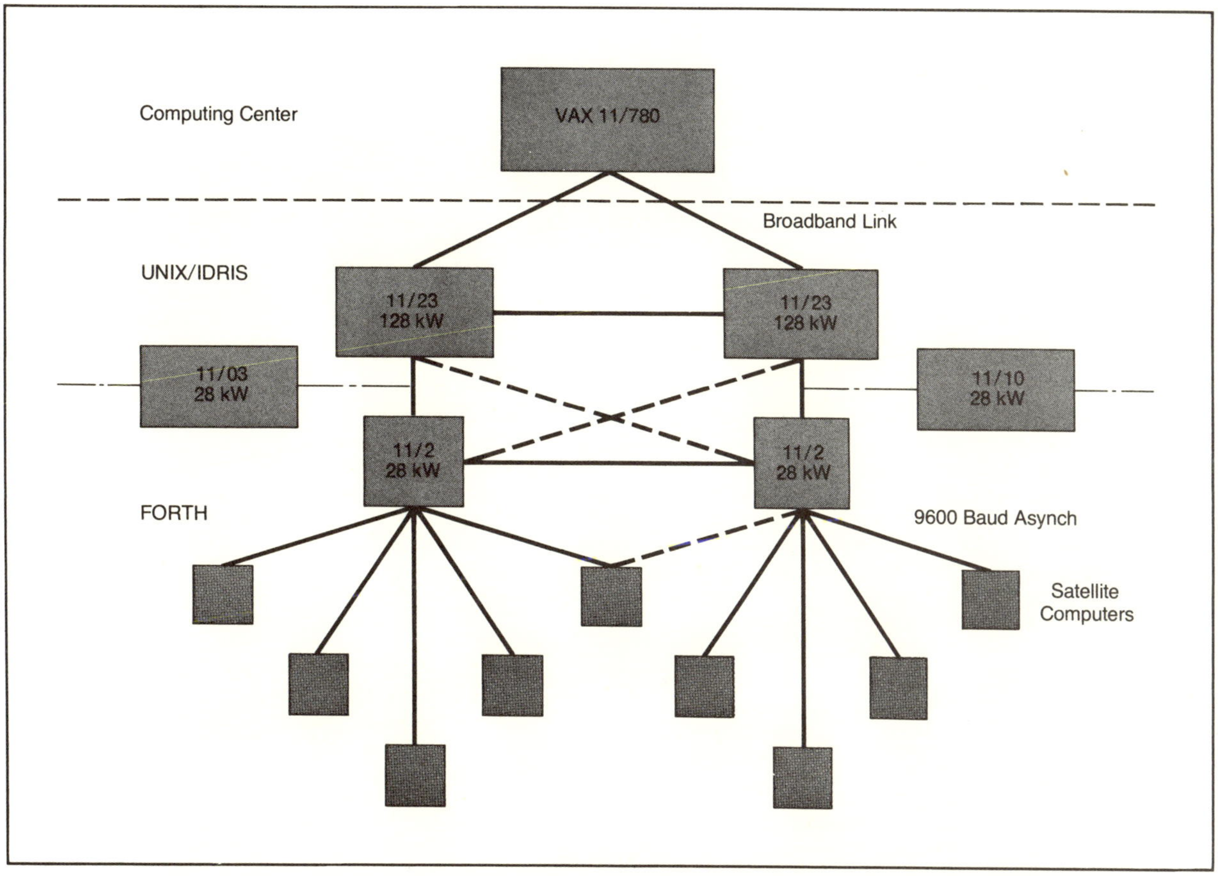

Figure 3. Star network coupled to a broadband net; VPI & SU Chemistry Department. Satellite computers (16 K word) are equipped with ADC/DAC subsystems, serial/parallel ports, real-time clock, and 60-Hz clock

porting technician handle the continuing growth of the system (installation of new distribution cables), routine periodic maintenance, and troubleshooting.

Pure frequency division multiplexing using fixed-frequency point-to-point dedicated links is used for real-time data acquisition and intercomputer communications. Contention access multiplexing using carrier sense multiple access/collision detection (CSMA/CD) techniques is used for interactive terminal support. This combination provides for up to 24 000 discrete links and occupies only 70% of the currently available cable bandwidth. Adequate bandwidth remains for other functions including intrasite video transmission.

Reliability of individual components and overall system integrity have been exceptional after initial tuning. Addition of new users is easily accomplished and has no impact on existing users. A suitable level of inventory has been established to ensure timely restoration of a failed link. The single most common problem has been the infant mortality failure rate in modems. A "burn-in" period prior to installation has minimized the effect on users. The system installed has met or exceeded expectations in every sense.

Burroughs Wellcome Co.
Research Triangle Park, N.C. 27709

Contributors: J. P. Koontz, W. S. Woodward [a], *J. Hinderliter–Smith, and D. F. Smith*

The network at the Burroughs Wellcome Research Laboratories is designed to support a large number of microcomputers engaged in data acquisition and report generation at the bench level in the analytical laboratory. Applications include gas and liquid chromatography analyses, UV reaction rate studies, and automatic pK_a determinations, among others. The research center employs 716 people. The network currently services an area of 154 000 sq ft on four floors, with cable runs of 5000 ft.

Before the network was implemented, approximately 30 stand-alone microcomputers were in operation controlling experiments. The primary mass storage medium for most of these units was cassette tape. The microcomputers had a variety of printers and plotters attached to provide the analysts with hard-copy graphics in addition to video graphics capabilities inherent in the computers.

The goal of the network is to serve as a link to connect these microcomputers to a host minicomputer. The host provides each microcomputer with the ability to down-line load software into the microcomputer's memory for data acquisition and manipulation. Executing programs can run "stand-alone," or transfer data to the host minicomputer for storage and later processing. Data files stored on the host can also be retrieved for manipulation within the microcomputer. Access to minicomputer resources is also provided to standard terminals transparently connected to the network through a microprocessor-controlled interface.

The automation system developed to provide these services consists of three elements: a host minicomputer, a high-performance communications network, and the remote laboratory subsystems. The communications network synthesizes the best qualities of a ring network and a multidrop bus system into a very flexible conduit (Figure 2).

The heart of the network is a microprocessor-based manager that routes data transfers between the host and the remote microcomputers and ter-

[a] Chemistry Department, University of North Carolina, Chapel Hill, N.C.

minals. The manager inserts data or control packets onto a single coaxial cable. These packets propagate around a coaxial loop and are read or modified by the node to which they are addressed. The loop terminates back at the manager where the packets are received by a data sink.

Remote devices access the loop stream through taps into the ring. Regeneration of the loop-signal quality and timing are functions of the active taps. They optically isolate the remote devices from the main loop. There may be 50 of these taps, each with a multidrop bus attached. Up to 16 remote devices may be connected to a multidrop bus and thus be served by that single tap. The maximum multidrop line length is limited to 30 m. This is the result of the two-level network (main loop plus radial multidrop buses), which requires a round-trip signal propagation from the tap to the local drop-line interface and back in one bit-time. Network communication speed is 500 K bits/s.

There are two types of local drop-line interfaces. The simpler kind of drop-line interface serves to connect microcomputers, with attached instruments, to the system. These interfaces provide differential signal interfacing and power to the drop line, packet addressing logic that recognizes one of 255 addresses, and access to the microcomputer's memory.

The second type of drop-line interface (gateway) provides an RS-232 connection that can be used by terminals, printers, plotters, and other compatible devices. These interfaces contain a dedicated microprocessor. Software is automatically down-line loaded to them at power-up, and thus the interface control program may be device-specific.

Connection/disconnection and power-up/power-down of user equipment does not interrupt service to other users over the primary coaxial cable. High-level error detection utilizes single-byte parity. Error correction is by retransmission.

These components provide a set of communication and network support functions that meets all of the original goals. Any microcomputer can access disk space on the minicomputer. This disk space is partitioned into individual user areas that appear to the remote microcomputer as small local disks, i.e., pseudo-floppy disks. Down-load time for a copy of the real-time BASIC interpreter is approximately 1/4 s. Transfer of a 7500-point data set (37 500 bytes) can be completed in 3/4 s.

A fundamental question is whether direct node-to-node communication is needed (or wanted). In the context of a laboratory network the answer is "no." Laboratory processors should not have to be concerned with unsolicited communications during data acquisition. All communication is between host and node.

Currently 41 microcomputers and 3 gateways are connected to the network. In addition to providing effective services for the scientists in the laboratories, the network also makes management of microcomputer software easier for the automation group. New software is distributed by storing the program on a read-only system pseudo-disk instead of making and delivering a cassette for each user.

Future enhancements of the system that will be available from remote microcomputers and terminals include communication with the corporate mainframe computer for archiving data and adding to the corporate data bases as well as access to a high-speed printer and a multipen plotter.

Virginia Polytechnic Institute and State University
Laboratory Automation and Instrument Design Group
Chemistry Department
Blacksburg, Va. 24061

Contributors: Raymond Dessy, Ian Chapple, James Currie, Steven Dueball, and Mark Thompson

The star point-to-point network serving the Laboratory Automation and Instrument Design Group at Virginia Tech provides computer resources to 16 people working on the development of new transducers and techniques in automated analytical chemistry. Laboratory rooms separated spatially by 150 ft are involved.

The network has grown from a small nucleus consisting of a floppy-disk-based 16 K word microcomputer servicing one satellite to its present configuration. Such expandibility is one of the key characteristics that make networking a rational approach to the needs of the automated lab.

Each scientist has his own microcomputer. A network approach was needed to meet the various requirements of each project. These needs centered around:

- the ability for continuous reconfiguration of hardware and software in the computers at the experimental level;
- convenient access to programming tools such as disk, printer, or debugging aids that cannot be cost justified at each control/collection node;
- access to computer resources for limited data management and calculation requirements; and
- access to a large CPU for the few experiments producing massive amounts of data or requiring major computation.

As Figure 3 indicates, the current configuration consists of 10 small satellite processors. Each of these 16-bit microcomputers (16 K word) is equipped with a multichannel analog-to-digital converter, multiple digital-to-analog converters, parallel ports, and serial ports. All real-time instrument control and data collection occur at this level. Each laboratory bench in the complex has the standard water, gas, and electric utilities plus a data plug that allows any satellite computer to be connected to the network near the equipment it controls. The communication path with the hosts is an RS-232C twisted-pair line supporting 9600 bits/s transmission. At the host level these lines terminate in a patch panel; thus, any satellite can be connected to any of the hosts.

At the host level, the network is bilaterally symmetrical. It is two-tiered vertically. First, there are identical 16-bit processors, each capable of supporting three terminals and four satellites concurrently. Multiple double-density flexible disk systems permit a number of users access to the system. Program preparation, data reduction requiring moderate math capability, printing, and plotting are performed at this level.

The programming language supported is FORTH. The network communication involves a "positive acknowledgment/check sum/request to retransmit on error" protocol. At power-up each satellite has a FORTH kernel from the host master disk down-line loaded into it. The advantages of centralized distribution of the lab software package have been described in the capsule report on the Burroughs Wellcome ring net. Software generation and testing are made easy, with both host and satellite hardware running the same machine language.

The hosts at this level are limited to 32 K word address space and "single precision float/double precision integer" math implemented in microcode.

To handle more demanding tasks, a duplicate set of larger host computers has been installed above them. These machines address 128 K words of memory and implement double precision float arithmetic in microcode. Rigid disks are attached to provide a large amount of bulk storage. The processors use the same machine code instruction set as the other two levels described. Thus, the three tiers involve computers with a common machine instruction set.

Except for the processors themselves, all of the other cards making up the computer systems may be exchanged freely from card cage to card cage. A complete set of spare cards for

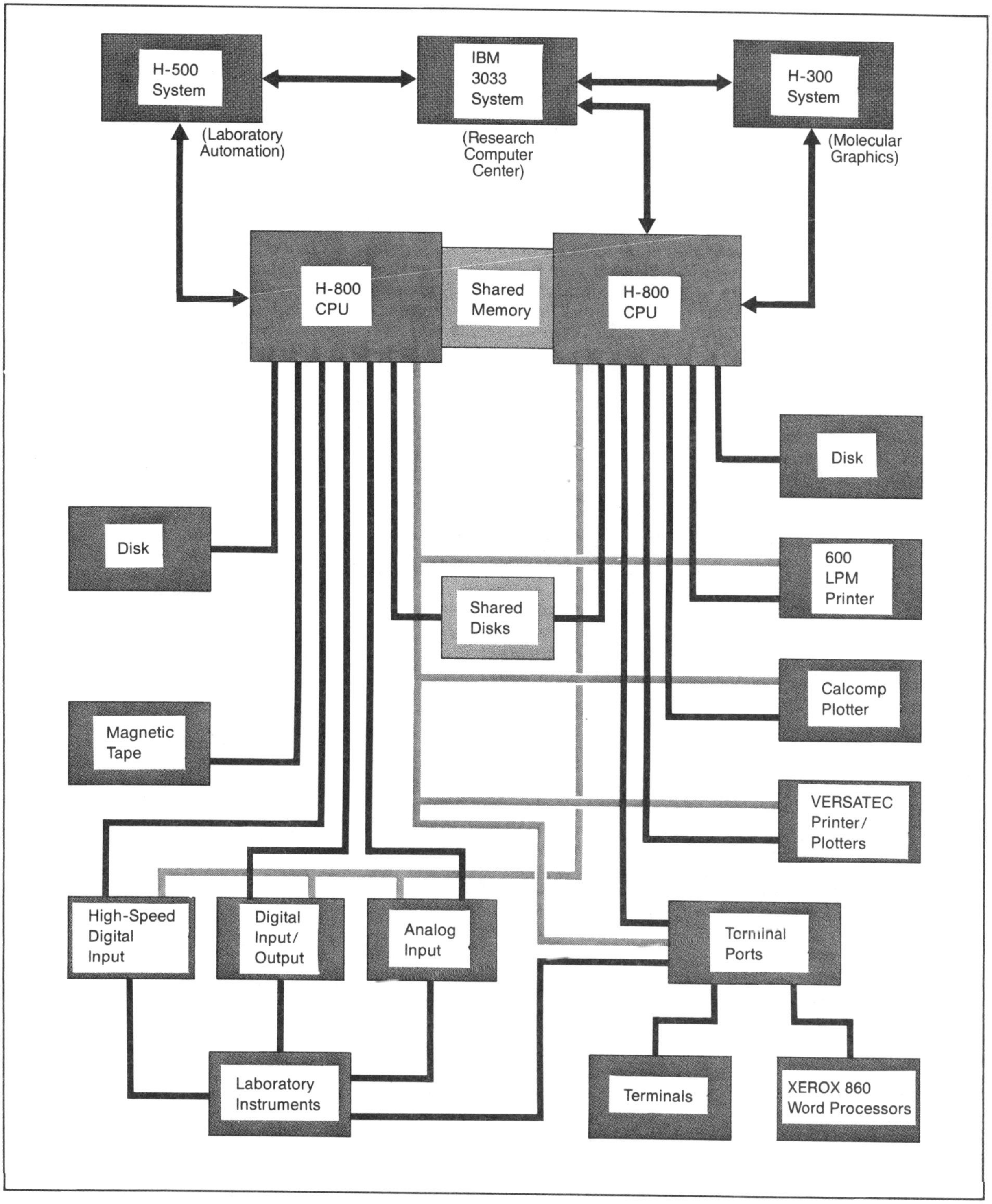

Figure 4. A single multiprocessor center; Upjohn Company, Physical and Analytical Chemistry Research. Darker lines indicate peripherals are switchable between CPUs

all installed units is maintained. Troubleshooting is therefore easily performed by board-swapping. The bilateral symmetry of the net allows the lab to continue to function during the brief downtime involved.

The bilateral symmetry also allows the system to use different operating systems. The larger hosts support classical operating systems such as RT-11, RSTS, or RSX-11M. They also support UNIX and a UNIX-like system called IDRIS, plus their associated language C. Languages and operating systems tailored to the research task can be easily called into play. The only alternative is a very large computer—which the lab cannot justify.

A fourth tier exists in this hierarchical network. The university has a variety of large 32-bit VAX computers as well as IBM facilities. A broadband link (two channels) connects the lab to these larger service units. Communication is over a frequency division multiplexed, frequency modulated coaxial line at 9600 bits/s.

The system has thus grown in two directions as the needs of the laboratory changed and more real-time capabilities as well as more powerful computing facilities became necessary. Networks are like trees. They can extend their root systems down and reach for sunlight with their branches.

The separation of responsibility at the broadband level has permitted the laboratory to continue to control how its chemistry is performed, and yet have easy access to the large computing resources it requires at a very low cost.

Acknowledgment

The network implementation at VPI & SU has been a group effort, to which Charles Knipe, Mark Wingerd, Jim Bergquist, Steve Conder, and two former members of the group, Michael Starling and David Hooley, have contributed greatly.

The Upjohn Company
Physical and Analytical Chemistry Research
Kalamazoo, Mich. 49001

Contributors: David J. Duchamp and Edward C. Olson

The Upjohn Physical and Analytical Chemistry System (UPACS II) supports the needs of Physical and Analytical Chemistry Research, a unit of about 40 research personnel, including 11 PhD scientists and 17 BS/MS research assistants. Approximately 60% of the efforts of this group are in organic structure proof via physical methods, 20% in applications of physical methods to drug design and drug interaction studies, and the remaining 20% in analytical method development and characterization of experimental bulk drug lots. The department houses much major instrumentation including two four-axis X-ray diffractometers, five mass spectrometers, three NMR instruments, two Fourier transform infrared spectrometers, three ultraviolet spectrometers, a circular dichroism instrument, an ESR spectrometer, several liquid and gas chromatographs, a thermogravimetric analyzer, a differential scanning calorimeter, an atomic absorption spectrometer, three CHN analyzers, an oxygen analyzer, and several balances.

The UPACS II system is required to provide real-time data acquisition, data reduction, on-line data storage and retrieval, and interactive data interpretation support. An earlier system, UPACS I, operational in 1969, was very successful in providing the data acquisition, reduction, and storage support. Experience with this system taught us that we needed to provide rapid and powerful interactive data interpretation support, essentially simultaneously with the data acquisition and reduction activities. Since our researchers frequently are collaborating on the same projects, much is gained by having them all on the same system, with access to centralized data files. The system is required to provide uninterrupted operation for long periods of time, 24 h/d, 7 d/wk. All experiments are initiated and terminated from the respective laboratories, via terminals in the laboratories. All interfaced instruments must operate simultaneously and asynchronously regardless of data acquisition rate, which varies from 80 K words per second for mass spectrometers, to less than a point per second for some of the slower analytical instruments. Some applications require that data be transferred to Upjohn's large mainframe research computer, an IBM 3033, for large-scale computations. Another requisite was the capability of processing chemical structures online to be able to effectively provide support for the interactive interpretation of spectral data and computer-aided structural elucidation. In addition, we required large amounts of online disk storage for spectral libraries to aid in interpretation and for archival storage.

After benchmarking the interactive requirements, a dual Harris S200 system with shared disk was installed in late 1977. The present system consists of a dual CPU Harris H-800 system with shared memory and 900 megabytes of shared disk (Figure 4). Presently about 20 instruments are interfaced to the system. Users communicate with the system via about 50 terminals, including 13 Tektronix 4014 high-resolution graphics terminals with graphic input tablets. Four Versatec printer/plotters located in laboratories and a Calcomp plotter in the computer room provide hard-copy graphics output. Graphics programming is done using a locally developed graphics subroutine package, based on the SIGGRAPH standard, which allows for device-independent graphics programming, i.e., the same program can be written for any of the four different graphics devices currently on the system. A set of sophisticated operating system extensions has been developed to support real-time data acquisition, including the digital input, analog input, and digital output functions used in most of the data acquisition activities.

The system is currently supported by one systems programmer, one applications programmer, a resident Harris service engineer, and part-time efforts of two other people. There is no computer operator. For application-specific programs, users usually do their own programming; for subsystems that span across applications, programming support is provided.

Roughly half of the plans that we have for the system are currently implemented and operational. We have found the system to be very powerful and very reliable, with a high uptime. It performs all functions very well. Building on the success of our dual CPU Harris system, we have recently installed a Harris H-500 system, using the same software as the system discussed above, to support the needs of our Drug Metabolism Research Unit (about 50 research personnel), requiring two additional programmers. This system, located in a different building, supports several dozen gas and liquid chromatographs, including many with automatic injection systems; several balances; and several liquid scintillation counters. The transfer of software between the dual H-800 system and the H-500 system was accomplished very smoothly. Since balances and chromatographs are supported on both systems, the savings on software have been great. We have also recently acquired a Harris H-300 system to support high-performance color graphics for use in drug design and molecular modeling activities. Again the same operating system is used.

We estimate that the H-500 and the dual H-800 laboratory automation systems are equivalent to approximately 50 additional personnel in terms of additional output from the units supported. Using a central super-mini approach to laboratory automation has saved us much in shared peripheral devices, ease of programming, cost of programming, and maintenance support. We have a very powerful system, which each user looks to as his or her own. Major goals for the system have been to provide good software documentation and acceptable audit trails for data and software manipulation, as well as to provide simple and reliable backup. These are tasks difficult to handle adequately on distributed computers. The disadvantages of the central approach are the high initial cost, the need to have a very capable systems programmer to make it work well, and the fact that all instrumentation is dependent upon a single computer system. In our experience, however, these disadvantages have been far outweighed by the advantages mentioned above.

Originally published in
ANALYTICAL CHEMISTRY **1983** *55*(1) 70A
and **1983** *55*(2) 277A.

Laboratory Information Management Systems
Part I

It all starts with a datum point derived from a microprocessor-controlled smart instrument. It is joined by other data from automated GC, LC, mass spectrometer, and infrared instruments to form a small data stream. This merges with related data streams created by ICP and AA facilities, as well as manual test methods. Emerging from the laboratory is a torrent of data that threatens to engulf the laboratory technician and manager.

The previous pair of articles in the A/C INTERFACE series (*Anal. Chem.* **1982,** *54*, 1167–84 A and 1295–1306 A) dealt with local area networks (LANs), the conduits by which instruments, computers, and terminals can synergistically interact to channel data flow. Intimately related to this hardware technology is the software-dominated area of laboratory information management systems (LIMS). Together they are providing the tools to harness these data more effectively.

LIMS are the subject of the present pair of articles. This article is a tutorial dealing with the nature of LIMS, why they are needed, and how they are implemented. The second article will feature capsule reports from several different types of analytical laboratories, each presenting a different solution to the LIMS problem.

Information management systems involve some of the most complex software in existence. However, there is no excuse for the poor communication that exists between the end users of these systems, the programmers who create such systems, and the vendors who sell the developed products. The typical laboratory manager, already inundated by a flood of data and frustrated by legal and profit imperatives, is being smothered by jargon like a torpedoed sailor under a blanket of oil. In an attempt to understand what system should be implemented in the lab, the user chokes on a stream of acronyms such as ISAM or PSAM. He or she cannot see through the films of schema definitions and cannot hear because of competitive system vendors shouting "B-tree," "plex," "hierarchy," and "relational."

An information management system is logical and highly structured. One suspects the obfuscation may be deliberate. Data General couches its business-oriented package (CEO) in terms of storing data electronically in "folders," "drawers," and "cabinets." Deletion of data is described in terms of "waste baskets" (material thrown in them can be recovered) and "shredders" (terminal destruction). The Xerox STAR video terminal actually displays replicas of documents, file folders, and drawers on the users' "electronic desk top." It should thus be possible to describe LIMS concepts in terms familiar to the analytical chemist. However, before exploring that realm, let us become convinced that we need the technology.

What Does a LIMS Do?

It is possible to list the activities for which machine assistance would be useful from the time of sample entry into the analytical laboratory to the time of sample release.

Log-in procedures bind the sample's identity and analysis requirements with a number that will aid in retrieving information about the sample (Figure 1). These sample numbers can be printed automatically on labels that also provide more human-oriented data about the sample and that aid in distribution of the sample throughout the laboratory (Figure 2). Since aliquoting is common, multiple copies of the label can be produced by a printer/plotter attached to the computer. Bar-coded labels can be printed and affixed, or preprinted bar-code labels can be purchased. Figure 3 shows how the vertical bars encode numbers. Labeling and bar-code identification techniques are often among the most beneficial aspects of an information management system since identification errors are drastically reduced.

As each day's samples are logged in, it is possible for a LIMS to prepare receipts for the submitter, indicating due date and estimated cost, and work lists (Figure 4), indicating a reasonable sample load and sample sequence for a given instrument. Nominal analysis times ensure that operators and instruments will not be overloaded. Even sample delivery lists can be generated. The work lists can then indicate sample location (cold room, testing rack, etc.) to expedite access. Realistic systems permit establishment of sample priorities to accommodate items that must be expedited.

The data from each analysis, as it is performed, can be captured electronically or entered manually from a video terminal. Computer-generated printed forms may be provided, each analyst filling them in and passing them to terminal operators for entry. Or video terminals can be made available to analysts for their personal entry of data. With such manual data entry it is important that the keystrokes required be minimal and user-cordial. Video-

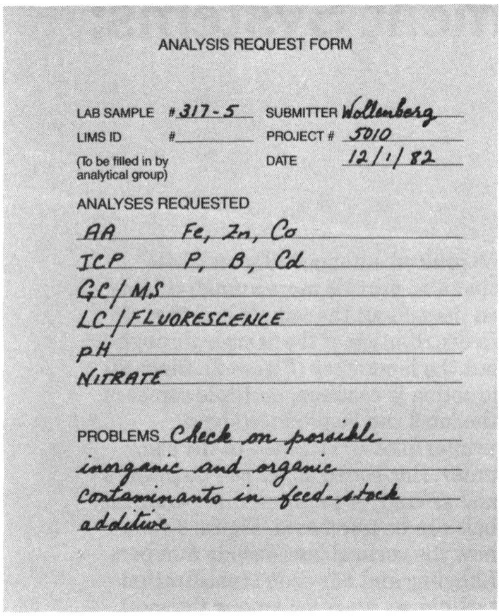

ANALYSIS REQUEST FORM

LAB SAMPLE #317-5 SUBMITTER Wollenberg
LIMS ID # PROJECT # 5010
(To be filled in by analytical group) DATE 12/1/82

ANALYSES REQUESTED
AA Fe, Zn, Co
ICP P, B, Cd
GC/MS
LC/FLUORESCENCE
pH
NITRATE

PROBLEMS Check on possible inorganic and organic contaminants in feed-stock additive

Figure 1. Analysis request form for a LIMS

displayed forms with highlighted areas for data entry locations are one viable approach (Figure 5). A menu or list to prompt the operator is another. Bar-code readers associated with the intelligent instrument or terminal may be used to identify the sample. The analyst and the analysis he or she is performing can be recorded the same way, improving the ergonomics.

Range information on normal limits permits the operator to be alerted if results are abnormal. The alert can be overridden if the operator suspects an unusual sample. Or the procedure may be rerun if an analytical error is suspected. These editing steps are essential to keep erroneous data out of the data base. Once there, it will be almost impossible to expunge it from the system.

Along with the data, the analyst may enter free-form comments that may be useful to the requester. Privileged comments may be added that will appear only on internal laboratory reports.

The laboratory manager can request operations information about a sample at any point while it is being tested. Submission dates and completion dates are available. This sample tracking includes statistical information on all samples submitted, in-test and completed, as well as backlog profiles (Figure 6). Such operational information can be broken down into instrument, analysis type, sample type, operator, or any other category that is meaningful in running the facility.

Although each analysis carries with it the name of the analyst performing the test and the instrument used to carry out the work, it is important to have one person authorize the release of the total data set associated with a single sample. It may be advisable to resubmit samples when normal ranges have been exceeded or results from different analytical methods suggest further work. It has become common to include audit trail features in laboratory data management systems that keep track of what alterations have been made, when, and by whom. This permits legitimate changes to be made, but makes it possible to retrace the alteration steps if legal or scientific mandates occur.

Many firms have found it inappropriate to release fragmented analytical data (sometimes even to the requester). It is too easy to jump to improper conclusions from incomplete analyses or to bias future work. Thus the access control to the various levels of information needs to be carefully considered. *Who* has access to *what* data or information *when*? Passwords or locks on data structures can easily define *no access*, or *read*, *create*, or *modify* status for individual users.

A well-designed system will back up (record an entire copy of its data) at appropriate points, so the system can be restored if failures occur. These copies should be kept in safe places away from the computer area. The most careful installations have backup copies stored at a different geographical site.

Archiving refers to the long-term, off-line storage of data and is a separate function. It may be desirable to store only condensed information (peak area) rather than raw data in an on-line disk to conserve space. However, the raw data can be archived for subsequent retrieval by storing it on magnetic tape. Alternatively, once a report has been generated, only sample number and identity information might be maintained on-line, all

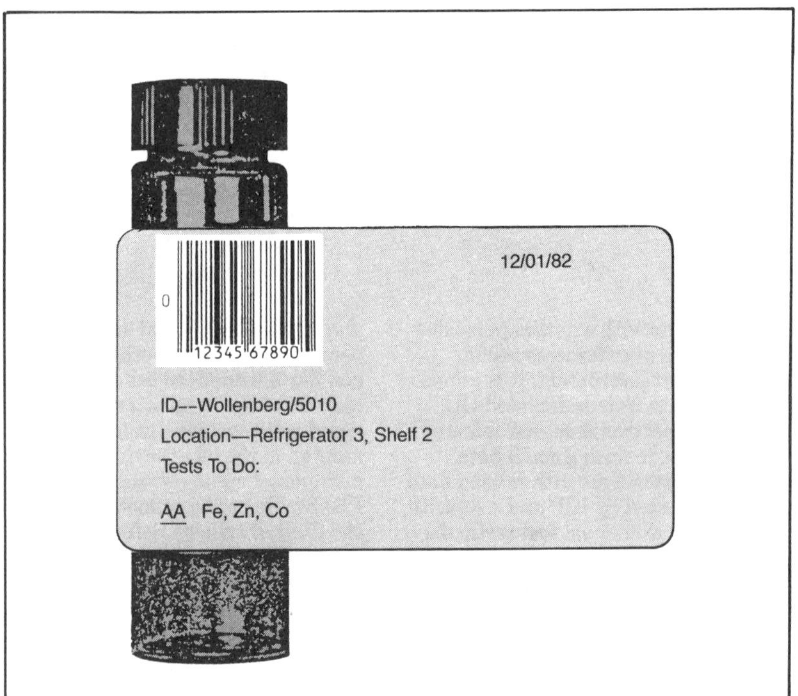

Figure 2. Computer-generated sample label (adapted from Radian's SAM)

Figure 3. A bar-code is a set of thin bars that represent binary ones, and spaces that represent binary zeros. A series of these is used to encode a sample number. Each number is expressed as seven bars and/or spaces. There are six digits expressed to the right of a center guard space and six to the left of the guard space. Digits to the left of center begin with a binary zero and end with a one. To the right of center they begin with a one and end with a zero. The bar-code sequences used to express the same character to right and left of the center guard are the complement of each other. An optical wand may read from either direction, the attached microcomputer determining the direction of the movement from the scanned pattern. The bits are then interpreted as an Arabic number. The last digit to the right is a check sum to confirm that the other 11 digits have been read correctly. If the calculated value and the read value for this digit disagree, an error is announced at the reader

chemical information being archived. In some systems, once a report is generated, the entire data set associated with a sample is archived. Each laboratory has its own use pattern for data.

The most important aspect of a LIMS is the report generation facilities it provides (Figure 7). Without a combined LAN/LIMS the analyst is doomed to become an error-prone, expensive secretary, copying data from one place to another. A well-designed LIMS provides a report program generator to allow the user to prepare programs that generate properly formatted data and text. Page headings, column headings, and columnar data formatting should be possible with a short learning curve to master the language elements involved. Word processing (text preparation) and list processing (sort, merge, and select) capabilities are built in. A math package allows the user to calculate intermediate and final values from stored raw data.

Some labs, and all quality assurance programs, require the presence of a higher level statistics and graphics package. This permits the quality assurance manager to use data from spiked samples and standards placed in the analysis stream to evaluate trends due to instruments, reagents, or personnel. With this as a base, developing trends in raw materials, intermediates, or products can be ascertained and the statistical results presented in various graphic styles. Simple modeling programs aid this goal, allowing data to be fit to mathematical models.

What Operations Should Your LIMS Perform?

All of the above is possible with existing software packages. Some are inexpensive and can be used by the interested analytical chemist with minimal training. Others require extensive hardware, expensive software, and data management personnel to coordinate activities. The proper choice requires a clear understanding of the problems. This is where most laboratories fail. Some binary-type questions may help clarify the point:

- Do you need an operations information and/or chemical data management system? Operations information deals with sample input, output, and backlog. It is the information needed to control the day-to-day operation of the laboratory. There is no need to carry analytical data along. This simplifies the data set involved and makes searching faster. Since updating operations management systems is much faster than adding to chemical data management systems some LIMS choose to update the latter by creating transaction files during the day and updating overnight. This presumes that instantaneous access to completed data is not necessary. Other laboratories demand concurrent updates.
- Do you require storage of analytical results or data? Some laboratories require raw, or nearly raw, data to be available at all times. Others require only condensed information derivable from the data. The latter reduces storage requirements and search times for the system. The complexity and cost of the data management system are lowered.
- Do you need to store procedures or analyses? Many production facilities and quality control laboratories need to store and update the procedures and tolerances they use for acceptance/release of material. As limits on incoming or outgoing chemicals or goods change because of supplier fluctuations, changes in regulations, or internal conditions, it is vital to keep track of these and associate the changes with a lot number.
- Are you involved with production, quality control, or research and development? The production environment is quite different from the R&D laboratory. In production, preset series of analyses are performed on batches of samples. Often the tests must be conducted repeatedly over long time periods. Sample entry into the system may need to be menu driven to match personnel capabilities. Automatic prompting is mandatory in time-scheduled reanalysis. Often the observations made on samples should be entered using menu selections to match the limited spectrum of allowed entries. The associated retrieval, statistics, and security elements must be well developed to meet the demands that will be made on the system for fiscal or legal reasons.
- Is detection or prevention of quality failure important? In some instances it is enough to detect that a product has failed. In others it is important to prevent a failure. Predictive capabilities are much more difficult to build into a system. They are a goal that will become more common as managers understand the use of computers and have access to proper modeling programs.
- Is it anticipated that changes will occur in how the data is accessed? Some data management systems permit only questions of the data that were envisaged as the data base was created. Others permit changes. There are important price/speed tradeoffs as the flexibility of a system increases.
- Is user cordiality (ease of use) or system capability important? Unfortunately these features are often inversely related.

INSTRUMENT WORK LIST

INSTRUMENT AA-#4 DATE 12/1/82 ANALYST J. Bergquist

SAMPLE #	LOCATION	ANALYSIS DESIRED	PRIORITY
1234567890	Ref. #3, Shelf 2, Tray 5	Fe, Zn, Co	Expedite
1756376520	Ref. #3, Shelf 2, Tray 5	Zn, Co	
1654329109	Ref. #3, Shelf 2, Tray 5	Fe	
1336198251	Ref. #3, Shelf 1, Tray 4	Ca, Cu	
3241117812	Ref. #1, Shelf 2, Tray 1	Pb, Cu, Zn	Expedite
7169119040	Ref. #1, Shelf 2, Tray 2	Cr, Fe, Zn	
1346118220	Ref. #1, Shelf 3, Tray 4	Mg	
1882643009	Ref. #2, Shelf 1, Tray 1	Ca, Cu, Pb, Cr	

Figure 4. Instrument work list

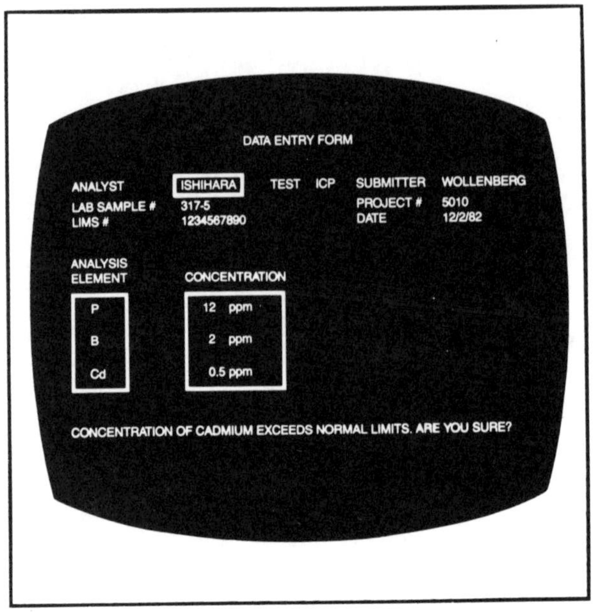

Figure 5. Video-displayed form with highlighted area for data entry

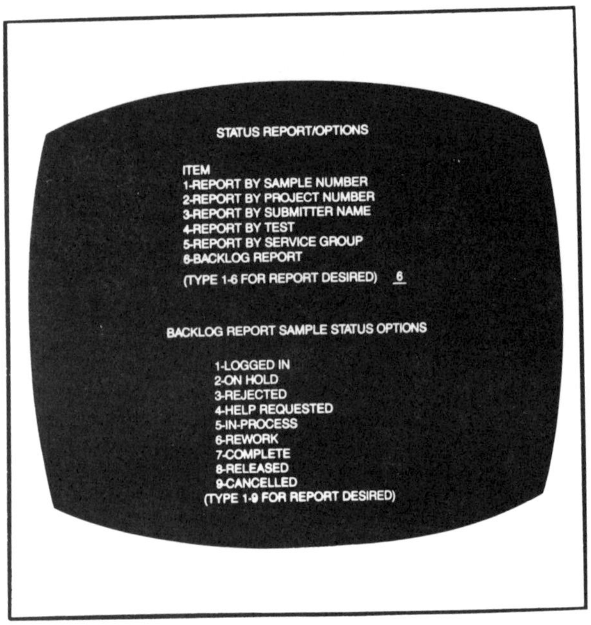

Figure 6. Typical operations report menus (adapted from Perkin-Elmer's LIMS-2000)

ANALYSIS REPORT FORM

LAB SAMPLE # 317-5 SUBMITTER WOLLENBERG
LIMS ID # 1234567890 PROJECT # 5010
CERTIFIED BY PETERSEN PHONE 5822 DATE 12/2/82

ANALYSIS				ANALYST
AA				BERGQUIST
	Fe	3	ppm	
	Zn	1	ppm	
	Co	0.06	ppm	
ICP				ISHIHARA
	P	12	ppm	
	B	2	ppm	
	Cd	0.5	ppm	
GC/MS			NO POLYBROMINATED BIPHENYL FOUND	DUEBALL
LC/FLUORESCENCE			12 ppb OF AFLATOXIN (B_1, B_2) FOUND	THOMPSON
pH			4.8 USING STANDARD SLURRY TECHNIQUE	KNIPE
NITRATE			NONE DETECTED	CONDER

Figure 7. Chemical data report

• Is update of the data structure important, or is speed of retrieval vital? Unfortunately these features are always inversely related.

How Do LIMS Store Data?

Faced with these questions and the need for a viable system the laboratory manager is confronted with two massive structures: the vendors of hardware and software and his or her own corporate computing center. The best defense is a working vocabulary.

The concepts in data storage resemble methods traditionally used on the printed page to store and retrieve information. Samples are submitted along with a form similar to that shown in Figure 1. The individual tests requested produce data with a structure specific to each method. Classical data storage involving work sheets in drawers or cabinets and sequenced by analytical method, sample number, date, or project number is being replaced by electronic storage on disk (Figure 8).

How can such data be stored and retrieved? Separate records could be kept for each analysis type. This approach is represented by the video display screens in Figure 9. The bench chemist sees data in this format. To prepare a report of data specifically related to sample 317-5 the separate records can be pulled and results merged. Alternatively, a single file on a sample could be kept; printing a report would entail pulling only that sample's file. That seems simpler. But suppose production asked if data were available on trends in heavy metal content, using ICP, for all samples associated with a certain project. Or suppose complaints were received

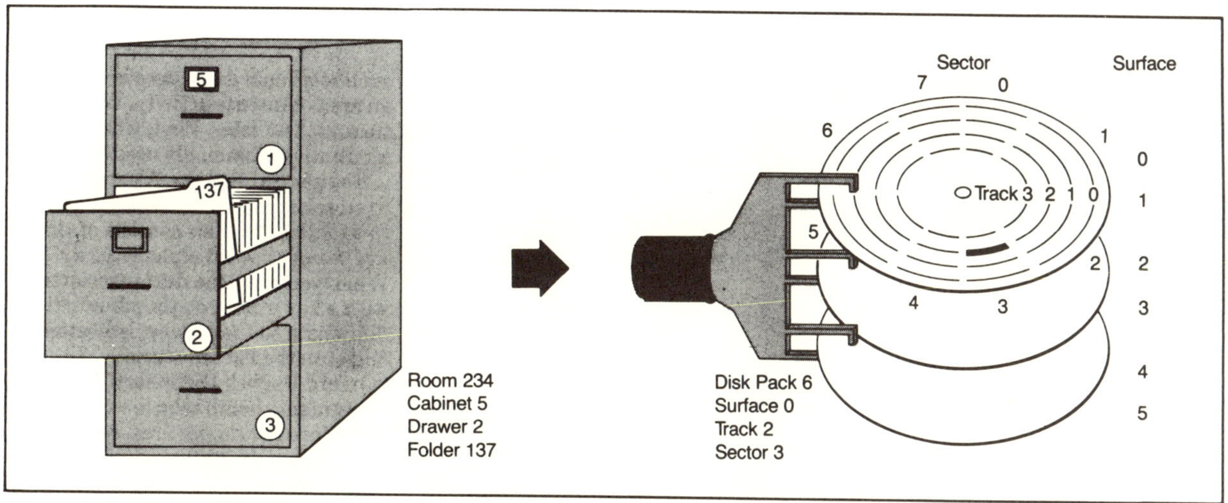

Figure 8. Classical data storage in cabinets is being replaced by electronic storage on disks

about the turnaround time for all samples submitted by a certain analyst. Appropriate structuring of the data is necessary to answer such questions and avoid long search times.

When data are stored in simple tables both status information and analytical data can be associated with each sample (Figure 10). Such tables are human-oriented and represent the classical paper approach. Information can be accessed in very different ways depending on one's function as analyst, lab manager, or lab director. This approach posed problems in the formative years of data management because the tables often were not flat. Instead of each intersection being a

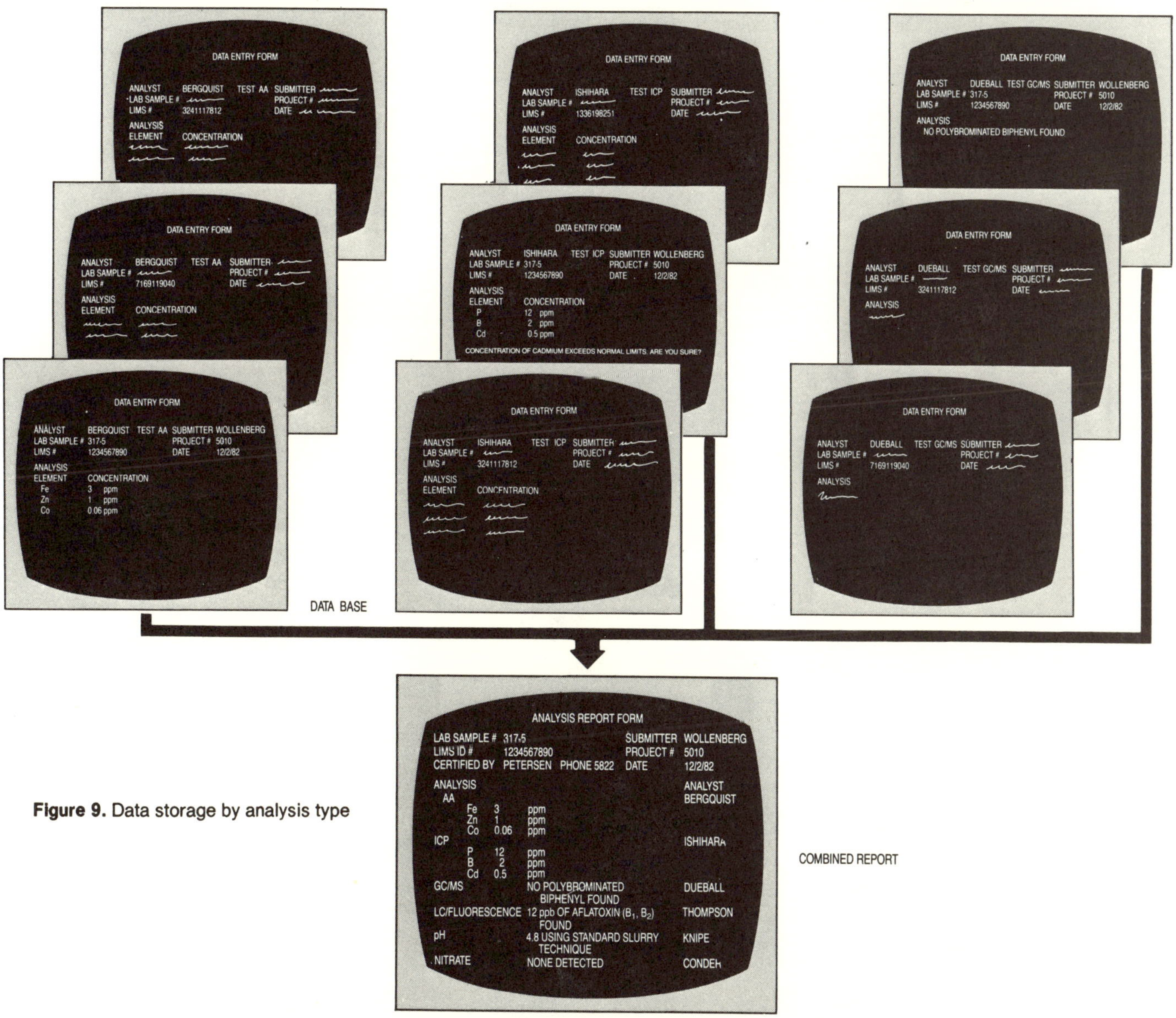

Figure 9. Data storage by analysis type

ID #	SAMPLE STATUS: PENDING COMPLETE RELEASED REPORTED DEACTIVATED	ANALYSIS	ANALYSIS STATUS: PENDING HELP RE-QUESTED REJECTED COMPLETE	SAMPLE IN	TESTS DUE	TESTS OUT	DESE-LECTED	AUDIT TRAIL	DATA
1234567890	PENDING	AA	PENDING	12/01	12/03		NO	NO	
		Fe							
		Zn							
		Co							
		ICP	HELP REQUESTED		12/05		NO	YES	
		P							
		B							
		Cd							
		GC/MS	PENDING		12/07				
		LC/FLUORESCENCE	REJECTED						
		pH	COMPLETE		12/02	12/02	NO	NO	4.8
		NITRATE	COMPLETE		12/02	12/02	NO	NO	NONE
0147598632	COMPLETE RELEASED REPORTED DEACTIVATED								

Figure 10. Data stored in tables can be either status information or analytical data

Figure 11. Partitioned sequential access methods (PSAM) and indexed sequential access methods (ISAM)

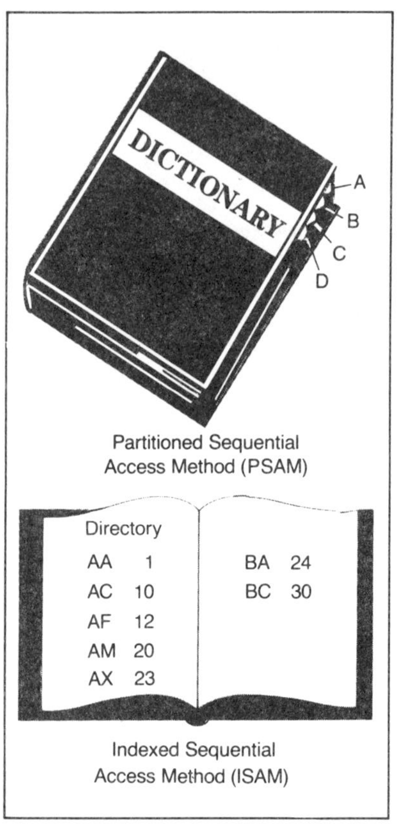

simple number or phrase, a complex relationship existed. Beginning in 1970, methods were developed to convert such files into flat, two-dimensional charts involving simple relationships. Storage of information in such relational data bases is currently an area of intense activity. We will return to them later. First, what are the alternatives commonly used?

Sequential Access. A common way to store data in an ordered manner is in a dictionary. This consists of the key words ordered alphabetically. When you want the data associated with a key word, you thumb into the dictionary to the appropriate letter and scan the list sequentially. The "Oxford English Dictionary" provides convenient thumb tabs to expedite entry into the proper area. Some dictionaries might have a small index to indicate the pages where first and second letters change. These are partitioned sequential access methods (PSAM) and indexed sequential access methods (ISAM) (Figure 11). They allow you to use sample number or date to locate the disk area where pertinent data are stored sequentially. You go to that location and read sequentially until the desired sample is recovered. If a word needs to be inserted or deleted it is necessary to move all of the dictionary entries beyond that location point.

Ordered Index Access. A better solution involves structuring the data base in a more complex way at the outset. Indexes can be kept in which the submitters' names are alphabetized. Next to each name can be an ordered list of sample numbers submitted by him/her, and the location of that data on the disk. This ordered

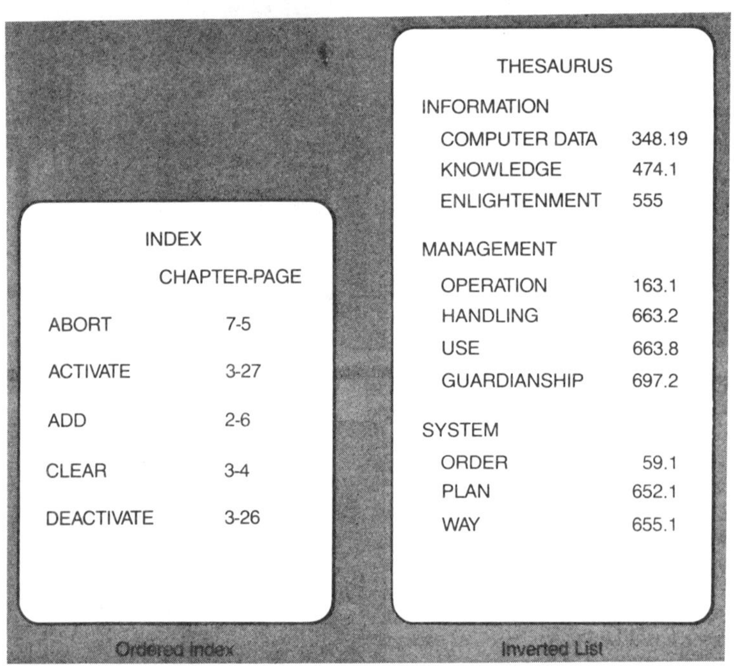

Figure 12. Ordered index and inverted list

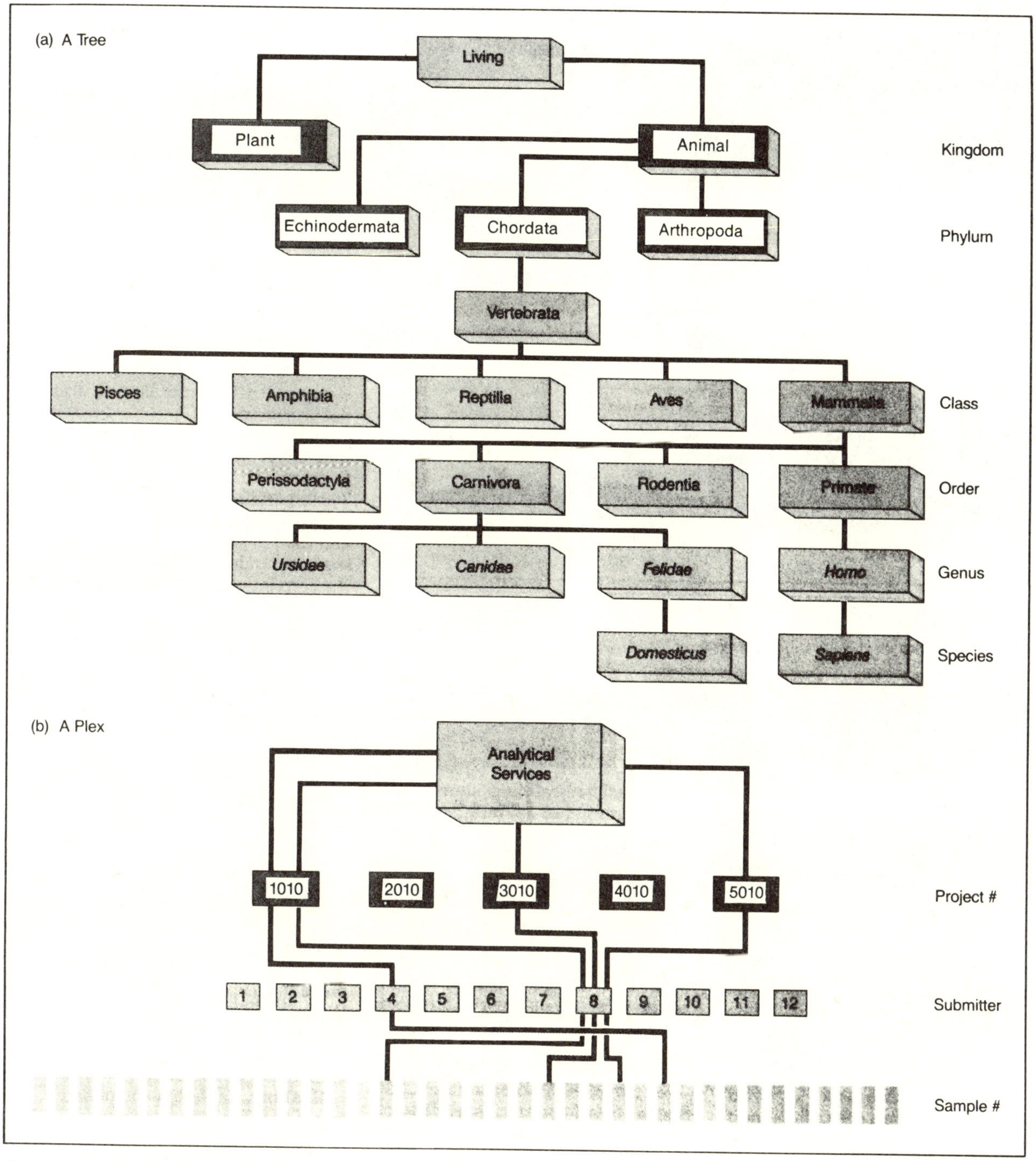

Figure 13. A tree and a plex

index access method involves rewriting only the index as changes are made (Figure 12). The main data files need not be altered. A good analogy is a three-ring notebook to which pages of information are added as required Appropriate additions are made to an index. The main file is unordered. Only the index is ordered.

Inverted List. Multiple indexes might be involved. These might include lists of *submitters*, their associated sample numbers, and location of related data; *project numbers* and their associated sample numbers; *analysts' names* and the samples they have run. We have moved away from storing data in a way that used a key word (sample number) to access a file containing information on project number, analyst, and results, to a method that has separate files for each of these attributes. These separate files for submitters, project numbers, and analysts are associated with lists of sample numbers and disk location where analytical results may be found. A thesaurus is an example of such an approach (Figure 12). Each word has associated with it a list of related words and the location where information on them can be found. This method makes it easy to respond quickly to questions like: What analyses have been run by analyst A on project 5010? What samples have been run for submitter B on project 5010?

This method is called an inverted list or inverted file approach, since it reverses the normal procedures. These systems retrieve information very rapidly, but update slowly. As information is added to the data system, each inverted file must be updated appropriately.

Trees. The Linnean classification is a hierarchical approach to classification. Each level becomes more specific

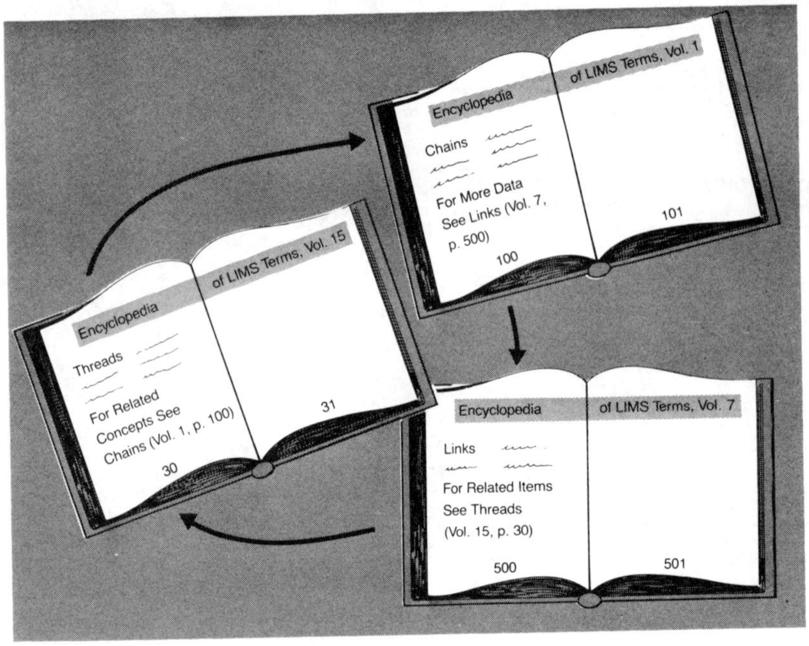

Figure 14. Threaded list

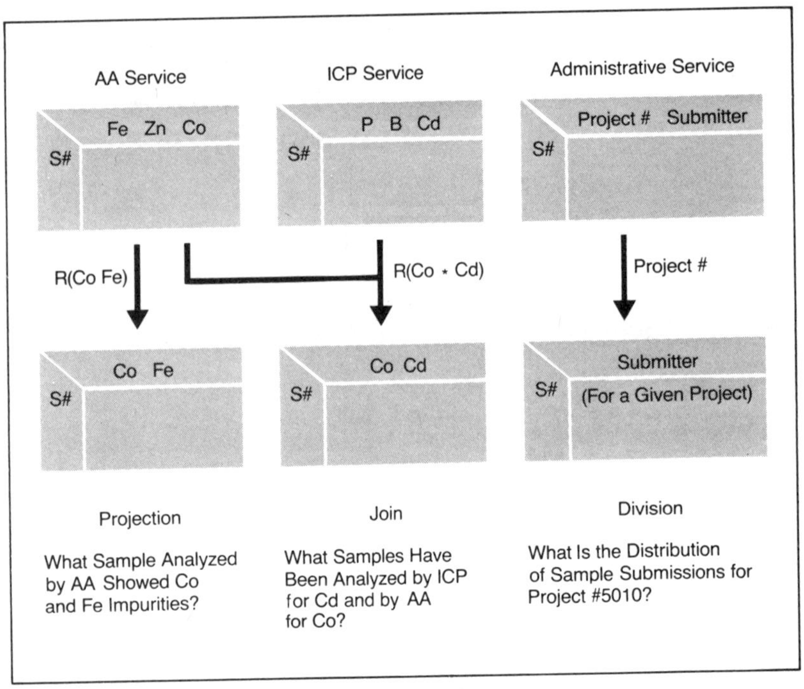

Figure 15. A relational data base

about the nature of the organism being identified (Figure 13a). Consider also a geneological tree where only the male side is presented. Laboratory data can be handled in tree form where there is a parent/offspring relationship between classification elements. It is easy to add things. Search times do tend to become long as the number of levels in the tree grow.

Plex. A genealogical tree in which both mother and father are considered, and in which intermarriage/divorce/remarriage occurs makes the simple tree more complex. The relationship is called a plex (Figure 13b). A laboratory information management system that stores data in a hierarchical approch, i.e., project number, submitter, and sample number, can actually be a plex since each submitter can have many projects as a parent. Handling plex structures requires storing the data in quite a different form, often invoking a complex set of internal pointers to thread the data.

Threaded-List Data Storage. An encyclopedia is a threaded set of data. Embedded in each entry on a subject are pointers to other volumes and subjects that have related data (Figure 14). To find all of the data associated with a given subject requires beginning at one end and following the thread. Threads update quickly but retrieve slowly. Some threads run in multiple dimensions, forward and backward, so searches can be in either direction.

Relational Data Bases. The data structures described up to this point allow the user to ask only those questions of it that were conceived at design time when the data base was structured. Relational data bases are not so constrained. Techniques of relational algebra and calculus have been developed that make it possible to create and use such data bases. These permit the user to imagine the data to be comprised of flat tables, and then to join, project, and otherwise manipulate the tables to produce the correlations or comparisons needed (Figure 15). These new relations, often called VIEWS, did not originally exist; however, they were present as virtual entities. Once abstracted they can be manipulated or stored as real entities. This encourages "what-if?" type questions. The current implementations are for large machines or for use on personal computers. The former speaks to the interest the professional community has in the concept, the latter to the fact that it could be done on laboratory-sized machines with proper pressure from the scientific community. Growth in the laboratory area can be expected. It is possible to take many of the storage methods previously described and make them appear "relational-like" to the user. With such approaches prospective users should investigate the flexibility of the final system. How relational is a relational-like data base?

Hashing. As the amount of data grows, another data management problem develops—managing the indexes to the data. Indexes to data, often called directories, can appear like our telephone directories, i.e., ordered lists partitioned for sequential search (Figure 16). Indexes can also appear as tree-shaped directories. Another approach is like the Chinese telephone directories. Family names are made up of a series of brush strokes. This collection is called an ideogram. The names in the Chinese directory are arranged by the number of brush strokes in the ideogram (Figure 16). It is possible to convert key words like submitter, project, etc. to numbers. This is done by a process called hashing. All alphanumeric expressions that "hash" to the same

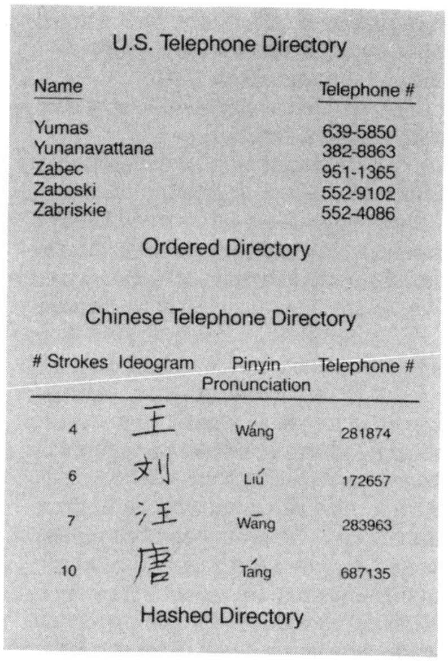

Figure 16. An ordered directory and a hashed directory

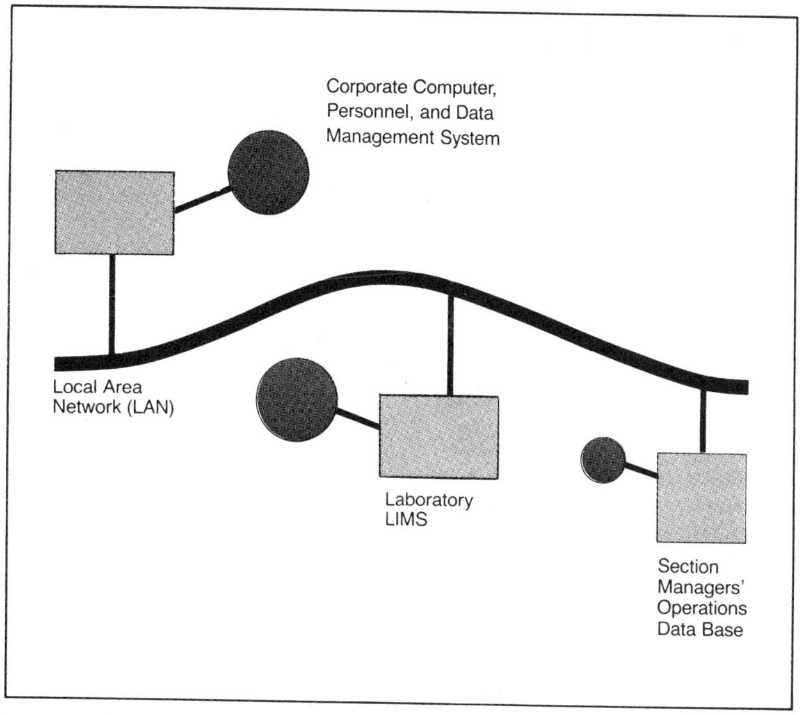

Figure 17. Distributed data management

number are stored in a separate area along with the location on the disk containing the relevant information. Hashing is commonly used to speed up access to information.

Who Supplies LIMS?

A variety of commercial LIMS packages are available from mainframe vendors and systems hours. The view of LIMS presented at the beginning of this article is actually a synthesis of various products currently in use in many analytical laboratories: LIMS-2000 (Perkin-Elmer), System-2000 (Intel), CALS/LBMGR and LBDMS (Computer Inquiries Systems), IMAGE (Hewlett-Packard), Datatrieve and MUMPS (Digital Equipment Corp.), SAM (Radian Corp.), RS/1 with RDE and Sample Tracking Options (Bolt, Beranek and Newman), RDM (Interactive Technologies), and LAB-MAN (Spectrogram). Very recent additions to the LIMS market place are packages from Varian/Digital Equipment, Hewlett-Packard, and Philips. They variously use simple trees, inverted files in conjunction with multiple threaded lists, a tree-structured ISAM approach, and a hashed access to a multikeyed tree structure. Installation costs of these systems range from $50 000 to $500 000, but pay-back periods of 18 months are often encountered.

What Should You Expect and Avoid in a LIMS?

With the fundamentals in place, what are reasonable expectations for a LIMS?

- Some prime rules: No more than a month should be required to become a skilled programmer of the system; no more than a day should be required to learn to use the system; response to moderately complex queries should be a few minutes; response to simple form and status requests should be a few seconds.
- Elaborate, multilayered log-on procedures should not be required.
- Repetitive entry of name, analysis type, and sample ID should not be required. The manufacturing industry now uses bar-codes routinely in production; so should the laboratory.
- Long queue lines for data entry at a few terminals must be avoided.
- Manual data entry should have good HELP functions available on the video screens so that recourse to an operating manual is not necessary. Some systems provide multiple-level HELP commands to provide the novice with detailed examples and experienced users with brief reminders.
- For experienced analysts, alternate data entry pathways should be available to avoid repetitive waiting for menus to be searched for and printed.
- Data should be capable of being selected/deselected for use in statistics or reports. Entire records should be capable of being activated/deactivated. This does not throw data away; it does allow the scientist to use only appropriate data.
- With a LIMS users should be able to *back out* of situations they may have created that are incorrect. The system should *back* itself *off* from contentions that have locked the system, and *backup* (copying) should be periodic and ritual. Required archiving must be automatic. Many of the current LIMS fail in these vital areas.
- Since most users will not know what they want to do with a data system until they begin to use it, the ability to implement change quickly and frequently is a necessity. Nonprocedural, user-cordial languages are available to permit this feature. Currently we face a conservatism and a hesitation to accept this change among computing center personnel and laboratory instrument designers. The language Nomad (National CSSS), running on large machines, has demonstrated the ability to create user-driven software in environments where applications development has been largely without programmers. The classical approach involving a long, detailed specifications study, with early signoff by the prospective user and subsequent implementation by isolated computing center personnel, must be replaced by a more flexible pilot development/change approach. Resistance to these new languages can be overcome with the simple realization that maintenance costs of a piece of software rise with time, as does rewriting the application in a more easily maintained language. The slope of the latter is lower. At the point where the costs of the systems cross, it is better to switch rather than fight.
- As LANs develop and it becomes possible to support distributed data bases, even greater changes in our use of the virtual file cabinet will be seen (Figure 17). Distributed data bases in

the business sector already exist, e.g., Datatrieve. In such networked data bases it is possible to concatenate relational requests made on data in the local computer with other relational requests that involve data on remote machines. Only relevant material is transmitted along the net to the requesting unit. Many managers find the combination of the "what-if?" capabilities of relational data bases and the local security of distributed data management compelling. Intel has announced an iDIS hardware/software system that supports multiple users enjoying full-function relational access to System-2000 operating on a larger computer. Two time-share vendors of production-oriented data management systems already support small microcomputers at the individual user sites (General Electric, Comshare). If true distributed data management develops in this sector, smaller labs may find this approach attractive.

How Do You Implement a LIMS?

With a working vocabulary in hand how does an analytical chemist go about implementing a LIMS?

• *In-house analysis of the problem and user preeducation.* Time spent analyzing the informational and psychological needs of users should focus on the normal flow of laboratory information. Laboratory managers' views of detailed laboratory operation may be oversimplified or erroneous. The technicians and section managers should be heavily consulted. The time to co-opt user support is at this point, making the user part of the decision-making process, or at least aware of the factors involved. This intimate contact with the user should continue during the entire development phase. Ego-less decisions and solutions that do not utilize the ability of LIMS to be centers of power structures are facets of successful implementations.

• *Turnkey vs. customized.* Unless you are very lucky, or you have a very simple operations information system in mind, some customization of an existing system will be necessary.

• *In-house or systems house?* Packages are available that provide in-house programmers with a set of powerful tools that allow development of applications programs in a reasonable time frame. It is possible to employ a systems house to customize software for you, developing it in-house or delivering and installing the system after generation elsewhere. Finally, computer vendors will often customize software for you. In any case, it is essential that some of your personnel develop expertise in any installed system so that your organization is capable of extending and supporting the system. As part of any purchase agreement insist on source code for sections of the system that you have any thought of altering. Sign nondisclosure agreements if necessary. Where work is performed by outside personnel, it should be made clear initially what freedom the vendor has to use the developed software elsewhere.

• *Choosing hardware.* Although the ideal installation involves software that would run on any reasonable machine, most LIMS will run on only one type of computer. Datatrieve, MUMPS, RS/1 (RDE/STO), RDM, and LABMAN run on Digital Equipment Corporation hardware. LIMS-2000 runs on Perkin-Elmer Interdata equipment, IMAGE, CALS/LBMGR, and LBDMS on Hewlett-Packard systems, and SAM on Data General computers. Documented assurances that new generations of hardware will not engender expensive software relicensing costs are essential. As the data base grows, it may be necessary to install not only more disks but different types of disks, terminals, printers, or other peripherals. The better information management systems are hardware-independent at this level. Another important feature to evaluate is field service support.

• *Compatibility.* LIMS that run under standard operating systems provide many software development tools that are essential if additions and modifications are implemented. Data files created can be accessed by programs written in other languages, and the files can be exchanged between computers relatively easily. For example, LABMAN and RDM run under DEC's RT-11 or RSX-11M operating systems. Some systems houses provide software front ends or handshakes between traditional data management systems from computer vendors and their own LIMS. For example, Bolt, Baranek and Newman is developing access methods to Datatrieve for RS/1. Where users will be responsible for applications software development the drop-off between the powerful data management command language and the next level of programming language should be small. Thus, although RS/1 is written in C to take advantage of the full power of the DEC VAX computer, the user sees either the high-level command language or a cordial RPL (research programming language). Many scientists would find a direct, precipitous drop to languages like FORTRAN, C, or APL shattering.

• *Networking.* Networking software should support communication from the LIMS host to larger computers. The LIMS should also support satellite computers for real-time data collection or as ancillary data entry points. Terminals at any level should have transparent access to computational facilities above them.

• *System benchmarking.* Comparison shopping is essential. Actually working with a system already installed at another location or, less satisfactory, interacting with a demonstration system at a vendor sales center is essential. Benchmarking is difficult at best. Try to compare systems with similar CPU capabilities, disk size, data base size, and number of active users.

• *Cost/benefit analysis and planning.* In preparing cost/benefit analyses it is seldom advisable to use arguments based on reducing staff. Sample throughput, the quality of data produced, and the ability to abstract information previously lost in manual manipulations are cogent arguments. More efficient use of instruments and analysts is achievable. Realistic PERT charts prevent prospective users and management from losing their faith in the project.

• *Start-up.* Credibility is most easily lost during the start-up phase. Good documentation and training of initial users characterize successful installations. These implementations have been done in stages so the problems that develop are localized, small, and quickly fixed. Manual entry services usually precede the more difficult real-time data capture services. Parallel manual backup is suggested during any initial phase. Eugene Schneider of Ralston Purina, a pioneer in the area, suggests "the project team must be highly visible and responsive to the requests and needs of the lab personnel during this time... then ... be prepared for requests for new and increased functions for the LIMS.... It is a natural outgrowth of the acceptance of the system by its users."

The area is just maturing. Vendor interest is high. The informed chemist has an opportunity and a responsibility to affect the market place by articulate input to prospective vendors. This tutorial was intended to give the analytical chemist a vocabulary and philosophy. The next A/C INTERFACE will contain a series of capsule reports describing how various chemists have successfully implemented a laboratory information management system.

Acknowledgment

Many of the philosophies developed in this article are the result of a series of discussions with David Hooley of the Standard Oil of Ohio Research Laboratories, Cleveland, Ohio. Szu-Lin Chen, a graduate student in the VPI & SU Chemistry Department, provided the information on Chinese telephone directories and the ideograms.

Part 2

Last month's tutorial (**1983,** *55,* 70–80 A) developed the vocabulary of laboratory information management systems (LIMS). The following capsule reports explore implementations of LIMS by eight different industrial firms, each unique in its problems and solutions. The prospective LIMS implementor may draw from each example those aspects applicable to his/her environment. The gamut of approaches spans resources from big mainframes to personal computers. The staff and time required for implementation range from tens of programmers and 25 man-years of internal effort to a single person purchasing hardware and software from external sources. Software has been developed in-house, installed by systems houses, or purchased from instrument vendors. Languages like APL, BASIC, Fortran, COBOL, and C, running under standard operating systems or turnkey stand-alone specialized systems are used. The next series of A/C INTERFACE articles will therefore address the subject of languages, operating systems, time, and human cordiality.

Ralston Purina Company
Checkerboard Square
St. Louis, Mo. 63164

Contributor: E. L. Schneider

The R&D Information Systems Group of Ralston Purina provides computer-based services in support of various activities at research facilities located in St. Louis as well as at branch laboratories. A total of over 300 professional and support personnel are involved. Together, these labs perform over 450 000 individual chemical analyses yearly across a spectrum of some 600 different procedures. In addition, they conduct biological research projects that result in a considerable amount of data.

The handling of the large amount of data arising from these activities is accomplished on a real-time interactive computer system. It consists of three software subsystems utilizing a common data base. The basic LIMS allows the user to enter, store, manipulate, retrieve, and/or print data about samples or sample loads in the laboratory as well as schedule new work. DACS (data acquisition control system) supports data acquisition from a variety of laboratory instruments. EAS (experiment analysis system) provides comprehensive statistical analysis and special report generation capabilities. The data base provides a single point of storage and allows any subsystem access to data generated by itself or any of the other modules.

The LIMS provides work scheduling for the bench chemist, and ensures that high-priority samples get preferential treatment. When an analysis has been completed, the chemist can report the results from his laboratory terminal. In addition to the data, the chemist must supply a disposition code that specifies what is to be done with the data and the sample. The sample may be declared as complete for a given assay or scheduled for a rerun on the next worksheet request. Optionally, the system can be made to override the chemist and reschedule a sample for analysis when the chemist attempts to report a single out-of-limit result for a particular assay.

Printed reports for all completed samples are generated each evening. This is done through the use of a program that scans the data base for newly completed samples after the close of operations. Reports suitable for direct mailing to the client are created with multiple and/or individually addressed copies produced when more than one user is involved. In addition to the data entered during the laboratory operations, other data can be appended to the report by the computer. Such items might include method references for the various assays performed, disclaimers, and/or signature blocks. In cases where long-term archiving of analytical data is desired or where hard-copy reports are not immediately necessary, sample data can be transferred to magnetic tape.

With information on all of the assays available in the data base, it becomes feasible to automatically generate management reports. Each of the laboratories is divided into one or more groups. Each group in turn is divided into from one to four sections. At the section level, management is interested in detailed information on workloads for the section. At the group level, production data, potential bottlenecks, and service times become important. The laboratory director is concerned about all of these parameters, as well as financial matters. Such specifically tailored reports are generated each day.

DACS supports a wide variety of laboratory instruments including chromatographs (LC and GC), Technicon AutoAnalyzers, electronic balances, and spectrophotometers. Instruments are connected to the computer through a specially designed interface that allows the chemist full control over the acquisition and storage procedure.

Chromatographic units are supported in an "on-the-fly" mode in which peak detection is performed as points are acquired, with no long-term storage of the chromatogram. This mode is most useful for routine analyses in which the processing parameters are well-known. For research or special applications, digitized chromatographic data can be stored and subsequently processed in the background. This mode allows the chemist to vary the processing parameters while analyzing the chromatogram and thus arrive at the proper algorithm for treatment of the data. In either case, detected peaks are transmitted to the data base for future use.

Data from electronic balances and other instruments that present single observations are sent directly to the data base.

With the proliferation of "smart" instruments having an embedded computer, an extension to DACS intercepts the normal printed output from the instrument. Thus, data from these instruments become available to DACS for processing and storage.

EAS utilizes the same type of video display terminal employed by LIMS but presents a different menu of programs to the user. Software consists of both Ralston-written and purchased packages. Because of the long-term storage requirements for some data elements, a separate section of the data base is maintained for use by EAS. Input can be from a terminal or from the LIMS-DACS data base. An almost endless variety of analysis schemes and report formats is available through the EAS system.

The total system described above did not emerge as an entity, but has evolved continuously since the start of the project in 1968. Six man-months alone were spent defining the data flow in the laboratory. In view of the size of the effort required to bring up all three systems, the implementation report recommended that the LIMS be done first, followed by the establishment of the DACS and EAS facilities only when the first phase was truly operational.

The current configuration is shown in Figure 1. A staff of 10 in St. Louis handles all of the day-to-day operations, applications programming, maintenance, and new system development. To date, 25 man-years have gone into system development. Estimates indicate that the system saves 15 to 20 man-years of effort per year.

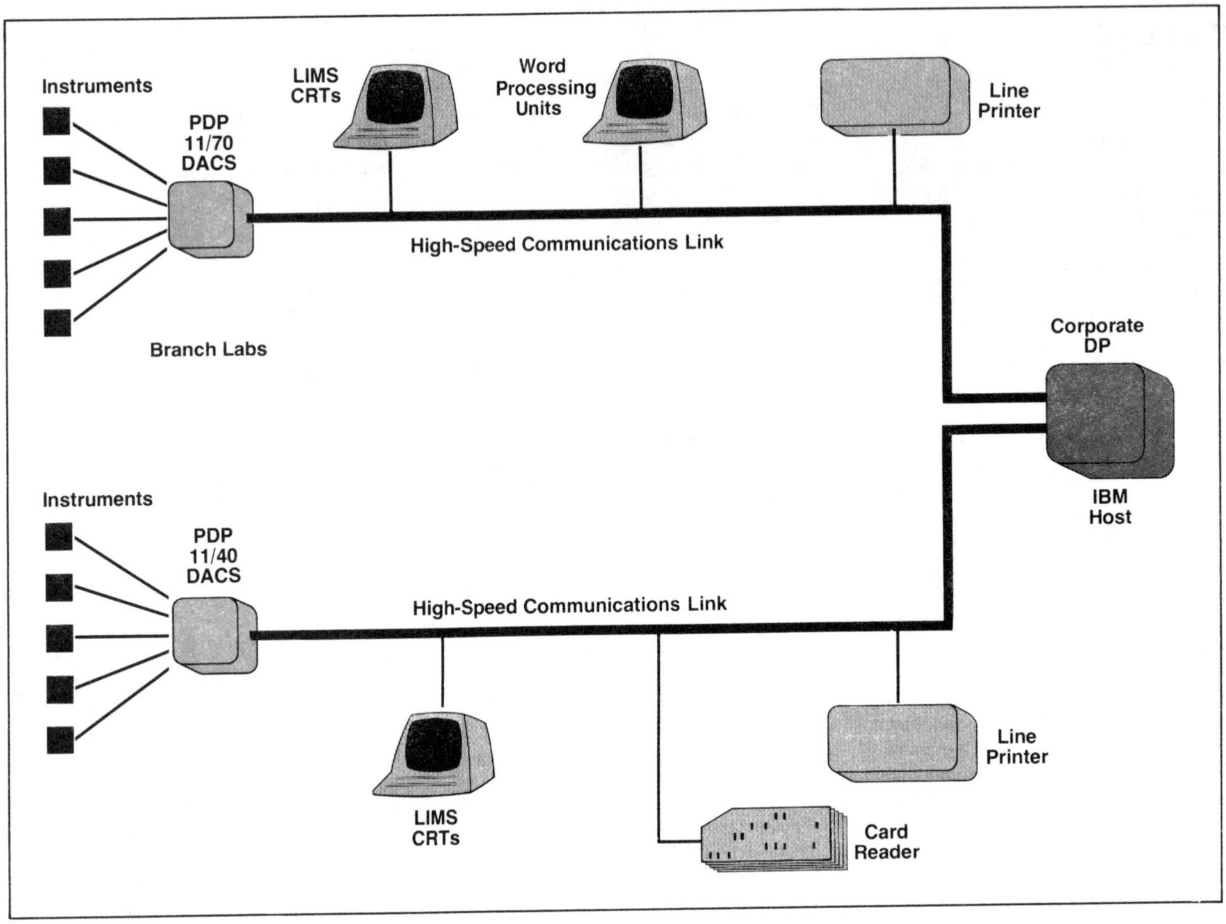

Figure 1. Ralston-Purina LIMS hardware

Mobil Research and Development Corporation
Paulsboro Laboratory
Paulsboro, N.J. 08066

Contributor: R. J. Kobrin

Fifteen years ago a review of the operation of the analytical section at Mobil's Paulsboro Laboratory resulted in a decision to use computers both as a management and an automation tool. Since the technology was not readily available commercially, the program relied heavily on developing in-house expertise to implement the system.

The automation/computerization has evolved through four distinct phases. Because excessive analysis time inhibited research, the laboratory was automated to improve throughput and reduce costs. This automation provided so much additional data that many researchers were unable to use the data effectively. The second phase was use of APL by researchers to facilitate assimilation of the data. As a result they began to demand faster, more reliable data transfer, resulting in the third phase, networking (Figure 2). The fourth phase, now being implemented, is to integrate word processing into the network to improve speed and accuracy of report preparation.

The analytical section provides analytical service to a research and development laboratory with nearly 1000 employees. A diverse repertoire of more than 800 tests is available. About 60 000 samples are submitted each year, and more than 200 000 tests are performed on them.

The system to improve control of the analytical laboratory began with a laboratory status report software package (LABSTAT) that mirrored the functions of existing tab cards associated with each sample. LABSTAT has evolved from a relatively small Fortran program that tabulated card decks to itemize test requests/completions to a large APL package that provides test result reports, status reports, cost accounting, and data communications to researchers. It still uses off-line data entry on tab cards for much of its input, although a fully on-line system is planned for the near future. An update is performed every morning to prepare status reports for distribution before work begins.

This daily LABSTAT report, which analytical managers and group leaders receive, has two main parts. The first consists of a listing of the work of each group, itemized for each test the group performs. Each test entry lists the number of requests and completions recorded for the previous day and the number of samples presently in backlog. A very important feature is that both the number of tests and the workload in man-days are listed for each item, based on a standard analysis time for each test. The second part of the report lists manpower availability—vacation, sick time, special assignments. This can be compared with the total required man-days of work in the group to flag the need for overtime work or staff reassignments.

Another report received by group leaders lists backlog samples, sorted by submitter's name. This report provides more information about specific samples and indicates where work is coming from. Sudden surges in requests from certain individuals cannot be allowed to adversely affect response owed to others. LABSTAT thus provides a means of finding problem areas as well as a basis for negotiating work schedules with major users of analytical services.

The automation program increased the volume of data available to researchers to the point where many were unable to effectively assimilate their results. Some individuals were receiving as many as 1500 analysis reports per month! The solution to this problem was found when a process research group at the Paulsboro Laboratory began using APL, a programming language uniquely suited to the data reduction problems encountered in a research environment. With that group providing training and support, APL use has expanded to displace virtually all other computer languages in our laboratory.

Researchers in this laboratory generally set up individualized data bases that are interrogated interactively with APL. For example, an engineer can extract an array of data from a file by typing a few lines of APL. The in-house APL graphics package is then invoked in a single-line command to scale, identify, and plot results against temperature or any other process variable. Various curve-fitting functions are available on-line, and can easily be experimented with until a suitable graph is prepared.

Once the analytical laboratory was automated and researchers began using APL, the time required to distribute results and enter them into the user's data base became a significant part of the research process. Data communications is an obvious step to connect the automated researcher with the automated analytical laboratory. The most direct way to do this is to use a single computer for laboratory automation, data-base operations, and computations. A more recent concept is to distribute these functions in a network of interconnected computers. Unfortunately, there are a number of problems with such approaches in some laboratories:

- Many laboratories already have a variety of data systems.
- Users may require data on a specific host computer.
- Few laboratory systems have communications capability.
- Most commercial systems are impossible to modify.

The solution to these problems interposes an intelligent data concentrator between the data systems and the host. The analytical data concentrator (ANDAC) system has the following characteristics:

- It requires no modifications to the satellite data system.
- It requires no modifications to the host computer.
- It provides local data storage during host downtime.
- It requires analysts to approve reports before transmission.

These features are achieved by placing essentially all system intelligence in ANDAC. ANDAC is designed to capture the data being sent to it by a satellite data system, to scan for features that indicate report start and end, and to extract pertinent information to be stored for subsequent transmission. Furthermore, it is capable of emulating a terminal on the host system, and logging in and running programs necessary to transfer, check, and store the data.

A single LISTENER task receives data records from all data systems and forwards each to a PROCESSOR task dedicated to each instrument. A PROCESSOR task is a small program that extracts specific data items from incoming reports and outputs them to a disk file in a standardized format for background processing. A number of utility programs are provided to handle the resulting data files. Analysts

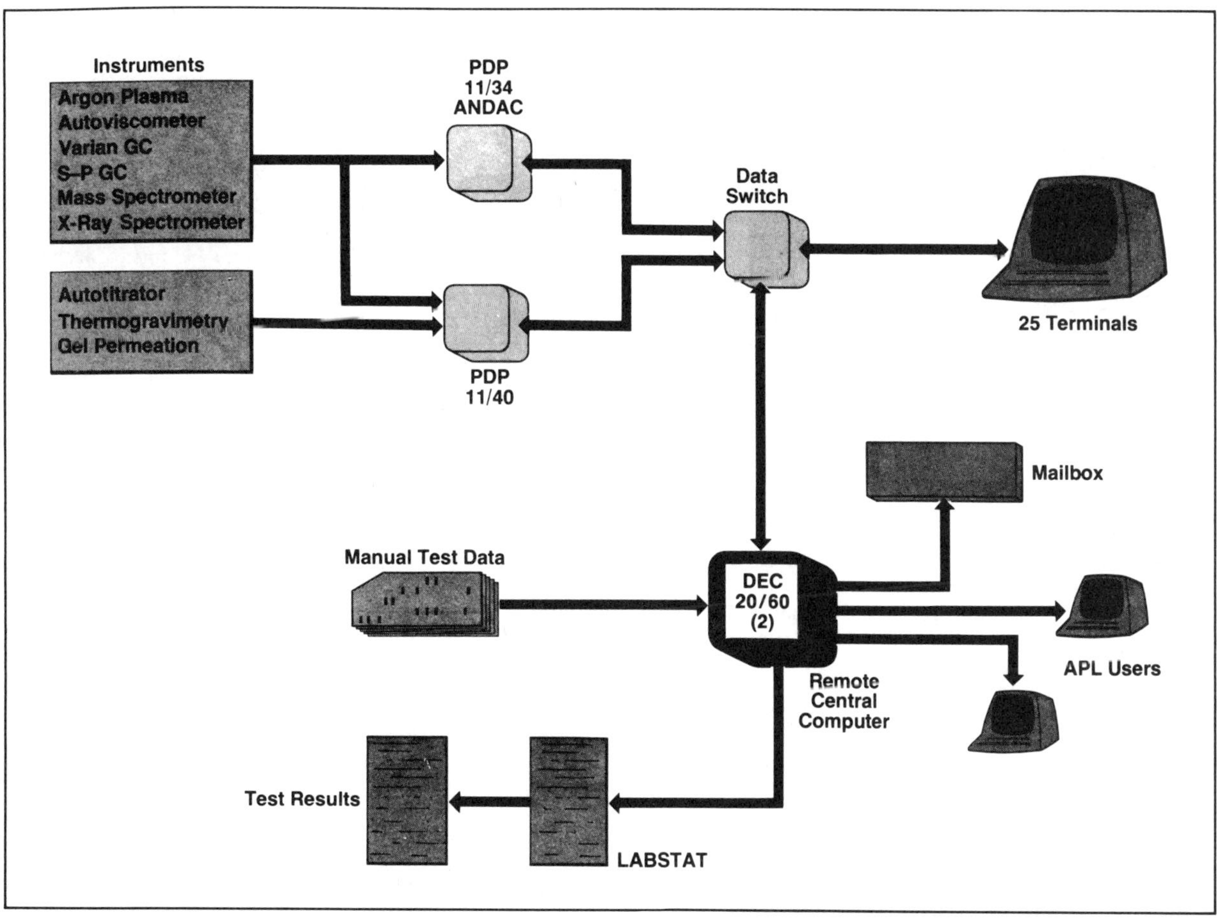

Figure 2. Mobil LIMS hardware

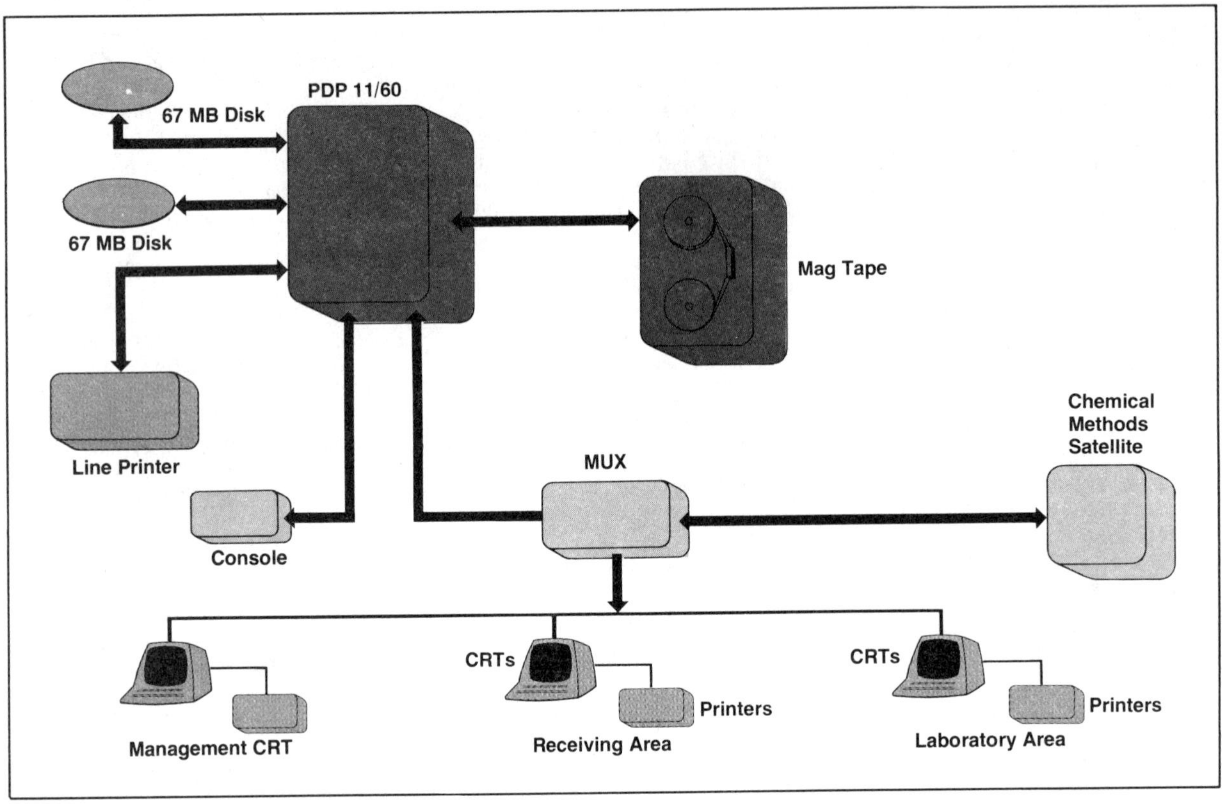

Figure 3. Phillips Petroleum LIMS hardware

use APPROVE to examine reports and declare each one as either approved for transmission or rejected. Approved reports will be transmitted by an automatically scheduled SEND task. Old reports are stored for about 3 days to allow retransmission of data if necessary, then are automatically removed by DELETE. Another task, KEY, is used for manual data.

MAILBOX is the output interface between the analytical laboratory and its users. It is an APL workspace containing functions necessary to access a large indexed file of analysis reports. This file contains all results reported by the analytical laboratory during the past 25 working days. The MAILBOX workspace includes tables and functions provided to extract an individual's results, translate codes, and copy the data into his own files. This arrangement decouples use of the data from its preparation. Application software, file structure, and maintenance are the user's responsibility. The nature of APL makes this process remarkably simple.

The Analytical Section is itself a user of APL and MAILBOX for many applications. One example is monitoring the precision of results. This vital program involves selection and resubmission of samples for analysis at least once a week for 81 different tests. Samples and standards are entered under new identification numbers, results are gathered and plotted, and discrepancies are investigated. A quality-assurance program of this size would be impractical without computer assistance; the clerical time required with the present system is about one man-day per week.

This automation and modernization program has increased productivity by about 64% since 1969. The equivalent of 49 technicians (allowing for overtime) perform routine analyses, an absolute reduction of 15 from our 1969 staff. With the observed 25% increase in workload since 1969, a staff of 80 technicians would be required had the program not been implemented.

Phillips Petroleum Company
Research and Development
Bartlesville, Okla. 74004

Contributors: P. R. Gray and T. V. Iorns

The Analysis Branch of Phillips Petroleum Company provides analytical services to the Research and Development Department and other parts of the company on request. In 1981, a staff of 89 persons performed over 200 000 tests on nearly 100 000 samples. The several hundred different test procedures performed range from routine ASTM methods, titrations, and simple chromatography to complex structure determinations and characterizations of alternate fuels, such as shale oils and coal liquids.

The benefits of laboratory automation have been recognized for many years. Presently 46 computers and many automated instruments are in use. Since 1976, productivity savings of several million dollars annually have been realized through automation. Data are also more reliable (i.e., accurate and precise), because automation facilitates the routine analysis of more standards and quality assurance samples.

A few years ago it became obvious that progress was being made toward automation of laboratory activities but that there was also an opportunity to achieve significant savings in the areas of management information and information transfer. Supervisors, analysts, and clerks were hindered by the daily handling of paperwork; any automation of these activities would yield large productivity gains.

In 1979, the AB/LMS (Analysis Branch/Laboratory Management System) was initiated. The first step was to identify the basic features needed. It was also recognized that the project was large and should be completed in stages so that some benefits could be obtained quickly and control of costs could be assured.

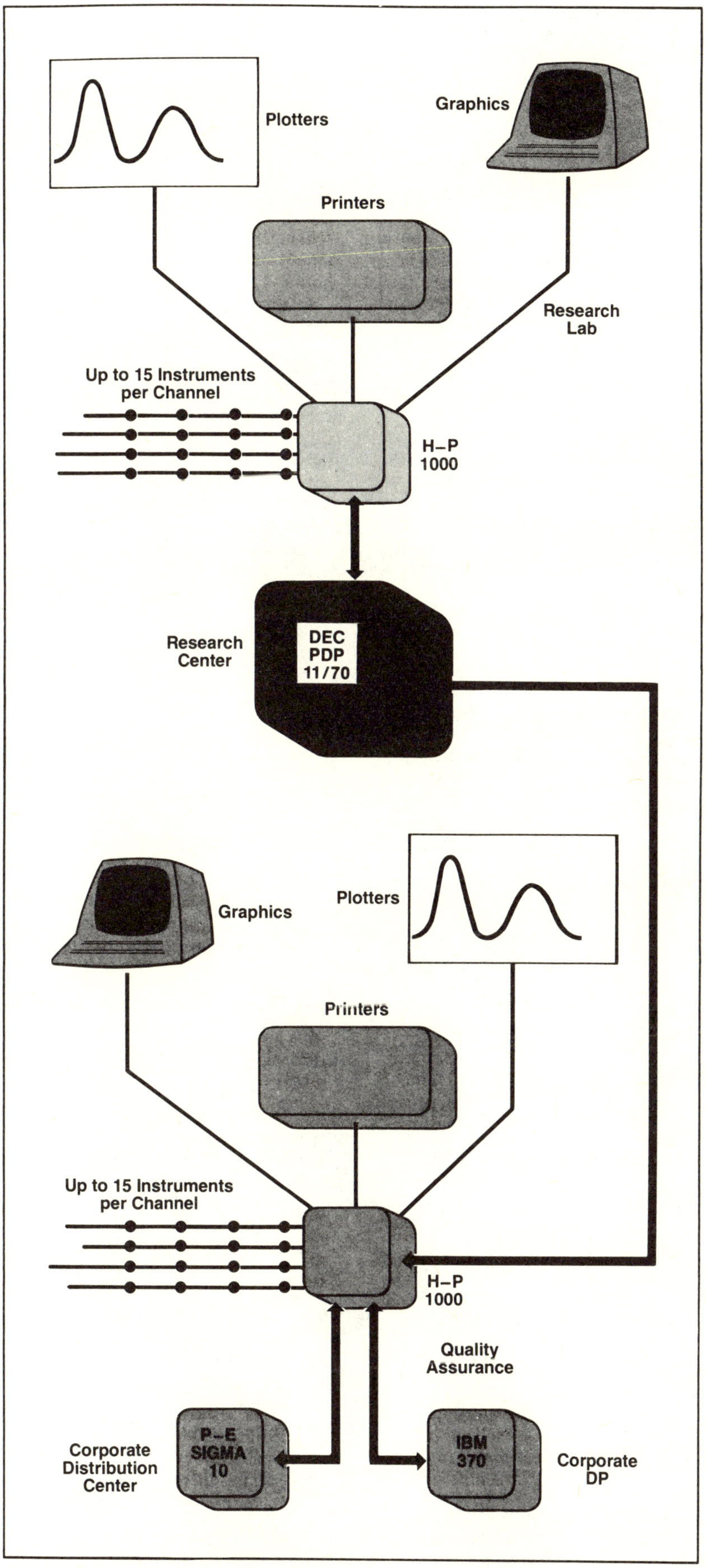

Figure 4. A. H. Robins LIMS hardware

Although the Analysis Branch had considerable expertise in laboratory automation, no one had experience with LIM systems. The decision was made to contact Purvis Systems, Inc., a software systems contractor.

After several meetings, a preliminary specification was written. This was followed by a more detailed document outlining specific requirements for data, reporting, and hardware. At this point it was recognized that insufficient internal manpower was available to complete the system in a reasonable period of time and a contract was given to Purvis to write the necessary software.

The computer operating system and programming software were selected on the basis of programming ease, functionality, and ease of program modification. Since the system was needed for lab management information and not detailed calculations, a simple data-base system, Digital Standard MUMPS (DSM-11), was chosen.

It was anticipated that a DEC 11/60 would satisfy corporate needs for about 5 years. Therefore, this computer, a line printer, three 67-Mbyte disk drives, a magnetic tape unit, bar-code readers and printers, 36 video terminals, and 32 tabletop printers were purchased (Figure 3).

After about a year of software development, the entire system was installed in April 1981. A few months of debugging, experimentation with limited data sets, and operation in tandem with our manual system demonstrated that the system met specifications, and in September 1981, conversion to the new system was completed. The system has been in operation since then with only minor problems. Significant additions to the original software have been made. The full potential and capabilities were not realized until the system had been put into use.

In the sample receiving area, all the information supplied by the customer is entered into the computer. This includes basic information such as the customer's name, address, phone number, and charge number. Sample information includes a sample number, description, physical form, priority requests, and hazardous handling instructions. The tests requested are entered and the computer utilizes a matrix of routing parameters to assign the tests to various work groups. For example, a request for sulfur analysis may be sent to any of several locations depending on the physical form of the sample, the estimate of the sulfur content, other tests requested, or identity of the customer or project. Manual routing is also possible. The computer generates a sample receipt for the customer detailing sample information,

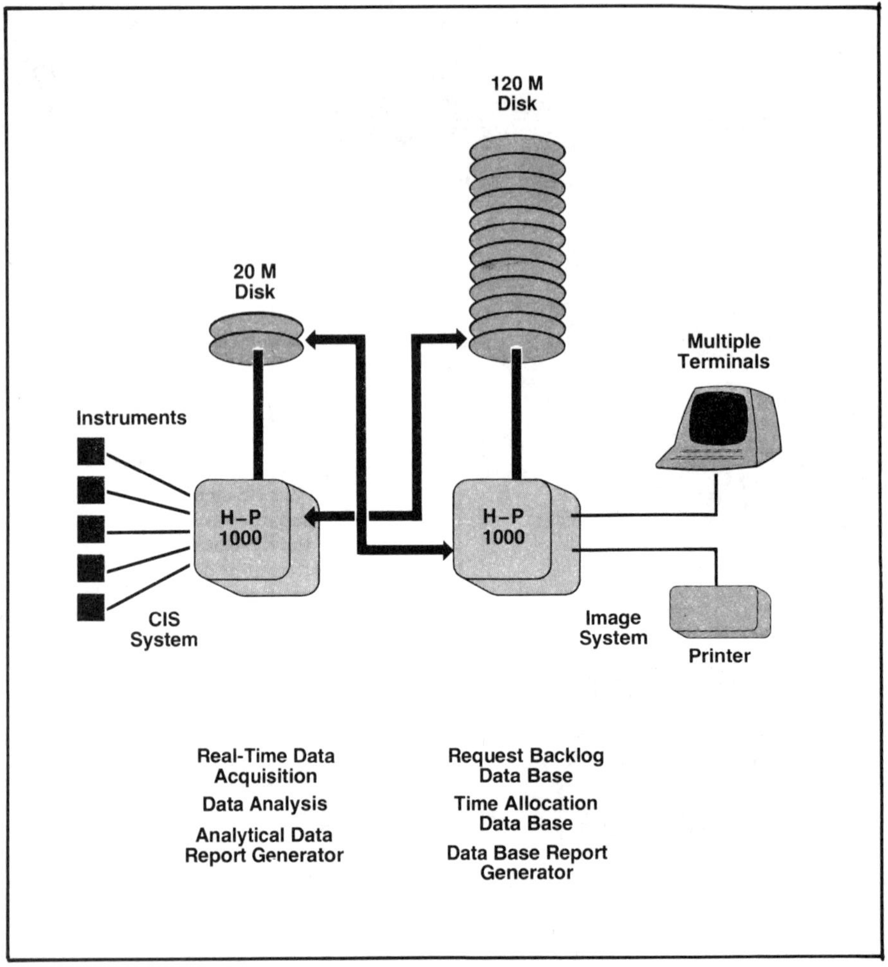

Figure 5. Dart & Kraft LIMS hardware

tests, and priorities, as well as the cost of the analyses and estimates of the completion data. Such receipt generation has reduced significantly the number of nonessential requests.

In the laboratory areas, individual queue lists contain information about samples assigned to that area for specific tests. Analysts can select priority samples, the oldest samples, or samples of similar types to analyze. The analyst can verify the location and availability of each sample and can request delivery of the sample from the central receiving area. In the near future, the computer, using analysts' criteria, will assist in sample selection during the night, and samples will be routinely delivered to the analytical laboratories before normal working hours.

Data entry into the system is accomplished in the sample receiving area using routines that allow the operator to work interactively. All the information in the system is accessible through any of the numerous video display terminals on the system. If a hard copy of the data is required, this can be obtained immediately at one of the attached tabletop printers.

Managers, supervisors, and group leaders no longer have to spend hours compiling reports on backlogs, productivity, and other management information. These reports are prepared by the computer routinely. Typical reports include a daily backlog for the group leaders, weekly backlogs for the supervisors, summary backlogs for the managers, monthly activity reports for the supervisors and manager, accounting reports for managers of other areas, and sample turnaround times associated with each test and customer, allowing supervisors to reassign analysts to optimize service. Custom reports can also be prepared. Reports frequently needed in the laboratories are generated automatically at tabletop printers during the evening.

The installation of the system required three persons, including one from Purvis. Since that time, two persons have responsibility for the system, one working on maintenance and minor problems, the other primarily occupied with major changes and enhancements. In addition, one extra clerk is required to enter data in the sample receiving area.

Our estimates of a 10% productivity increase in analytical services have been surpassed and are expected to grow even further as more of the system's potential is realized.

Looking ahead, the next challenge is in the area of information transfer. A customized local area network, ETHERNET, to connect computers and work stations into a comprehensive system is being developed by Purvis and Phillips and will be installed soon. This will allow more efficient transfer of information, preparation of reports, graphical interpretations, and archival storage of data. In the future, connection of this local area network into the Phillips R&D network will allow transmission of reports directly to customers via electronic mail, and the chemists and engineers will be able to interrogate our computers to determine the status of their samples.

A. H. Robins Company
Scientific Information Department
Richmond, Va. 26609

Contributor: R. J. Barbie

In September of 1979 a task force was formed at Robins to investigate available alternatives for laboratory automation. Task force representatives were from Quality Assurance, Drug Metabolism, Analytical Research, the Corporate Computer Center, and Scientific Information.

The task force formulated the following system requirements:

- Sample management functions, such as sample initialization, scheduling, and reporting, must be present.
- Hardware and software must support data acquisition from varied laboratory instruments in real time.
- Software must permit both standard and customized reports to be produced for laboratory personnel. This includes graphic capabilities.
- The system must archive all instrument-produced data. Audit trail and access control features are required.
- The system must be expandable to support additional instruments.
- The system must have the ability to transfer data within a network.

Among the vendors responding to these needs was Computer Inquiry Systems, Inc. Their Computer Automation Laboratory System (CALS) with Lab Manager System (LBMGR) came remarkably close to meeting the above requirements.

CALS consists of a Hewlett-Packard 1000 minicomputer with instrument interface equipment and computer terminals placed in the laboratory areas (Figure 4). For instruments having an analog output (gas chromatographs, spectrophotometers, etc.), the software provides for peak area integration, automatic baseline recognition, detection of shoulders, and tangent peak skimming. Relative response factors, area normalization, standard sample, and internal standard modes of calculation are implemented. Plotting of instrument data is supported, and instrument runs may be overlaid for comparison. CALS can also handle single reading instruments, such as nonscanning spectrophotometers and analytical balances. Any instrument having a digital display or a printer-type output can have its data transmitted to CALS for processing.

All subsequent processing of the data is via the terminal in the laboratory area. Files can be accessed by standard CALS software and the data can be plotted and/or reported. The chemist can also construct programs (in BASIC or Fortran) to access, calculate, and report data in a manner of his or her choosing.

Results of tests run on instruments interfaced to CALS are automatically posted to LBMGR. The LBMGR package is a comprehensive set of programs that allows each installation to customize the system based on the particulars of the environment. This is accomplished by storing customer parameters in what are called dictionaries. The ID dictionary identifies the users of the system and the functions that they are allowed to perform. The TEST dictionary defines each laboratory procedure or test. The SPECification dictionary defines which tests are to be performed on each product. A CALCulation dictionary can be used to define routine calculations associated with a particular test procedure.

Using the information in the dictionaries, the system users can initialize samples, schedule tests, and report sample status or results on request. Completed sample results are stored in the LBMGR data base for long-term reports for management. All these facilities are available to the chemist on interactive terminals in each lab.

The Standard Oil Company
Research and Development Department
Cleveland, Ohio 44128

Correspondent: M. K. Snavely

Dart and Kraft, Inc.
Analytical and Testing Department
Paramus, N.J. 07652

Correspondent: F. Scardiglia

Aluminum Company of America
Alcoa Technical Center
Alcoa Center, Pa. 15069

Correspondent: R. S. Danchik

3M
Central Research Laboratories
St. Paul, Minn. 55144

Correspondent: W. L. Stebbings

Four other quite different approaches to a LIMS installation will be of interest to the reader attempting to synthesize a LIMS solution appropriate to his or her own environment.

The Standard Oil Company of Ohio has chosen to implement a Sample Logging and Reporting System in its analytical laboratory. Incoming samples are "bound" to an ID number. User-cordial MAINTENANCE, COMPLETION, and REPORTING functions are available to the analyst. Users can access only those functions of the menu-driven system for which they have authorization. As tests are completed, data and/or status information can be entered by first specifying a test number or a mnemonic equivalent to the test (VISBRO equals Brookfield viscosimeter). Data can be accepted as free-form textual material, by prompted fixed-form entry, and by use of headers created at sample log-in time.

When a sample is completed and released, INQUIRY functions permit the user to fill in known information on a preformated video form. After the form has been completed, a custom program searches the data base and displays information about one or more samples and related data. With the QUERY/REPORT program the user asks relational questions like "Show me all of the samples with priority of ONE that are due next month." This mode services the ad hoc type of questions that occur on a day-to-day basis. Good HELP and TRAIN features are easily accessible to new or infrequent users.

The present system is capable of handling some 3600 samples/month (12 000 tests), maintaining detailed sample information for 1 year on-line, and statistical information for 5 years. The system is constructed around DEC's Datatrieve. Sytek, Inc. (Ohio) has developed for Sohio front end as well as stand alone modules for data maintenance and data reporting. These provide for normalization of the data and the ability to access information in a relational manner. The relational JOIN operation allows access to specific information on a sample. DIVISION operations can yield specific subsets of status information. These VIEWS of data are virtual records. They did not exist in the original structuring but have been created by the relational calculus invoked by the access commands present in relational-like systems. The LIMS tutorial described how much pseudo-records can be used in distributed data bases to exchange relevant information between computers connected by a local area network. Sohio is implementing such technology.

Yet another approach is used at Dart and Kraft. Dual Hewlett-Packard processors are employed (Figure 5). One runs Computer Inquiry Systems' real-time data acquisition software described in the A. H. Robins example. The Dart real-time system has

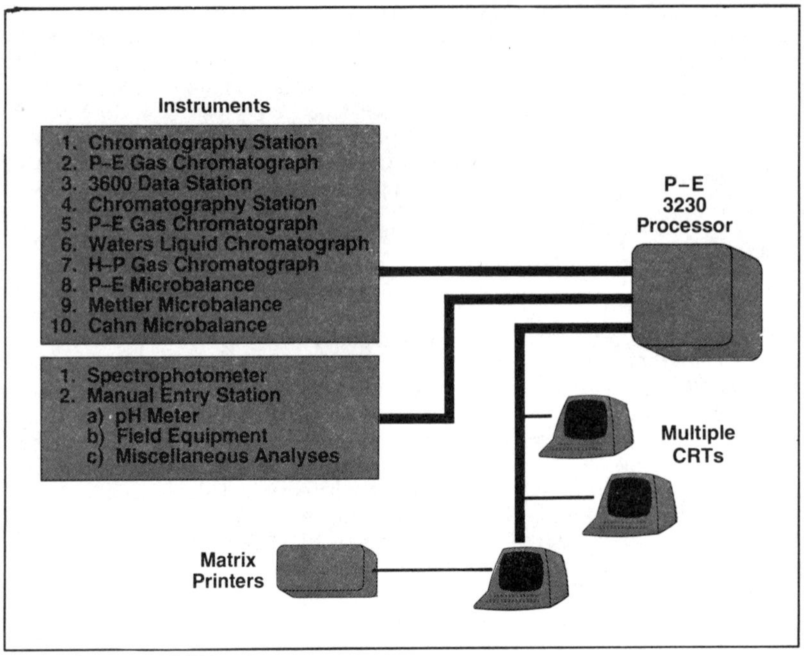

Figure 6. Alcoa LIMS hardware

the standard capabilities of peak area integration, baseline recognition/correction, and peak area normalization. Special application programs for size exclusion chromatography, high-speed impact testing, hydrostatic pipe testing, rheology, X-ray diffraction, and differential scanning calorimetry have been written by outside software houses or in-house personnel and incorporated into the CIS system. Twenty-five different instruments are currently being handled. The captured data are stored on dual-ported disks and are used to generate analytical reports. The second disk port is used by a second H–P Computer System, which runs H–P's IMAGE. This software package allows the user to correlate status information on the analytical work load. No experimental data is retained on this part of the system. Ready access to the data can be obtained by identifying the analytical work sheet request number involved. In this way the data base is kept relatively small and quick to access. With this system it has been possible to accommodate 11 000 requests per year, comprising 25 000 samples and 60 000 individual tests.

At Alcoa's Environmental Health Laboratory, a centralized analytical laboratory is responsible for development of sampling and analytical methodologies, analysis of samples, and maintenance of industrial hygiene data records. Over 3200 samples (10 000 determinations) are involved each year. The services provided include GC/LC, selective-ion electrodes, microscopy, atomic absorption, UV/VIS spectroscopy, ICP, and X-ray diffraction.

A LIMS was needed that provided two separate data bases coresident on one disk; one for management (operations) information, the other for scientific data. Good Methods-Files had to be supported. The stand-alone computer facility needed to be capable of future expansion and allow the development of totally new application programs in BASIC, Fortran, PASCAL, and COBOL.

This end has been met using Perkin-Elmer's LIMS/2000 software package running on an Interdata 3230 computer (Figure 6). The various instrument hardware and software interfaces (Fortran VII) were provided by the manufacturer.

Finally, at the 3M Central Research Laboratories an unconventional approach is being implemented. Approximately 25 Apple personal computers are being networked together using the Corvus Omninet. This is a CSMA protocol, using RS-422 twisted-pair wires as the communication medium. The network allows the individual microcomputers to interact with pseudo-disk areas on a large Winchester disk attached to the network. Backup is provided by a video tape cassette.

Each analyst will be able to enter data and status information at his own computer terminal. Each Apple computer has its own flexible disks to allow local storage in case of network failure. Since each microcomputer can function in a stand-alone environment, the analyst will have access to local word processing, math packages, etc., all of which can be down-line loaded from the Winchester disk. The software package, still under development, will have only prescribed and limited access methods to a tree-indexed ISAM. That software will be generated from BASIC source code, partially compiled, with the remainder running interpretively. The software is being written by a small systems house in the area.

Originally published in
Analytical Chemistry **1983** *55*(6) 650A
and **1983** *55*(7) 756A.

Languages for the Laboratory

Part I

It all starts with a bit, a binary one or zero, controlling an electronic gate. A collection of these bits, called a computer word, controls an ensemble of logic gates to effect the execution of a computer instruction. At the end of every execution phase the computer's central processing unit (CPU) begins the fetch of the next instruction. These alternate fetch/execute phases comprise the execution of a computer program.

The computer is a superb servant, capable of executing millions of operations per second. But it is common to find chemists faced with a computer that cannot take more than 20 data points a second from an instrument successfully! Two tutorials will explore time and the computer, focusing on languages and operating systems. These are the software components of the ménage à trois represented by the instrument/computer/program ensemble so vital to today's laboratory.

Evolution of Languages

The language of the machine, a string of 1's and 0's, is not human cordial (Figure 1). Expressing the string in a different base numbering system, such as octal or hexadecimal, improves the situation. Still, the simple numeric syntax makes it difficult to extract the semantics of a message. Artificial languages that use short mnemonics to represent repeatedly used numeric code phrases have been developed. The simple item-for-item code translation required to convert the mnemonics to machine code is called *assembly*, and the mnemonic structure *assembly language*. This machine-oriented approach produces an instruction sequence for the computer that requires minimal space and executes very rapidly. But the approach is not human oriented, and programmers have developed higher level languages that are structured more along the lines of our natural languages to express algebraic, algorithmic, logical, or control concepts. These various new languages are used to create program statements that are converted to the required machine code by programs called compilers or interpreters (described later). These labels have been used to classify a language as compiled or interpreted. Such languages usually require a coordinating facility fitted with commonly used utility programs—an *operating system.*

The artificial languages used to program computers have varied strengths and weaknesses. Many scientists refuse to consider alternate computer languages, just as they often resist learning other natural languages. This leads to situations where the chemist is trapped without the vocabulary or grammar to precisely and concisely solve his or her problem.

Rather than compare languages at the outset, let's first list some of the attributes that a good laboratory language should have.

Attributes of a Good Laboratory Language

User/Computer Communication

Operators. A set of arithmetic (+, −, ∗, /) and mathematical operators (log, exp, trig, etc.), relational operators (=, <, >, <=, >=, etc.), Boolean functions (AND, OR, XOR, NOT), and binary-number-oriented commands such as SHIFT, ROTATE, and one's and two's complement operations are all essential. Also necessary are procedures to handle the comparison and concatenation of character strings.

Variables/Constants. The ability to define named single- and multiple-precision variables and constants with support for the various types of numbers, such as integers, fractional numbers, and floating-point formats, is vital. Options should include 16-, 32-, and 64-bit format. Bit-for-bit, integer and scaled fractional arithmetic have better resolution and execute much faster than floating-point math, except on machines with microcoded math packages, coprocessors, or array processors (hardware math support facilities). However, the dynamic range for integer/fraction arithmetic is smaller, and it is harder to write general programs because the data must be properly scaled. The system should also support single- and multiple-precision number literals and alphanumeric literals. There should be no restriction on the length and characters used in any label.

Named Procedures. The capability to define new named operations that issue commands or accept and return values must be included. Parameters should be capable of being passed from the calling operation to the called operation, and values returned to the calling operation. Various languages label such activities procedures, functions, packages, or subroutines. Confusion results if we use such names interchangeably. Therefore this article will call the concept a *procedure.* In many systems these procedures can be threaded together to form complete programs.

Conditionals/Repetitives. A good set of conditionals and repetitives that allows versatile control over the program flow is necessary.

Conditional expressions such as IF . . . THEN . . . ELSE . . . test a counter or variable value, and depending on the result, alter the flow of the program. The IF process should allow named numeric, Boolean, and character-testing operations.

A CASE statement compares a numeric or string variable against a list of predetermined possibilities and passes control to the appropriate procedure.

Repetitives allow an operation to be performed a certain number of times. The usual approach sets up a LOOP counter that starts at some value, adds a fixed or variable amount at each pass through the loop, and terminates when a fixed limit is reached. *While* and *until* conditionals embedded in

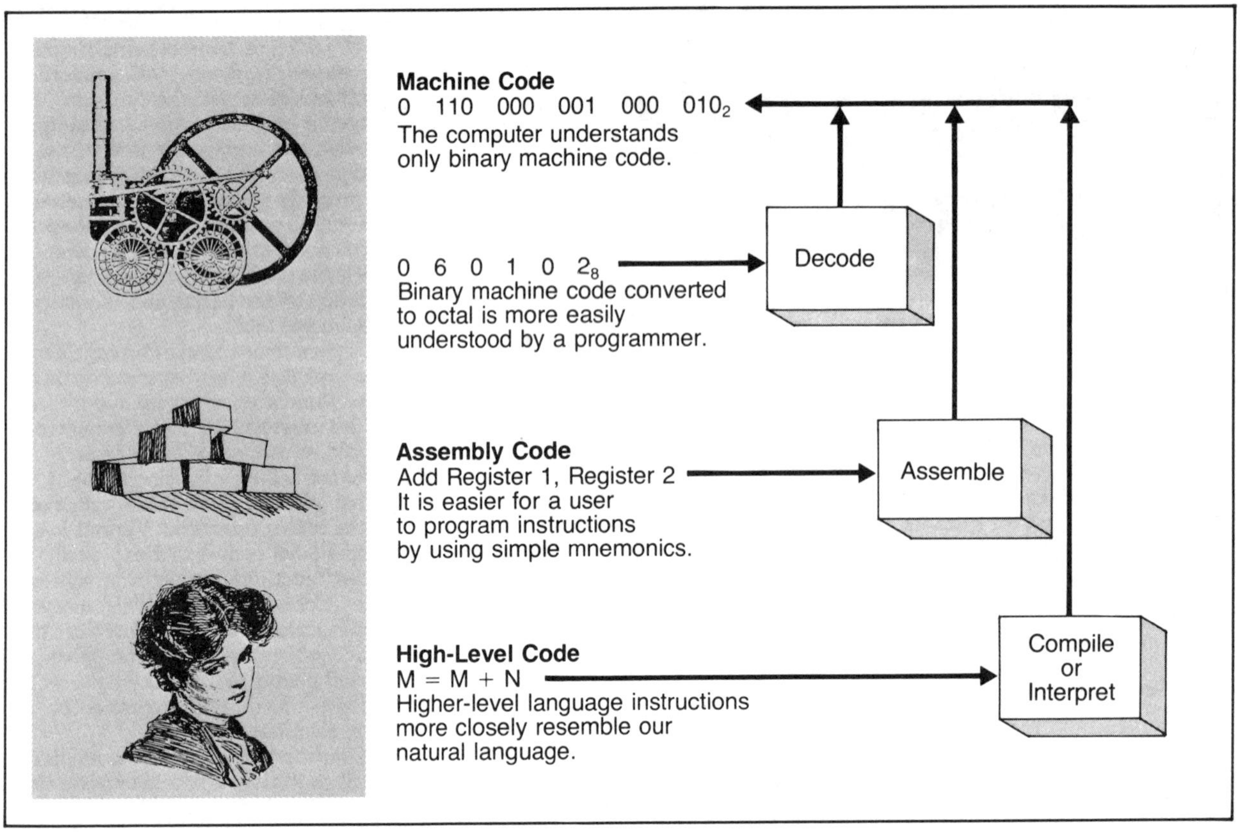

Figure 1. Human/computer communication

the loop structure permit logical control of exit from the loop. An embedded *continue* can force execution of the next iteration of the loop, bypassing some of the loop code. Positive and negative increments in the loop should be permitted.

An indefinite loop repeats an operation until a certain condition is met. No counter is involved. A typical structure is: BEGIN . . . UNTIL . . . END.

All of the above loop control structures should be capable of being nested to any degree required by the situation.

A GOTO . . . statement transfers control of the program to another section. Many programmers consider a GOTO statement improper because it tends to create spaghettilike code. It is useful in handling error conditions and creating jump-table situations. Used sparingly, with named arguments, it is a useful tool.

Many language dialects developed for the hobby or personal computer fail to meet these minimal criteria. In addition, there are other sets of criteria that must be met if a language is to be useful in the laboratory environment.

Reentrancy. It should be easy to write routines that can be shared among various users. This requires that none of the executing instructions modify the procedure itself. Any transient values belonging to individual users should be kept separate. Multiple copies of the same code should be avoided to conserve valuable memory space.

Recursion. It should be possible for procedures to call themselves. For example, calculating the factorial of a number can be done iteratively or via a recursive call to a routine.

Pointer Variables. It should be possible to request the *value* of a certain item (call by argument), as well as its *address* (call by reference). This address access permits pointer variables to be used to construct and manipulate complex data structures.

Extensibility. The language should allow user implementation of procedures that have all of the power and flexibility of the primary functions built into the system by the software engineer. For example, if DO . . . LOOPs need to be nested one level deeper than the system provides, such an implementation should be easily accomplished. Extending a CASE control statement to allow execution through a default procedure if no match is found (THRU-CASE vs. CASE) must be easy. User-defined generic procedures, capable of handling the input of a variety of data types, should be easy to implement. The different data types are then handled in context.

Overload. It should be possible for user-defined procedures to locally supercede the primary defined functions. For example, a new user-defined procedure, F*, that implements a user-oriented floating-multiply operation should be allowed, even though a basic F* already exists and even though the new F* uses the already-established F*.

User-Defined Data Types/Records. The user should be able to define new data types, with control over the creation and conditional filling of the substructures within each member of the class. A primitive data type is a finite ordered set of values, for example:

```
TYPE DATE = RECORD
DAY: 1 . . . 31
MONTH: 1 . . . 12
YEAR: 1950 . . . 2000
END
```

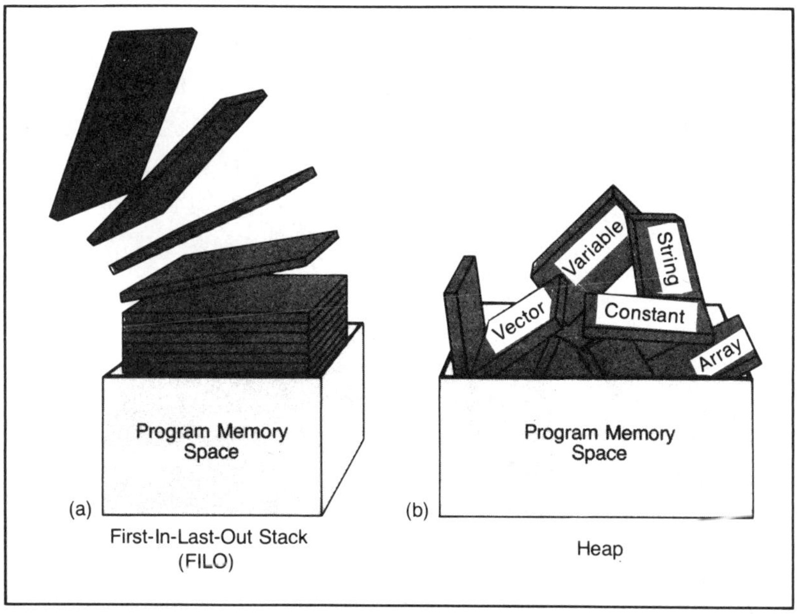

Figure 2. Memory space allocation

(The m . . . n structure creates the required sequence, such as 1, 2, 3 . . . 29, 30, 31.) A laboratory sample record would require a more complex structure that could be conditionally filled.

Interactive. All of the above functions should be available in an interactive mode. This permits the user to define new constants, variables, and procedures at the keyboard, a line at a time.

Immediate. Most of the primary functions of the language should be available by keystroke entry from the keyboard. Thus "*5* + *3* =" should execute and produce the value *8*.

Multidimensional Arrays. Some languages support only serial lists of associated data, in the form of a vector. Using pointer variables these vectors can be accessed randomly—a one-dimensional subscripted variable approach. Viewed as "slices" (a row or column in a matrix) the vectors can be used to manipulate information involved in a two-dimensional relationship and are sufficient for much lab work. Other languages allow true two-dimensional array processing, providing operators for manipulation of elements and entire rows and columns. Another view requires multidimensional data accessing—and even demands fully supported matrix manipulation techniques.

Virtual Arrays. The amount of memory space that can be allocated to a task is finite. Many data manipulations require array sizes exceeding that limit. Under these circumstances it is desirable to treat rotating disk storage as an extension of memory. In the computer world, *virtual* implies that something is not real, but appears to exist; *transparent* implies that something is indeed real, but appears not to exist. A program manipulating an array can determine if the array segment it desires is in memory. If so, it uses it; if not, the disk is accessed to bring the desired segment into memory, permitting the user to access it. This is virtual array handling. Its implementation is completely transparent to the user.

Storage Allocation. Manipulated items can be handled in a space-saving manner on a stack. Items are "pushed" on the stack and "popped" off in a first-in/last-out (FILO) order. The area may constantly be reused in this orderly operation (Figure 2a). For longer tenancy some systems allocate storage space statically, allocating space for variables, constants, or arrays before execution. These static declarations tie up such space. A dynamic storage allocation uses a "heap" of unallocated space that is assigned to user items as needed (Figure 2b). Space is recovered after use by a "garbage collection" routine. Memory is conserved since space is only used as needed.

All of the previous criteria affect the speed and facility with which a user and program can interact with the computer. In the laboratory we must consider how the instrument and computer communicate.

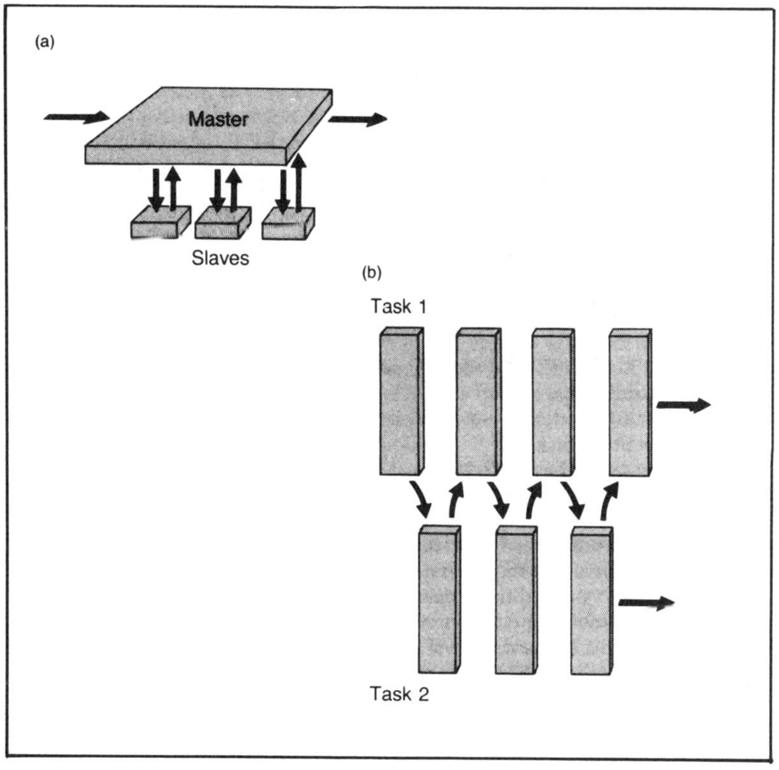

Figure 3. (a) Two-level tree; (b) coroutines

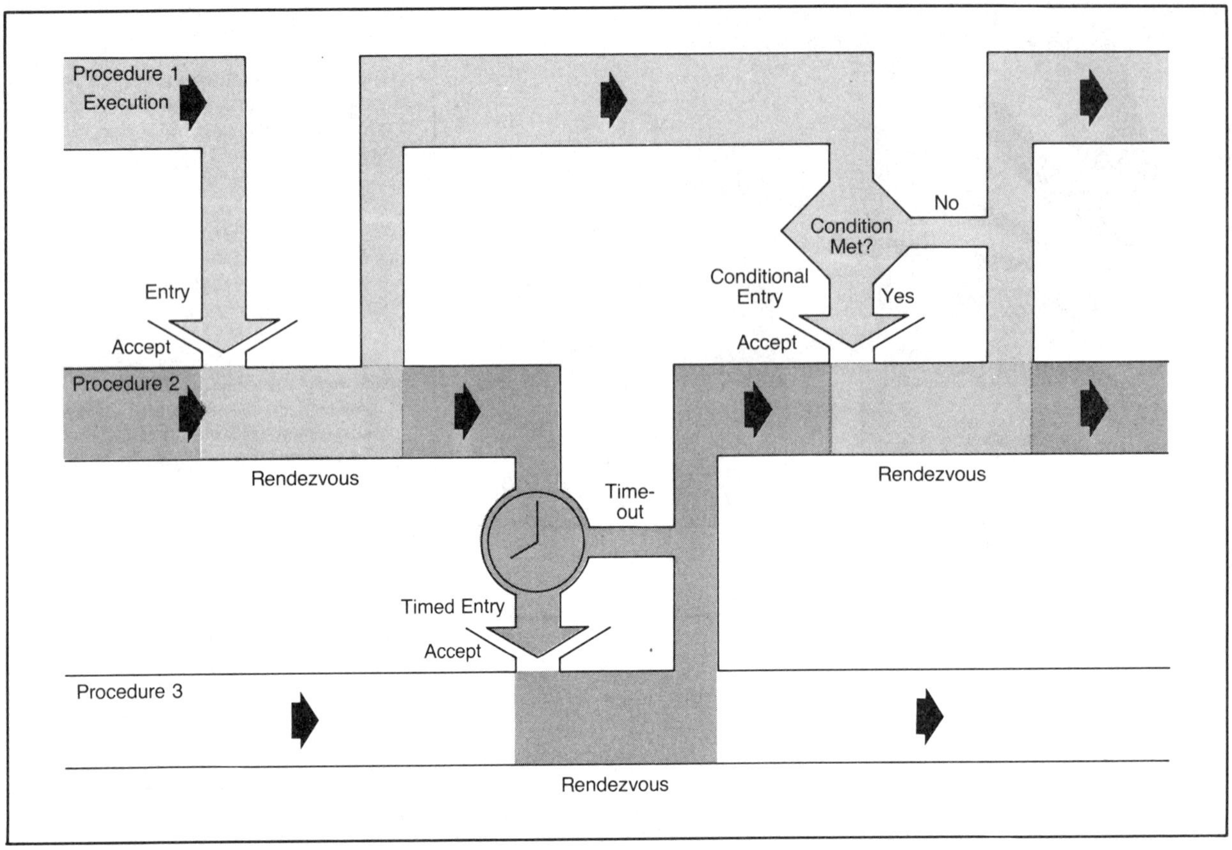

Figure 4. Rendezvous

Instrument/Computer Communication

Multitasking. Even if the computer is dedicated to a single task, it is essential that multitasking capability be supported. Many instruments and processes involve an overlapping set of synchronous and asynchronous activities. It is possible to write a single monolithic program to effect control over such an operation, but it is much easier and less expensive if the following features are available:

- ACTIVATE, ABORT, and TERMINATE permit a task to be started, aborted upon error or request, and successfully terminated. Tasks should be able to activate other tasks in a treelike fashion.
- WAIT, PAUSE, and SLEEP all allow an individual task to stop executing for a period, independent of other executing tasks.
- Coroutines. In simple tree approaches to intertask coordination only a master program exists, with slaves that can be called from it (Figure 3a). Where equal importance is given to each task and they make alternate calls to one another, it is possible to achieve a better degree of synchronization (Figure 3b). This relationship and implementation is called a coroutine.
- Rendezvous. In more complicated cases it is desirable for one task to desire ENTRY into another, at a certain ACCEPT point. When the ENTRY is invoked the calling program is suspended until the called program executes to the ACCEPT point. A common set of code is then executed; subsequently both tasks proceed to execute concurrently. This is termed a *rendezvous.* Some more advanced coordinating schemes permit *conditional* entry, where the rendezvous is entered into only if some condition is met. Or a timed entry may be requested, specifying the maximum delay that will be tolerated for a rendezvous to be established. At the end of that elapsed period, if the rendezvous is not met, the first task proceeds to execute independently. (Figure 4).

Exception Handling. Most systems have error detection mechanisms built into them. For example, division by zero is detected. Should the executing program be aborted? In many systems the user has no control over this fatal response and the program is aborted. It is preferable to separate the occurrence, or *raising*, of an exception condition, and the *service* of that exception. This allows the user to decide what should be done upon error.

Finally, there are a number of features that serious programmers *do not* agree upon. These are related to the form and documentation of software. The lack of agreement does not lessen the importance of the decision.

User/User Communication

Typing. Some languages require explicit declaration of data types in a special area. This binds the labels to a type (character, real, integer, etc.) and carefully monitors any operations involving these entities to make certain that no mixed-mode operations are being attempted. These highly typed languages contrast with those that have default conditions: If you do not specify the type, the system will automatically assign a preset standard type to the label. Finally there are languages that give the user total responsibility over the ways numbers and characters are used. With typing, mistakes are more readily detected and corrected. Better program documentation and maintenance are assured. Some default-type languages allow mixed-mode arithmetic by automatically converting one type into another. The untyped languages actually encourage mixed-mode arithmetic, for example, allowing two single-precision numbers to be multiplied together, producing a double-precision result.

Page Layout. Some languages are line oriented, i.e., expressions must begin and end on a line. Others are paragraph oriented. The physical structuring of the "sentences" in the paragraph may be free-style. Or it may be rigorously defined, with inden-

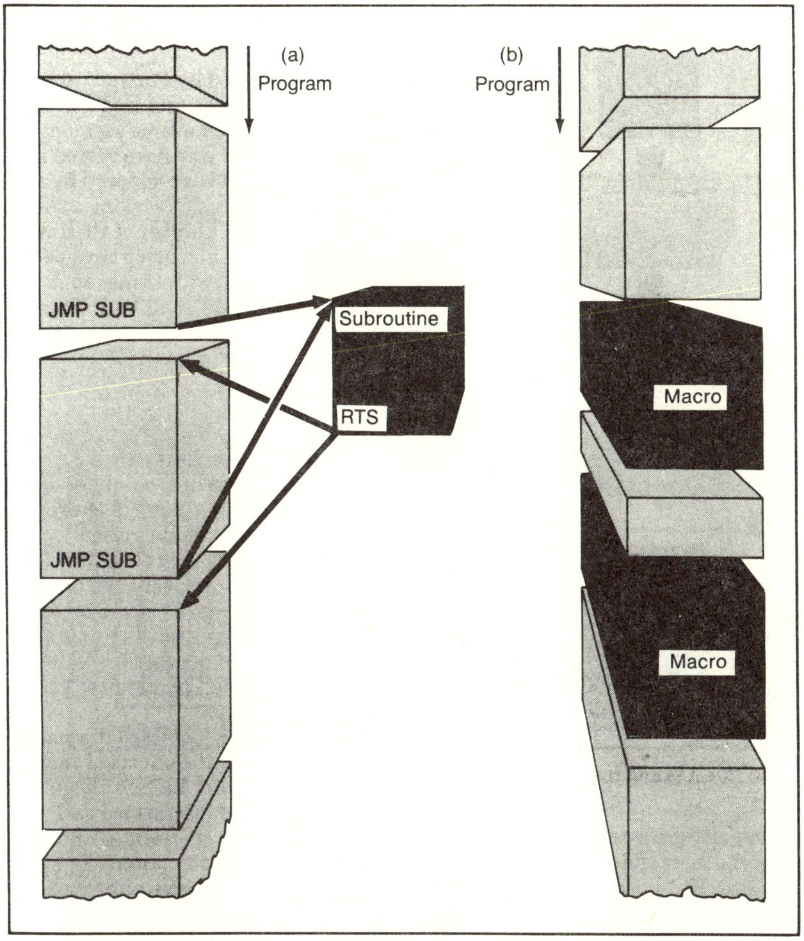

Figure 5. Assembly language. (a) Subroutine; (b) macro

tation profiles and character font and case forced on the programmer. Advocates of the latter support the strictures by claims that the maintenance costs for the program's life cycle are reduced. Free-style advocates suggest that the program should dictate the page format, with as much flexibility as possible allowed the programmer.

Structured Programming. There is general agreement that a structured programming language does drastically reduce production and maintenance costs. One view of structured programming sees it as a modular approach to software engineering. The function of the various named procedures required can be specified in the beginning, and the actual implementation programmed at a later time.

Some languages permit both conception and coding of the various modules from the top down. In this approach the body of the procedure contains the actual code that implements the operation. The specifications for the procedure are separated from this body. The specifications section contains all of the information necessary to use a procedure. This includes the number, names, and types of all passed arguments, as well as the logical control requirements. As various procedures are combined into a user program, the compilation process would use their specification segments to link them together. It would not matter that the code for some of the bodies did not yet exist.

Other languages allow design conception to be top down but require that the procedure be coded from the bottom up. In this approach the compilation process builds up more complex functions from simpler entities and requires prior existence of the latter. The code for specifications and body implementation is more intimately connected.

In both approaches the goal of structured programming is to prevent changes in implementation within a given procedure from affecting other procedures that use it as long as the interface specifications are maintained. In the best systems individual procedures may be recompiled without recompiling the others associated with it.

Let us look briefly at some laboratory languages, exercising the terms just introduced.

Laboratory Languages

Assembly language is specific to a particular machine. It is not portable, in the sense that it cannot be transferred to another brand of computer. A simple assembler contains only mnemonics for the primitive machine operations. However, it is possible to augment an assembly language and provide a mechanism for faster code generation. One approach uses a collection of commonly used subprograms or subroutines that are called from the main program (Figure 5a). The machine code for each subroutine exists in only one place in memory space. Jump-to-subroutine and return-from-subroutine movements are used to enter and exit the subroutine. Although this saves space, it is time-consuming, and it generates code that is often difficult to follow. An alternate approach creates a library of commonly used procedures that may be employed by another program. At assembly time the code for a required library program is laid down in-place in the main program each time it is required. This in-line code is space inefficient, but executes rapidly. Since one line of assembler code results in the in-line deposition of many machine code instructions, the technique is called *macro programming* (Figure 5b).

When more than one program is required to occupy memory at the same time, each must reside in different space. The most primitive assemblers produce only *absolute code*, which will run only in the place the programmer designed it for. Other assemblers produce an intermediate *object code* that a linker program can relocate into available space. This requires modification to referenced addresses. Finally, some machine architectures allow generation of *position-independent code.* Such code will run wherever there is space to put it. All three capabilities are essential to a successful laboratory package (Figure 6). They allow individual modules to be prepared independently, even by different programmers, with no worry about where to position their code in memory. The resulting object modules are fused into a single run-time module by the linker.

As we look at the high-level languages in a real-time environment, it is essential to remember that they occasionally will have to use assembly-written code. The macro library concept will also be used by them for commonly used procedures.

BASIC is a language originally intended to teach programming to novices. Its form is extremely variable, ranging from an interpreted language to a compiled form. In the former, an executable machine code module is

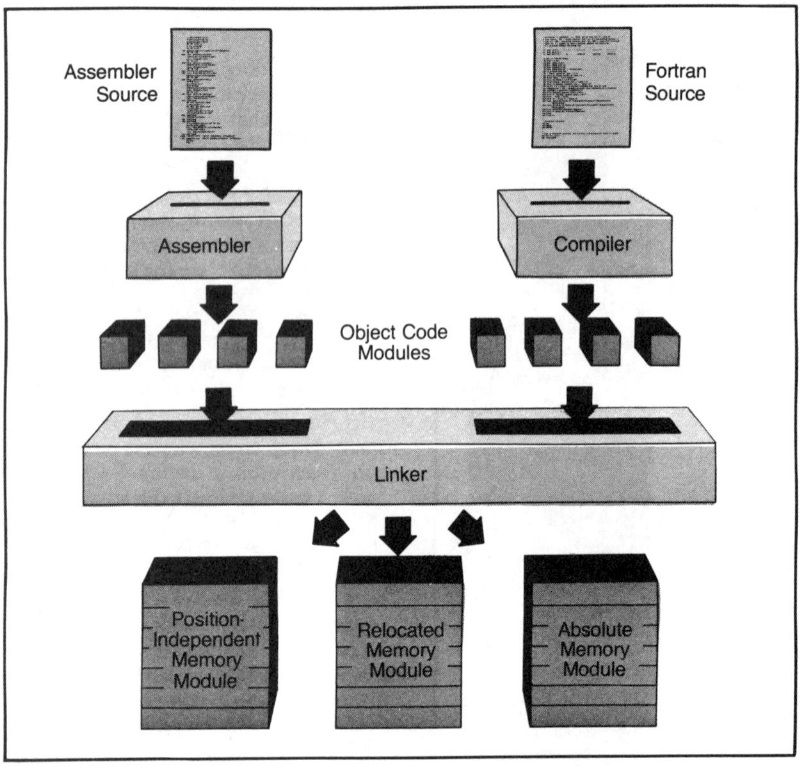

Figure 6. Code from various sources can be linked into a single run-time module

not created directly from the source; rather the common operators invoked in the user code are replaced by tokens that condense the program strings. At execute time each program statement is interpreted in a way that leads to an executable set of machine instructions. This results in very slow execution speeds; however, it does allow detection of syntax errors interactively. A compiled BASIC language produces an entire executable module of machine instructions from the user's source code.

The strength of an interpreted BASIC language lies in its simple, interactive capabilities. There is no need to invoke highly complex operating system functions such as separate editors, assemblers/compilers, and linkers. However, the simpler versions of BASIC do not meet the minimum criteria set out above. Developed as a teaching tool, the personal computer has pushed BASIC into a larger role, not because it is necessarily easy for the programmer, but because it is easy for the machine. Its memory, disk, and architectural requirements are minimal. Small problems are easy to code; large or complex programs become impossible.

Fortran-77 is an ANSI standard (American National Standards Institute) that upgrades the ubiquitous Fortran-66 ANSI standard (Fortran IV) to a useful compiled language. Moderately typed, with default, its portability, available libraries, and extensive software availability make it one of the most widespread scientific languages. It is not really suited to structured programming and contains a number of special cases that are often frustrating. The language does not permit recursion, synchronization, or complex data structures.

The compiled code can be linked with assembly-written subroutines that can perform real-time tasks. Such linkage steps require that the Fortran compiler also produce the intermediate form of machine code called object code. To pass arguments and data between separate modules requires the declaration of GLOBAL labels that will be used commonly. Each object module knows what it will operate on, but does not know where these items will be located as the object modules are merged. Therefore, address resolution is performed by the linker, creating a single run-time module. With such linking, we can have the best of both worlds: a good high-level language, yet access to assembly programming when speed and intimate bit-on-bit control are required.

APL is an interactive, interpreted, untyped, extensible language that was developed for use in blackboard demonstrations to math students on how to manipulate arrays, strings, and other data constructs. Not constrained by the type font of a typewriter keyboard, it created new powerful operators by chalk strokes. It therefore requires a special keyboard to run on a computer today. APL is extremely useful in math and statistical environments.

PL/1, developed by Fortran users, includes some of the required features. It has not attracted a large segment of the Fortran users who have vested interests, nor has it been adopted by the more progressive users seeking a successor to Algol. A number of PL/1 dialects for smaller machines have become widespread, with names such as PL/M, PLZ, PLMX, and PLI. These differ greatly in their characteristics, and some begin to appear Pascal-like.

Pascal, a highly typed, highly structured, compiled language was designed to teach computer programmers structured programming. Pascal, like BASIC, comes in a variety of dialects. They include many of the required characteristics described above. Dynamic storage allocation and pointer variables are available, but many versions are weak in concurrency and real-time control. The expressive power of the language lies in its ability to create and manipulate variable data structures.

Pascal's use has increased dramatically as a result of the efforts of the University of California/San Diego team. A variety of interpreted and compiled versions of Pascal are now available for small computers. The UCSD approach has also pioneered the concept of a compiler that produces machine code for an ideal, hypothetical central processor: p-code (pseudo machine code). A target machine merely needs to emulate this machine, and the language becomes highly transportable.

In some computers, the instruction set that can be executed can be altered by changing the contents of a control read-only-memory (ROM). Instead of actual machine instructions setting up the gates and flip-flops of the CPU, the machine instruction word is used to address a location in the control ROM that contains a longer binary bit string. This string, in turn, sets up the CPU for execution of the desired operation. This is *microcoding* (Figure 7). It is possible with this technology to build machines that can directly execute p-code, providing incredible speed of execution.

LISP was developed primarily for the manipulation of nonnumeric data. It is interactive, interpreted, untyped and extensible. It is ideal for applications in artifactual intelligence and robotics. (Canada is north of Mexico; Mexico is north of Guatemala. Is Canada north of Guatemala?)

LISP uses pre-fix operation notation. We are taught an in-fix arithmetic in which the numbers in the expression 4 + 3 sandwich the operator. In complex cases, such as 3 * 5 + 7 *

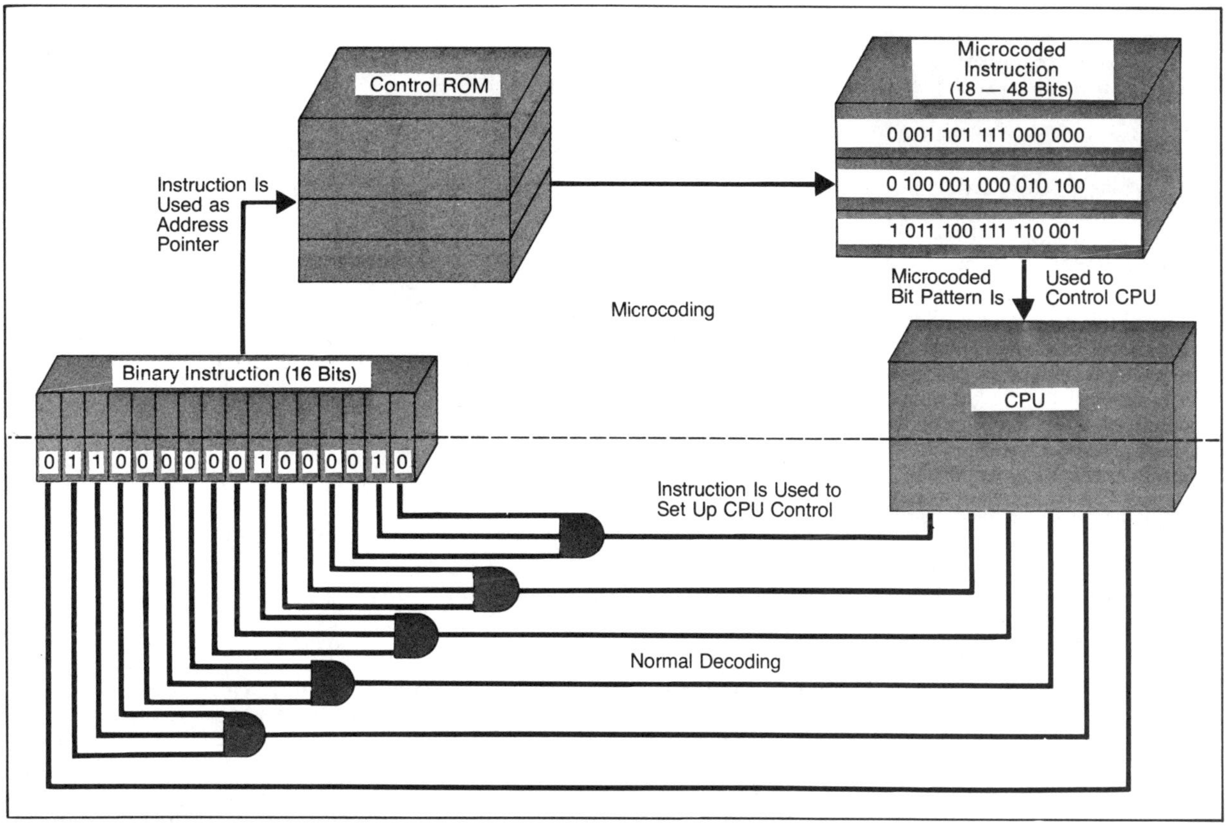

Figure 7. Microcoding is depicted in top half of figure; normal decoding is operation below dashed line

11 we have to carefully prioritize the operators + and ∗, or use parentheses to prioritize the operations, to avoid confusion between (3 ∗ 5) + (7 ∗ 11) and (3 ∗ 5 + 7) ∗ 11. In Polish notation the former would be written + ∗ 3 5 ∗ 7 11, pre-fixing the operators. A reverse-Polish approach would use 3 5 ∗ 7 11 ∗ + to express the same sequence, post-fixing the operators. The pre-fix approach of LISP allows multiple arguments to follow the math operator. It also avoids in-fix ambiguities, or the overhead of parsing required to avoid interpretation errors.

LOGO is a variety of LISP that incorporates some very desirable graphic features.

C is an intermediate-level compiled language developed at Bell Laboratories that is not as strongly typed as Pascal. It contains most of the desirable features cited above, but does not implement parallel operations, synchronization, or coroutines. It is pointer-oriented, permits direct access to all machine facilities, and is well structured. It does not directly implement slice array manipulations or dynamic storage allocation. However, its use of pointer variables permits easy addition of such functions.

One of its most distinctive features is a pipe. This refers to the ability of a command to pipeline the output, say, of a generation procedure to its report procedure. As output from the first routine occurs it is passed immediately to the second routine, without the need for explicit command by the programmer. One need not wait for the first operation to run to completion (Figure 8).

C has been used to write a new type of operating system called UNIX, whose architecture and vocabulary resemble that of C. There is a growing tendency for such development, where the language and the environment it works in begin to merge in the user's mind. The user is no longer faced with different rules, vocabularies, and command structures as he or she moves back and forth between the language and the operating system. Some people refer to an operating system as a collection of programs that helps in writing other programs. Others say it is a collection of programs that does not "fit" into the language. C and UNIX avoid the last criticism because of their intimate relationship. C and UNIX are beginning to be used extensively in laboratory environments and by instrument manufacturers. Many UNIX look-alike operating systems are becoming available, such as IDRIS and XENIX.

Forth is an interactive and immediate applications-oriented language designed to control scientific equipment. Ideal for computers embedded in instruments, it is truly extensible. It permits rapid generation of top-down-designed bottom-up-coded structured programs. Incrementally compiled, it has the ability to allow keyboard or disk definition of words that implement procedures. These words are built into a dictionary that can be called upon as new words are created using previously defined ones. All of the words can be concatenated into

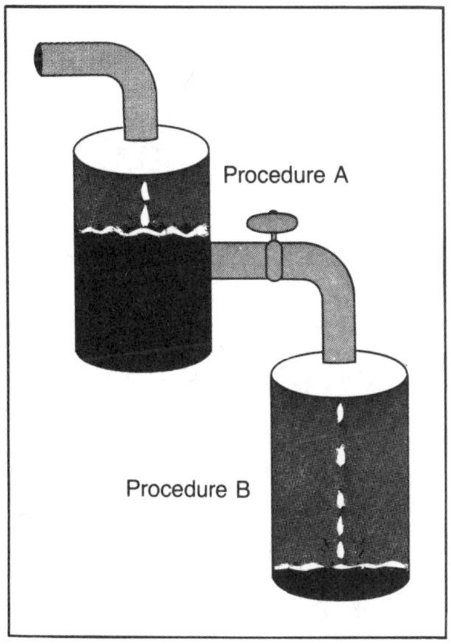

Figure 8. In a pipe the output of procedure A may be passed to procedure B immediately, before A has run to completion

sentencelike structures that, in turn, are given names. Reverse Polish notation is employed.

Forth contains most of the attributes of a good laboratory language. It can have task synchronization implemented by coroutines. Modification to include other desired features is trivial. Forth is inherently pointer-oriented, virtual, and reentrant. Exception handling is left to the discretion of the programmer. Forth is being used by a number of instrument vendors who are interested in a language that permits quick programming, produces rapidly executing code, and runs in minimum memory. Running on well-designed processors, such as the DEC 11 family or the Motorola 68000, Forth provides an excellent example of a systems language, where the language matches the machine it runs on. All machine facilities are fully available, and the language does not contain any nontrivial extensions beyond the machine capability. A machine is imminent that executes most of the Forth basic instructions in microcode. The best Forth language implementations are written in Forth.

Ada is a language developed under the sponsorship of the Department of Defense for use in embedded computers. It is highly typed and has perhaps the best-structured programming features of any of the languages mentioned. Ada has good array-handling capabilities and allows generic procedures, where the handling of various data types is done "in context." Its control structures and multitasking coordination facilities are outstanding. It supports true top-down development. It is just becoming available to smaller machines, using a microcoded approach. The Intel 432 chip is particularly fitted to Ada. It was designed specifically to execute Ada instructions. This mating of language and processor architecture is a major step forward.

BASIC and Fortran provide traditional language support and an extensive user base. APL and LISP have specific attributes that make them of particular importance to specific applications. Pascal and Ada furnish powerful structures in a framework that is rather rigid. Forth and C allow considerable flexibility in the use and implementation of features by the programmer.

Time/Benchmarks. It is very difficult to write benchmark programs that test fairly; each processor has its respective strengths and weaknesses. However, the integration of a number of sources indicates the following:

- If the processors considered are limited to the range 6502, Z80, 8080, 8086, LSI-11, and 68000, then running the same algorithm, on the same machine, in different languages, can result in an execution speed differential of over 500.
- Running the same algorithm, on the same machine, in dialects of the same language can result in a differential of 20.
- Running the same algorithm, in the same language, on different processors can result in a differential of more than 50.
- Float vs. integer arithmetic can cost a factor of five in execution speed.

Thus, the incorrect choice of language and processor can easily result in a thousand-fold drop in execution speed.

- There is no simple correlation between the speed of execution and memory requirements.
- There is no simple correlation between the speed of execution and compilation time (Figure 9) (*1*).

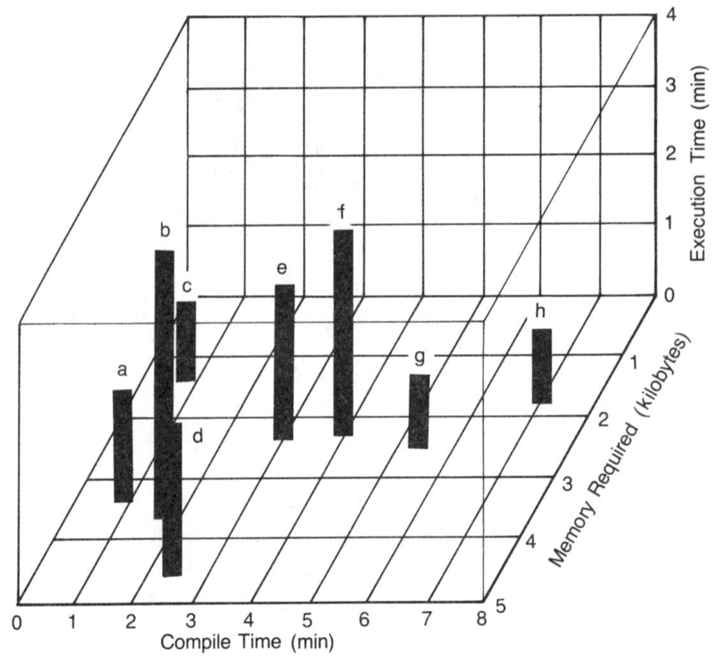

Figure 9. What can be seen from this graph? There is no simple correlation between compile time, required memory, and execution speed (*1*). Languages depicted are: (a) Pascal/MT; (b) Fortran (80); (c) Interactive Systems C; (d) PL/1; (e) Pascal/Z; (f) PLZ; (g) Whitesmith C; (h) PLMX

Compilation time affects the overall cost of program generation, since a successful program is always arrived at by iterative processes. The importance of execution speed and memory requirements is determined by the task involved.

Programming language selection should therefore not be performed in an environment characterized by hasty, intuitive, or close-minded attitudes.

In addition to these technical specifications there are important management criteria. These involve development time and cost and the life-cycle maintenance cost. The latter involves both correction and augmentation functions. It is difficult to evaluate software costs, particularly in environments where scientific personnel not classified as programmers generate their own code. In 1981 the value of shipped computer hardware was approximately $15 billion. The cost of programmers, according to the Bureau of Labor Statistics, was approximately $10 billion. The untold thousands of scientific programmers toiling in the cottage-industry environment of their offices and labs would swell the latter figure greatly. It seems safe to assume that software represents at least half of the hardware/software combined costs. For special-purpose microcomputer programs, with a limited number of users, the software/hardware ratio can approach 3:1.

Language choice is the first major step in determining the actual performance of our million-instruction-per-second computers. Operating system choice will further influence performance. This will be the subject of our next tutorial. Next month's capsule reports will focus on the use of Forth, Pascal, C, APL, BASIC and Fortran-77 in analytical laboratories and instruments.

Acknowledgment

During the development of this tutorial valuable discussions were held with Ian Chapple, Jim Currie, Ken Hinson, Mark Thompson, and Steve Duball.

Reference

(1) Anderson, Gordon E.; Shugate, Kenneth C. *Computer* **1982,** *15,* 29–36.

Part 2

```
VARIABLE SCAN_COUNT          ( number of scans to collect )
VARIABLE COUNTER             ( running scan counter )
VARIABLE GAIN                ( gain value )
VARIABLE I-BUFFER            ( pointer to input buffer )
VARIABLE A-BUFFER            ( pointer to averaging buffer )


( assembly language primitives: )
COLLECT    ( input data from spectrometer to buffer)
ADD_SCAN   ( add data in first buffer to second)
MOVE_DATA  ( move data from first buffer to second)
DO_FFT     ( perform Fast Fourier Transform on buffer)

: COLLECT_SCAN
        I-BUFFER @ GAIN @ COLLECT  ;
: AVERAGE_SCAN
        COUNTER @ O =
    IF     ( first scan )
        I-BUFFER @ A-BUFFER @ MOVE_DATA
    ELSE  ( not first scan, add to average )
        I-BUFFER @ A-BUFFER @ ADD_SCAN
    ENDIF   ;


: CHECK_COUNT?  ( return TRUE flag if done )
        COUNTER @  1 +  COUNTER !
        COUNTER @  SCAN_COUNT @  =   ;


: SCAN
        O COUNTER !          ( zero counter )
    BEGIN
        COLLECT_SCAN AVERAGE_SCAN CHECK_COUNT?
    UNTIL
        A-BUFFER @ DO_FFT  ;
```

Figure 1. Forth SCAN process. Comments in Forth are set off by parentheses

Nicolet Analytical Instruments
5225-1 Verona Rd.
P.O. Box 4508
Madison, Wis. 53711

Contributor: Joel V. Petersen

Forth and Instrument Automation

Forth originated as an automation language for control of large telescopes. It has since been used for control of processes of many types, from spectrometers to laser fusion research. It employs a structure that is comfortable and useful to both the system programmer and the system user.

The Nicolet 5-MX and 5-DX FTIR spectrometers are programmed entirely in Forth. The 5-MX is primarily operated through a touch panel, generating output on a plotter. The 5-DX is operated through a standard keyboard, which controls an interactive color graphics display. In both cases, the existence of Forth is not apparent to the user, but the language is easily accessible.

The decision to use Forth was primarily a result of three constraints: The program had to be small, fast, and user programmable. The only program storage on the 5-MX was ROM (read-only memory). A floppy disk is standard on the 5-DX.

Since the ROM-based instrument requires its own operating system, Forth takes over the tasks of device input/output (I/O) and disk file structure. On the other hand, the floppy-based 5-DX uses the operating system common to the other Nicolet systems; Forth interacts with the resident system.

In Forth, a stack is used to manipulate parameters, and the post-fix notation common to calculators is employed.

Simplicity is the rule in Forth in every aspect, including the definition of new commands. In the following, ⟨name⟩ indicates the label of the new word being defined; word1, word2, word3 indicate previously defined words: ": ⟨name⟩ word1 word2 word3 ;". A variable is created simply by using the phrase, "VARIABLE ⟨name⟩".

Abbreviations are used in Forth to reduce its memory requirements. The most obvious are the store and retrieve operators for variables. The exclamation mark (!) indicates *store* and

```
8190 CONSTANT SIZE
0 VARIABLE FLAGS    SIZE ALLOT


: DO-PRIME
  FLAGS SIZE 1 FILL                ( SET ARRAY FLAGS TO 1)
  0                                ( SET COUNT TO 0 ON STACK)
  SIZE  0                          ( RANGE OF DO)
   DO  FLAGS I + C@                ( GET FLAG)
      IF I 2* 3 + DUP              ( TRUE?, CALC PRIME, 2*I + 3)
         ( CR DUP .) I +           ( PRINT IT)
         BEGIN DUP SIZE <          ( I + PRIME GETS MULTIPLES)
         WHILE  0 OVER FLAGS +  C!
         OVER + REPEAT             ( AND SETS FLAGS TO 0)
         DROP DROP 1+
      THEN
   LOOP
   .  ." PRIMES"  ;                ( PRINTS # FOUND)
```

Figure 2. The prime number program in Forth. This program does only one iteration. Comments are set off by parentheses

the commercial "at" sign (@) indicates *retrieve.* "360 ⟨name⟩ !" puts 360 in the variable storage area; "⟨name⟩ @" retrieves the data. With simple definitions, the user could rename these functions as STORE and RETRIEVE. Similarly the colon (:) and semicolon (;), which are used to delimit definitions, are abbreviations for START_OF_DEFINITION and END_OF_DEFINITION.

Program execution in any language is controlled through conditional and unconditional jumps. Forth provides loop structures, ": ⟨name⟩ BEGIN word1 UNTIL ;", where the word(s) between BEGIN and UNTIL are repeated until a logical true is passed to UNTIL. Conditional branch structures are also available: ": ⟨name⟩ word1 IF word2 ELSE word3 ENDIF ; ", where the word(s) between IF and ELSE are executed when IF receives a true from word1 and the word(s) between ELSE and ENDIF are executed on a false.

The most persuasive aspect of Forth syntax is demonstrated in process control environments, where definition of new commands involves the simple linking together of primitives. Meaningful names can be given to various processes. For example, on the 5-MX and 5-DX, the most common process, collecting data, is called SCAN (Figure 1).

An important feature of Forth is that the parts of SCAN, such as COLLECT_SCAN, are equally available to the programmer (and user) for testing or use in other routines. In Fortran and Pascal, subroutines are testable only by use in a main program.

Like other high-level languages, Forth provides many benefits to the programmer over assembly language programming: system portability, structured programming, modularity, easier documentation, and easier training of other programmers. But unlike most of the other languages, Forth offers real-time interactive programming and debugging to both the system programmer and the user. A benchmark Forth program appears in Figure 2. (See box, "A Benchmark and Comparison Program," for information on comparative versions of the prime number program listed in each section of this article.)

Varian
611 Hansen Way
Palo Alto, Calif. 94303

Contributors: Robert S. Codrington and Stephen Smallcombe

Pascal

Background. In the early 1970s Varian recognized that it would have to develop a new NMR spectrometer based on advanced Fourier transform and superconducting magnet technologies. Analyses suggested that separate CPUs would be needed for pulse sequence timing and data acquisition and manipulation. A multitasking main CPU was essential.

The proposed new data system implied a much larger and more complex set of software programs than had been developed for any previous NMR instrument. The majority of software programs up to that time had been programmed in assembly language. The general programming efficiency was roughly 10 lines of released code per programmer day.

Definition Phase. Two considerations requiring immediate attention at the start of the project were the type of programming language and the operating system to be used. In view

A Benchmark and Comparison Program

For interested readers each capsule report on a language is followed by a benchmark program that illustrates its syntax. The program implements the Sieve of Eratosthenes, which calculates the prime numbers between three and 16 381. The algorithm involved creates an array of flags; the flags represent the odd numbers between these two values. A loop counter, or index, is used to find primes by the calculation, 2∗i + 3. As each prime is found, multiples of itself are struck from the table, by changing the flag value. The collection of programs is used with the permission and courtesy of Jim Gilbreath, Head, Computer Sciences and Simulation Department, Naval Ocean Systems Center, San Diego; and Gary Gilbreath, of the University of California at San Diego. The APL program was submitted by W. D. Griffith of Mobil. The Fortran 77 program was submitted by Charles Dumoulin of Syracuse University.

of the problems that had been experienced in previous software development programs, it was felt that the software should be modular and the operating system should force on the programmers a set of protocols that would prevent programming errors from propagating outside the module in which they were resident.

The choice of Pascal as the programming language was heavily influenced by ideas then being developed within the computer science community concerning structured programming.

One member of our project team was a trained NMR spectroscopist as well as a computer system specialist. He and members of the application laboratory staff developed a complete command/parameter list with definitions that were oriented to a spectroscopist/operator, not to a computer programmer. This list of command and parameter descriptions was a major factor in keeping the software development program on track and was later to form the basis of the software manual.

Development Project. The project engineers felt it would be wise to have the software as portable as possible in case cost and performance considerations should cause a change in the CPU hardware. It was decided to develop compilers that would convert the Pascal source programs into generalized pseudo-machine code, or p-code, which could be stored on the instrument's disk memory. The p-code programs could then be decoded, loaded, and executed by a software interpreter. One necessary modification was to provide an additional pass to link Pascal programs to assembly language programs, which were needed to speed up certain instrument operations.

A number of considerations contribute to the success of a software development program, including:

- Modularity: Keeping procedures reasonably short is helpful in debugging and allows the module to be used in other program segments.
- Transportability: The fact that the software has to be programmed so that it can be used with different hardware causes the programmers to produce more flexible and understandable code.
- Hierarchical structures: Hierarchical structuring in multitasking systems of directories, process communication, and resource allocation help to maintain the integrity of the system.

Several months before the new instrument was to be announced, the chief programmer left Varian. Ordinarily this would have been disastrous for a software project of any size. With only occasional consulting, the remaining personnel were able to meet the scheduled release dates. The programmers felt this continuity of effort would not have been possible if the software had been written in assembly language or Fortran.

Product Introduction. The software package introduced in 1978 permitted the novice operator to quickly master the procedures and commands for routine operation of the instrument without penalizing the more experienced operator with unnecessary comments and prompts. One advantage of using Pascal-like languages in dedicated instrument systems is that it is relatively easy to string several commands together to create a simple macro command. This makes it possible to produce a simplified and operation-oriented command set for the operator who does not wish to interact with the software, while at the same time keeping the generalized command set available to the operator who wishes to write specialized programs.

The initial reception of the software was not as positive as had been expected because some of the operator interface concepts were so new. Many operators were not used to multitasking systems and preferred to carry out all operations serially, as was done with older instruments. Operators having programming experience with assembly language as well as Fortran and BASIC were not very comfortable with the formalism of the Pascal program structures.

Fortunately, the development of interest in the Pascal language in the past five years resulted in many excellent publications and programs that were helpful to those operators who wished to use the programming capability of the new instrument.

The released software had its share of both major and minor bugs, but the structure of the programs made it relatively easy to correct them. Most of the programmers involved agreed that the formalism of the language played a major role in avoiding the catastrophic software errors that had plagued some earlier projects. The average efficiency of a programmer over the development period was about 42 lines of machine code per day.

In the five years since the software package was initially relased, it has grown from about 250 000 machine instructions to close to one million instructions, with much of the new code being contributed by instrument customers.

Did all the programmers in the project become Pascal disciples? No. Pascal was an ideal language for an instrument development program at that time because it combined the attributes of a high-level language with enforced structured programming and was supported by smaller operating systems (2000 machine instructions) and compiler/interpreters (8000 instructions). But since its completion, parts of the XL-200 software package have been adapted to other systems using newer, higher-level languages such as PLM, C, and Forth, all of which have advantages depending upon the application. A sample Pascal program appears in Figure 3.

Mobil Research and Development Corporation
Paulsboro, N.J. 08066

Contributor: R. J. Kobrin

APL and the Analytical Chemist

The first impression one gets of APL is that it is complex and hard to understand. It uses a strange character set, which may require a special computer terminal to use. This capsule attempts to demonstrate the effectiveness of APL as a personal tool for the chemist. It does not attempt to argue its applicability to major programming projects, or as a teaching language—although there are valid reasons for both.

Researchers in our facility typically use APL as a replacement for calculators, graph paper, and file cabinets; for quick, one-time calculations and plots; and to maintain personal data bases. A great many functions are available in on-line libraries to simplify storage and retrieval of data, statistical analysis, and graphics. A significant advantage of APL is the ease of using functions developed by others; a large multiuser computer environment encourages exchange of useful functions and data files.

Because APL is a conversational, interpreter-based language, users do not need to write formal programs for everything they want to do. They quickly become fluent enough in APL to "converse" with their data without spending a great deal of time in programming.

Suppose we need to calculate the molecular weights of a few organic compounds for use as elemental analysis standards. We first enter some atomic weights in a reasonable order: ATWT ← 12.011 1.008 16. This simply loads a list of the atomic weights of carbon, hydrogen, and oxygen into a variable called ATWT. In APL a variable containing a single value is called a *scalar*, a one-dimensional array is called a *vector*, and a multidimensional unit is called an *array*. Special sym-

```
program prime;
const
  size = 8190;

var
  flags : array [0..size] of boolean;
  i,prime,k,count,iter : integer;
begin
  writeln('10 iterations');
  for iter := 1 to 10 do begin                  {do program 10 times}
        count := 0;                             {prime counter}
        for i := 0 to size do
              flags[i] := true;                 {set flags all true}
        for i := 0 to size do
              if flags[i] then begin            {found a prime}
                    prime := i+i+3;             {twice the index + 3}
                    {writeln(prime);}
                    k := i + prime;             {first multiple to kill}
                    while k <= size do begin
                          flags[k] := false;    {zero non-prime}
                          k := k + prime        {next multiple}
                    end;
                    count := count + 1                {primes found}
              end;
  end;
  writeln(count,' primes')                      {# primes found}
end.
```

Figure 3. The prime number program in Pascal. Comments in Pascal are set off by braces: { }

bols, such as the left arrow, are APL operators. They replace the shifted alphabetic characters on an APL terminal. The unshifted alphabetic keys produce the upper-case alphabet; lower case is not a feature of APL.

To print the contents of vector ATWT we just type its name:

```
ATWT
12.011 1.008 16
```

To add nitrogen to our vector, we simply use the *catenation* operator, the comma. Then we print the vector to see if it worked:

```
ATWT ← ATWT, 14.008
ATWT
12.011 1.008 16 14.008
```

Benzoic acid has the empirical formula $C_7H_6O_2$. If we represent this as the vector 7 6 2 0 (the zero means no nitrogen) we can type:

```
ELWT ← ATWT × 7 6 2 0
ELWT
84.077 6.048 32 0
```

and instantly get the contribution of each element to the molecular weight in a new vector, ELWT. The entire atomic weight array was multiplied by the empirical formula array in a single step. This unique property of APL applies to arrays of any size or shape.

To set the molecular weight itself, we perform a *plus reduction* (+/) on ELWT. This has the effect of placing plus signs between the items in ELWT and performing the addition to get the molecular weight of benzoic acid:

```
+/ELWT
122.125
```

Of course this would normally be done in one step:

```
+/ATWT × 7 6 2 0
122.125
```

Note that APL evaluates expressions strictly from right to left, first doing the vector multiply operation, then the plus reduction. There is no precedence of one operator over another; parentheses are used to alter order of evaluation.

If we had a number of these to do, we could write a useful APL function called ELPCT:

```
    ∇ ELPCT FORMULA
[1]   ⍝ ENTER EMPIRICAL
        FORMULA AS C  H  O  N
[2]   ATWT ← 12.011  1.008  16
      14.008
[3]   ELWT ← FORMULA × ATWT
[4]   MWT ← +/ELWT
[5]   PCT ← 100 × ELWT ÷ MWT
[6]   MWT, PCT
    ∇
```

The del operators (∇) begin and end our function definition. The function name ELPCT is followed by its argument. When we invoke ELPCT the list of elements it needs will appear where FORMULA is. While ELPCT is running, FORMULA will take on the value of this list. Programmers will recognize this as a dummy argument used to pass data between software modules. Variables used within a function may be defined as local, so functions can be used freely and exchanged with other users without concern for naming conflicts.

Line [1] is a comment; it appears only when the function is listed. The characters following the lamp character (⍝) are ignored by APL. Line [2] enters the vector of atomic weights. Lines [3] and [4] are the same as previously shown. Line [5] divides the vector of element weights by the molecular weight and multiplies by 100 to produce a vector of percentage composition. Line [6] displays the molecular weight followed by the percentages of C, H, O, and N in the compound.

Our function can now be used to find the composition of acetanilid, C_8H_9ON, by typing:

```
ELPCT 8  9  1  1
135.168  71.0878  6.71165  11.8371
10.3634
```

This function is not very elegant. But ELPCT represents the kind of simple program a chemist might write for one-time use. The conversational nature of APL encourages programming by experiment rather than by plan. A structured language such as Pascal represents the opposite end of the spectrum from APL.

Most of the output of a research project is in the form of arrays of data: tables of starting materials, experimental conditions, and results. Viewed in this light, APL offers significant advantages for interactive computing in a research environment. APL can be obtained for all mainframe computers and many mini- and microcomputers—although the latter may offer limited implementations. A public domain version is available for PDP-11 computers from Decus. Terminals equipped with the APL character set are not common, but most type-ball and daisy-wheel hard-copy terminals accept APL typing elements, and some dot-matrix terminals offer optional APL character sets. Some APL implementations support special ASCII character sequences, allowing use of regular terminals. The prime number program in APL is very brief:

```
    ∇ R ← PRIME N
[1]   R ← (1 ≥ (−N ≤ 9) + +/0 =
      ⍉(⍳9)∘.|N)/N ← ⍳N
    ∇
```

Perkin-Elmer
Main Avenue
Norwalk, Conn. 06856

Contributor: Andrew R. Muir

The C Language

Application Area. The Perkin-Elmer Model 7500 laboratory computer is designed to operate with a variety of analytical instruments. A typical application is a spectroscopy system in which the computer connects to a spectrometer by an RS-232 communication interface. The functionality of such a system includes a menu-driven dialogue between the system and the human user; computer control of the instrument; acquisition of digitized instrumental data; calculations on the data; presentation of computed results both graphically and as printed reports; and archiving of methods with data and results.

Choice of C as the Programming Language. The choice of C follows partly from the selection of the computer's operating system, the Unix-like Idris, which is written in C. The system software needs to be augmented and maintained with supplemental software written in C. As it is desirable to have a common culture for systems and applications software, the general use of C is advantageous. Programmers can switch easily from systems to applications work, and sections of systems code may be used directly in applications software. In addition, C is increasingly popular as a standard language. This usage (together with related vendor support) makes it likely that the use of C will result in software transportability from other systems and into the future. Finally, C is such a powerful and flexible language that it is capable of meeting the requirements of demanding applications software.

Characteristics of C. Turnkey applications must be developed to run efficiently and to make efficient use of computer resources. C is well suited to such purposes because it is a high-level structured language, with all required computational facilities, and features such as data structures and address pointers. In cases where assembly code programming would otherwise be needed, such as for direct access to hardware facilities or fast processing of data arrays, C permits such operations to be performed directly. Thus C makes it possible to write an entire demanding application as an integrated package in a single language, without resorting to unportable assembly language.

Savings and Benefits. C can lead to statistically and dynamically efficient code, although this is, of course, also a function of programming practices and compiler performance. Experienced programmers report that they can often develop working code in a shorter time than is possible with other high-level languages. When C is used for functions that would otherwise need to be implemented in assembly code, the productivity increases can be an order of magnitude or more—mostly in reduced debugging time. The language naturally anticipates good programming practices, such as the use of meaningful names for variable and structure elements. This results in self-documenting code,

```
#define true 1
#define false 0
#define size 8190
main() {
    char flags[size + 1];
    int i,prime,k,count,iter;

    for(iter = 1; iter <= 10; iter++) {  /*do program 10 times*/
        count=0;                          /*prime counter*/
        for(i = 0; i <= size; i++)        /*set all flags true*/
            flags[i] = true;
        for (i = 0; i <= size; i++) {
            if(flags[i]) {                /*found a prime*/
                prime = i + i + 3;        /*twice index + 3*/
              /*printf("\n%d",prime);*/
                for(k = i + prime; k <= size; k += prime)
                    flags[k] = false;     /*kill all multiples*/
                count++;                  /*primes found*/
            }
        }
    }
    printf("\n%d primes.",count);  /*primes found on 10th pass*/
}
```

Figure 4. The prime number program in C. In C, comments are set off by a leading /* and a trailing */

which renders C programs readable, maintainable, and directly transferable. Minimally documented programs have been transferred between programmers, or even acquired externally, and modified or enhanced after reading only the source code.

In this time of substantial and increasing software development costs, the financial benefits of the C language are twofold. First, the rapidity of software development by fluent programmers helps to limit initial program development costs. Then, the improved software maintainability reduces future costs as programs are enhanced or transferred to other uses. See Figure 4 for the prime number program in C.

Nelson Analytical
10061 Bubb Rd.
Cupertino, Calif. 95014

Correspondent: Glenn Ouchi

Spectra-Physics, Inc.
Autolab Division
3333 North First St.
San Jose, Calif. 95134

Correspondents: Barrett L. Tomlinson and Carl Flarity

BASIC: A Programming Language for the Masses

The creation of BASIC was the start of the revolution that now allows nearly anyone access to and control over computers. Early forms of BASIC were written for machines with severe memory limitations and are rightly criticized. An ANSI X3.60 BASIC, unenhanced, is not suitable for laboratory control applications. This capsule will discuss the use of BASIC as a valuable language for both simple, single-use programs and for larger programs developed by project teams.

Interpreted BASIC has a distinct advantage over all compiled languages in rapid program or algorithm development. Each new version does not need to be compiled, linked, and loaded before each debugging test.

In some systems, program development can be iteratively pursued using an interactive BASIC interpreter. The finished program is then compiled with a BASIC compiler. Versions of BASIC variously support character string and file handling, extended-length variable names, and structured programming features.

BASIC is the language selected most frequently to implement chromatographic data systems. Our data systems divide the work load into two major parts. Because of speed requirements, acquisition and processing of data are performed using software written in assembly language. Final data processing and report generation are done in BASIC.

The products of our data system represent two different hardware strategies, but have similar software architectures. The Spectra-Physics line of computing integrators incorporates data acquisition, instrument control, and data processing and report generation into a single module supported by a single microcomputer.

```
5 DEFINT A-Z
10 dim flags(8191)
20 print "10 iterations"
30 for m = 1 to 10
40      count = 0
50      for i = 0 to 8190
60              flags(i) = 1
70      next i
80      for i = 0 to 8190
90              if flags(i) = 0 goto 170
100                     prime = i + i + 3
105 REM                 print prime
110                     k = i + prime
120                     while k <= 8190
130                             flags(k) = 0
140                             k + k + prime
150                     wend
160                     count = count + 1
170     next i
180 next m
190 print count,"primes"
100 end
```

Figure 5. The prime number program in BASIC

The Nelson Analytical System emphasizes data storage for postrun manipulation. It uses a networking scheme with separate modules for acquisition and control, and a central microcomputer for processing, storage, display, and reporting.

A BASIC That Was Developed "In-House." The principal design objective for the Spectra-Physics line of computing integrators was software that could be used and modified by chemists with little or no programming experience. Users wanted to alter the format of output reports as well as calculation routines. The system had to provide for interactive control of GC and LC equipment by user-written programs. Spectra-Physics enhanced the original Dartmouth version in the following ways to meet instrument needs:

- Concurrent execution of programs was implemented. This permits a user to edit a parameter file using an interactive dialogue while another program prints a postrun report.
- External switch closures trigger execution of specific BASIC programs. This gives the user programmable function keys and "interrupt" process handling capabilities.
- All background tasks are BASIC programs stored in ROM. A number of commands were added to BASIC that can initiate execution of the various reports and functions. Preprogrammed tasks, such as *reports* or *status*, can be used by an operator who knows nothing about the BASIC language.
- A feature was added that allows the user to redefine (overload) ROM BASIC statements by typing in new line definitions as normal BASIC statements.

A Commercial BASIC Adapted for Laboratory Applications. Nelson Analytical selected BASIC for its microcomputer products because programs can be rapidly developed, and the user has access to the code for custom programming. Hewlett-Packard Series 200 microcomputers are used, and their implementation of BASIC has been found to be excellent.

This BASIC has the capability to interrogate the current value of variables by simply pausing the program (either by inserting the command PAUSE in the program or pressing a PAUSE function key) and then typing in the variable name. A TRACE feature, which lists all variable changes, makes program debugging fast and simple.

Other features of this implementation of BASIC are:

- Multiple-character variable and array names. Instead of one or two letter variables like Al(I) or Z, one can have meaningful names like Area(Peak_Number) and Component_Amount.
- Program lines and subroutines can be referred to as named labels instead of just line numbers. Thus you can "GOSUB Print_Report" rather than "GOSUB 6000".
- Character string manipulation commands are available.
- Easily used formatting for output and input is available. Input can be checked for proper type (numbers or text characters) and value range.
- To increase the execution speed of a program, subroutines can be written in assembly language or Pascal and called from a BASIC program.
- Graphics and external instrument control capabilities are built into the language.
- Recovery from errors can be gracefully handled.

This version of BASIC is interpretive, and the source code is immediately available to the user. If the user wants specific formats for reports or additional information calculated and reported, they can be added easily. Such programs can be written by the user and automatically called by CHAINing. See Figure 5 for the prime number program in BASIC.

Syracuse University
NIH Biotechnology Resource for Multinuclei NMR and Data Processing
Chemistry Department
Syracuse, N.Y. 13210

Contributors: Charles Dumoulin and George Levy

Fortran 77

Since its genesis in the mid-1950s, Fortran has become the most widely used computer language for scientific and engineering applications. The Fortran IV standard, defined in 1966, is still in everyday use. Through the seventies manufacturers managed to add and extend their particular implementations of Fortran. In 1978 the American National Standards Institute publication X39-1978 established what is now commonly called Fortran 77.

Features. Fortran syntax is straightforward. Individual statements are extremely easy to read and understand, although segments of source code can become intractable if discipline is not exercised in writing the code. Each line does not have to be numbered; only those that are referenced by other source lines in GOTO and DO statements.

Several necessary features have been added to Fortran 77.

- The use of the IF . . . THEN . . . ELSE construct obviates the need for the GOTO statement and makes Fortran 77 a structured language in the same sense as Pascal.
- Another enhancement is the addition of a character variable type that makes the manipulation of text data much more straightforward.
- Error recovery has been added for all I/O statements. The error recovery mechanism is an embedded jump to another statement upon raising of an exception. Error codes can be stored in variables for further processing.
- I/O capabilities have been extended with the introduction of list-directed READ and WRITE state-

```
logical flags(8191)
        integer count,prime

        print *,'10 iterations'

do 92 iter = 1,10
           count = 0
           do 10 i = 1,8191
10            flags(i) = .true.
           do 91 i = 1,8191
              if (flags(i)) then
                 prime = i+i+1
C                print *,prime
                 count = count+1
                 k = i+prime
                 if (k .le. 8191) then
                    do 60 j = k,8191,prime
60                  flags(j) = .false.
                 end if
              end if
91            continue
92         continue
        print *,prime,' is the largest of ',count,' primes'
        end
```

Figure 6. The prime number program in Fortran

ments and an unformatted PRINT statement.

Although Fortran 77 is a fairly comprehensive improvement, the ANSI standard still has some room for growth. In some ways standard Fortran 77 is unnecessarily restrictive. For example, the use of quotation marks is not allowed, and the Fortran 77 standard does not have provisions for renaming files and for multitasking.

In our laboratory we make full use of Fortran's wide user base to disseminate software. ORACLE-1D is a powerful spectral data analysis system designed for the reduction of one-dimensional NMR data. ORACLE-2D has been developed for the processing of two-dimensional data sets. Both systems use color graphics output and are designed to be exported from our Data General (DG) computers to other systems.

Portability. We have successfully installed ORACLE-1D on several computers including the Digital Equipment Corporation (DEC) VAX 11/780. Several minor but insidious problems were encountered in these transfers. Fortran computational source code is very portable and is readily compiled on new hosts. Fortran I/O source code is not as portable. Most problems were due to different interpretations of the ANSI standard. For example, DG and DEC have chosen different default characteristics for OPEN statements. A worthwhile practice is to declare every file descriptor explicitly during opening. This solution is complicated by the fact that both manufacturers provide a variety of supplemental descriptors that are strictly incompatible. A particularly interesting case is the MAXRECL descriptor, which in DG Fortran is valid only for dynamic records, but in DEC Fortran is valid only for fixed-length records. Use of direct-access records is complicated by the fact that during file opening, DEC accepts the value of RECL in units of 32-bit words, whereas DG expects bytes.

Hardware and operating system differences also affect portability. The VAX 11/780 has a standard maximum floating value of about E 35, whereas the MV/8000 has a maximum floating value of about E 76. (DEC can provide an optional extended floating point format.) Thus any computations written for the latter processor have to be modified to run on the former if large floating point quantities are involved.

File name conventions are different in the operating systems associated with the two processors. Even if the source code compiles correctly, errors will be encountered at run time when file names are rejected by the operating system. Other differences between DEC Fortran 77 and DG Fortran 77 have been compiled in a list that is available from the authors upon request.

To enhance utility, each manufacturer extends its own version of Fortran. These extensions, however, are almost always incompatible between manufacturers and should be avoided if portability is ever a remote consideration in a program's design. The prime number program in Fortran appears in Figure 6.

Originally published in
ANALYTICAL CHEMISTRY **1983** *55*(8) 883A.

Operating Systems for the Laboratory

CP/M, MP/M, MS/DOS, p-System, Unix, Idris, RT-11, RSX-11/M, RSTS, RTE, RDOS. What are they? *Operating systems.* Most of the languages discussed in the previous tutorial (*Anal. Chem.* **1983,** *55,* 650–62 A) run under an operating system that allocates time and resources, such as disk and memory space, and provides powerful tools to protect the system and the user from errors.

A ticking clock paces the operation of the laboratory computer, approaching a million instructions per second. One instruction may start an analog-to-digital converter (ADC) that can digitize a signal in 25 μs. Yet the system can acquire only 25 data samples a second. Where does time go?

Time, Where Does It Go?

The assembly language routine to start the converter and wait for the finished operation involves a conditional loop. When data are available a pointer is manipulated and used to store the data. A counter is tested to see if all points have been acquired. These operations consume time. Because of this software overhead, a 25-μs ADC can really only acquire data at 20 kHz. If 50 μs is allowed for execution of integer arithmetic operations to determine if a peak maximum has been reached, data can be taken only every 100 μs; the acquisition rate is halved again.

Wait loops are not the only way to operate an ADC. The ADC can assert *interrupts.* These indicate to a running processor that an external device requires service. Interrupt service requires that the current task be stopped, current status stored, the request serviced, and the machine restored to its original state. These bookkeeping operations may take another 100 μs, reducing the acquisition rate to 5 kHz. A program written in an interpretive language or an inefficient high-level compiled language may reduce execution speed tenfold. Data can be acquired at only 500 Hz.

Finally, the operating system can reduce the data acquisition rate to 25 Hz or less. It may even preclude real-time response. The magnitude of its effect makes the choice of an operating system particularly critical. This tutorial will explore desirable and undesirable features in operating systems, current and future magnetic storage technology, and memory management techniques, which all affect the response time of a system. As an ensemble, these factors determine whether a computer system is applicable to a real-time laboratory environment.

Basic Elements of Operating Systems

An operating system (OS) is a collection of programs that allows users to create and use their own programs.

Kernel. The heart of the system is a *kernel.* A minimum kernel contains interrupt-driven software to control communication between the computer and the user's terminal, as well as between the computer and its disk. It contains a *command decoder* to interpret strings of characters input from a terminal into the required machine operations, such as:

COPY from: DISK2:DATA__FILE
to: PRINTER

The kernel also will control time, memory allocation, and disk space, using techniques to be discussed.

When the computer is started each day a "bootstrap" program, usually stored in ROM, is executed. This reads a more complex program from the disk, which in turn reads the required kernel into memory.

Drivers. Each peripheral associated with the system must have a software driver (or handler) that controls data transfer via interrupts. These are equipment specific. It is essential that the user insist on source code for all peripherals added to the system. As new OS revision levels are acquired it is common to find that drivers must be changed, recompiled, and linked to execute properly.

In some systems all of the drivers necessary to service installed peripherals are kept memory resident; others require only terminal/disk drivers to be permanent. The rest are read from disk as needed. Such a trade-off of required memory vs. access time is typical of the options facing a prospective OS user.

Compilers, Assemblers, and Libraries. See "Languages for the Laboratory: Part I," *Anal. Chem.* **1983,** *55,* 650–62 A.

Diagnostic and Maintenance Utilities. A good set of utility programs is essential. These permit files stored on disk to be examined in source code or object/binary code. Debugging programs execute programs a step at a time, allowing the user to locate logical errors in code. Maintenance programs also permit deletion and compression of the disk contents.

File Storage. As source programs in any language are compiled, object code and absolute, relocatable, and

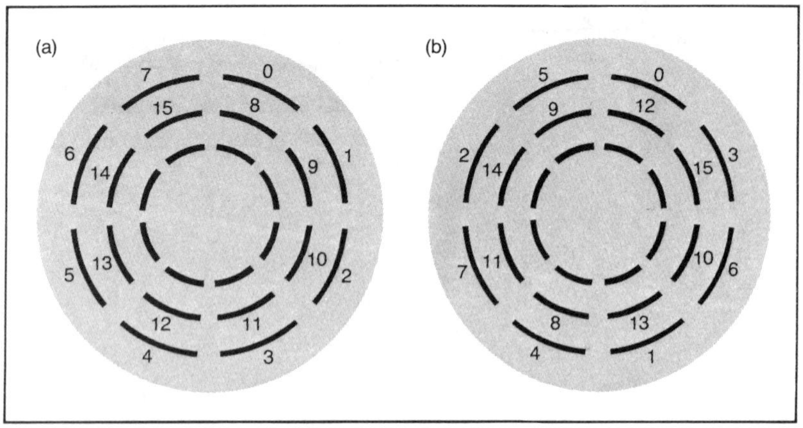

Figure 1. (a) Disk sectors. (b) Interleaved and staggered sectors

position-independent code versions are created and stored on disk. Each operating system has its own unique way of storing these files. Associated with each file is a core-control block that carries information specific to the program. The number and format of these vary from machine to machine, OS to OS, and revision level to revision level. The only common communication mode is the source code itself, transferred as ASCII files. The differences in language dialects and their associated compilers sometimes make even transfer of source codes a nontrivial task.

Finally, the floating point format used by different vendors, or even by different OSs running on the same processor, may not be identical. Again, protocols are important.

Choosing an operating system is therefore an important step that should be carefully analyzed. Many people purchase a piece of computer hardware and learn to live with the OS provided. If the CPU is from a vendor with a large installed base this will probably have good consequences.

Purchasers of turnkey systems with an embedded computer may acquire the same piece of hardware, but find themselves using a customized piece of software running under a nonstandard OS or running as a stand-alone unit. A nonstandard OS makes it extremely difficult to use the computer for other functions. Even with program listings, altering existing software may be nontrivial. In the latter case it will be impossible.

Disks

Operating systems provide access to and control of disk space. A disk is a plastic or metal medium upon which a coating of magnetic material is deposited. As the disk is spun, read/write/erase heads induce magnetic domains in the material using the same technology found in tape recorders. The heads are located on movable arms driven by digital stepping motors or voice coil assemblies. At each resting point in its radial motion, the head assembly can read/write/erase a circular path of magnetically encoded data called a track. Fifty to 1000 tracks per radial inch are found. Several inches of radial space form the recording area.

Electrical Protocols. The data are written onto the media by flux changes in the write coils of the head, and reading is accomplished as the newly created magnetic domains in the media pass under the read coils. Various technologies are used to encode the data, described by the terms NRZ (nonreturn to zero), phase encoding, or FM (frequency modulated) and MFM (modified FM). MFM records at twice the density [bits/circumferential inch (bpi)] of FM, giving rise to the terms *single* and *double density* recording. Densities of 2000 to 10 000 bpi are found. This is transparent to the user. However, the actual recording method will affect whether media written on one system can be read by another. Thus, the physical protocol, involving the size and housing of media, and the electrical recording protocol must be identical for media exchange to succeed.

Physical Protocols. Three basic types of physical protocols are found in laboratory computers.

• *Rigid* disks comprise a family that is characterized by a metal medium housed in an inflexible plastic case. The ferrite core heads are usually driven by massive voice coil magnets that allow fast, accurate positioning. The heads fly microinches over the medium surface on an air cushion. The medium is usually removable. The heads remain "unloaded" during the disk-mounting procedure. When the medium is up to speed, the heads move into position over the disk. The disks are about 14 in. in diameter and rotate from 1500 to 3600 rpm. Single- and double-density recording techniques are used.

• *Flexible* disks, popular because of their low cost, use a Mylar support base. The coated medium rotates in a plastic envelope lined with a dirt-removing material. The heads are usually driven by digital stepping motors and ride on the medium during read/write operations. This, coupled with the contact of the medium with the envelope, limits the rotational speed to about 360 rpm. Single-density recording techniques provide an almost universal standard in the form of IBM 3741 compatible media. Double-density recording techniques differ greatly from manufacturer to manufacturer. Some systems record on one side of the medium only, others use both (single sided vs. double sided). Diameters of the disks range from 3 to 8 in.

• *Winchester* technology uses rigid media. The heads, often fabricated from thin-film technology, ride on the disk when it is static. As the disk begins to revolve, the heads slide across the surface. Winchester disks therefore have very high capacities. They were originally made nonremovable to lower construction costs. This has led to problems related to the cost and inconvenience of maintaining backup copies (a duplicate copy of the Winchester's contents). Streaming magnetic tape (reel to reel), floppy disks, and various tape cartridges are used. The future would appear to lie with removable Winchester media. Packaging standards (size, shape, and head disposition) are necessary before this is a settled area. Winchesters are available in various sizes, ranging from a few inches to 8 in. in diameter.

Format Protocols. As disks are used, it is common to want to erase old information and add new. To accommodate this process, all of the tracks are subdivided into sectors that contain the same number of bits of information. This might be anywhere from 64 to 1024 bytes (1 byte = 8 bits). Each sector is separated from its neighbors by an intersector gap, where no recording occurs (Figure 1a). At the beginning of each sector a preamble is written by a hardware–software combination called a formatter. This preamble contains a series of characters that the disk controller electronics can use to recognize the beginning of a sector. The physical address of the sector is included, indicating track and sector number. At the end of each sector an error-detecting word (CRC check) is written. In subsequent read operations the CRC is calculated again and the two values compared. An error is reported if the two do not agree. Such disks are termed *soft-sectored* (software formatted). If holes or slots in the media delineate sectors the disk

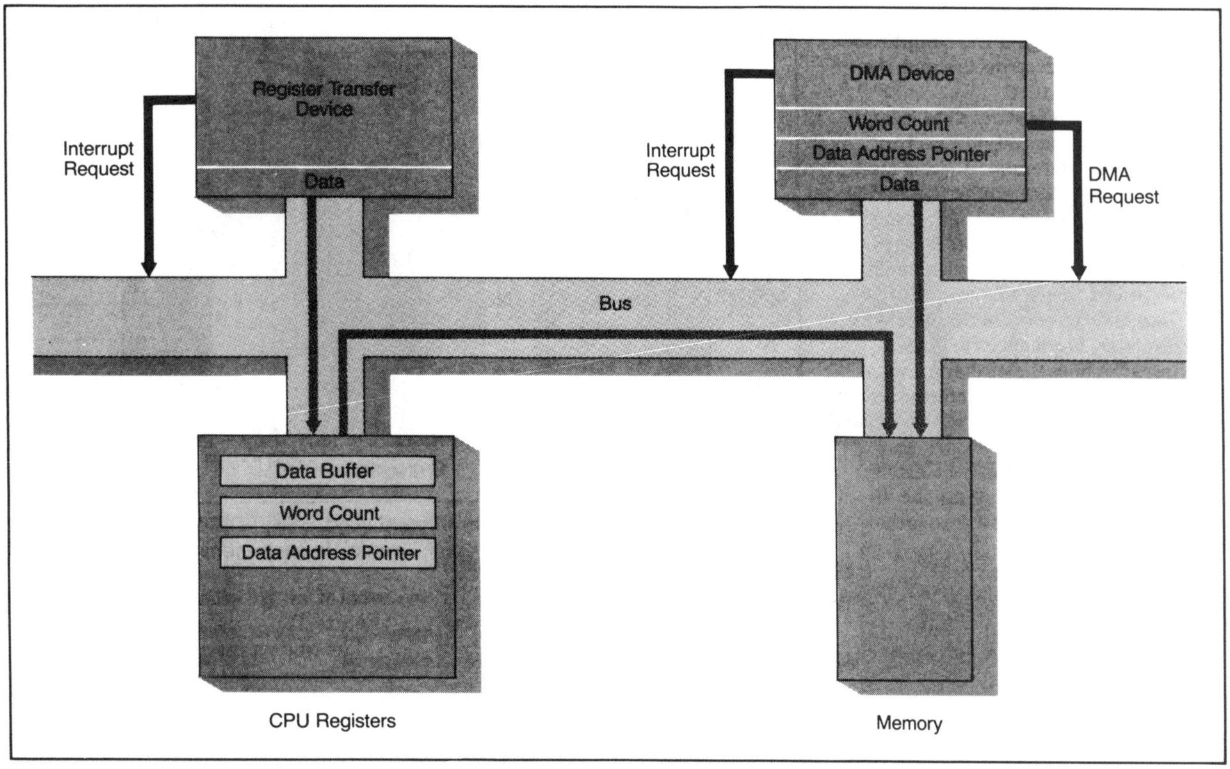

Figure 2. With interrupt requests data go "through" the CPU to memory. Data go directly into memory when direct memory access (DMA) is used

is *hard-sectored.* Another protocol to be met if disks are exchanged!

Access Latencies. The access time of a disk is fixed by the worst-case time required for radial movement of the head assembly to the desired track and the waiting period involved until the data appear under the head as a result of disk rotation. These are called head latency and rotational latency respectively. The size of the object being read, the writing density, and the disk rotational speed then determine the read or write latency (Table I).

Data transfer via interrupt control requires the update of memory or CPU registers that point to the next storage cell and count whether all of the data have been transferred. The software overhead is a few hundred microseconds for each datum transaction. If the counter and pointer registers are placed in hardware, a datum can be directly transferred to or from memory by stealing a few hundred nanoseconds of time from the CPU. This is called direct memory access (DMA). DMA facilities use the interrupt service only when they have completed the transfer of an entire block of data. DMA is used for all disk/memory transfers. Only physical motion now limits disk access (Figure 2).

To optimize disk access times, it is often desirable to interleave sector addresses, as shown in Figure 1b. This lets the computer process data from one sector while interleaved ones pass under the heads. With proper formatting, the next required sector will pass under the heads exactly when needed. Such separation of physical and logical assignments is common. Otherwise each new sector would require a seek operation, and the worst-case latencies would occur time after time. For a similar reason, as one "spirals" to the next track, the sector numbering assignment is often staggered to allow time for the head motion to occur without incurring the worst-case rotational latency each time. These format changes can affect speed performance of the system by a factor of five or more.

In operating systems (OSs) that support multiple users or multiple tasks many calls for disk service may be made concurrently. In some OSs the disk facility is grabbed by the first task and not released until done, a first-come, first-served arrangement. In managed systems, service requests queue up, and the head motions are optimized. Imagine a robot running up and down a series of aisles, depositing finished parts in bins and picking up new raw materials. It could handle requests for service as received. However, a more efficient way would be to sweep a repetitive path up and down the aisles in sequence, depositing and collecting, and then returning to a home base after each sweep for new orders.

Such considerations are vital. Although a disk seek may involve only a few hundred milliseconds, access to a required area may involve five to 25

Table I. Characteristics of Typical Laboratory Disk Systems

Disk type	Capacity (Mbyte)	Rotational latency (ms)	Head latency (ms)	Transfer rate (μs/byte)
Flexible				
Single density	0.25	160	400	32
Double density	0.5	160	400	16
Double sided	1	160	400	16
Rigid	5	30	100	3.2
Winchester	20	15	60	1.6

accesses, depending on the complexity of the data storage structure and the directory architecture.

Directories. OSs store data as logical blocks. In some systems, sectors and blocks are the same physical size. In others many sectors may make up a block—yet another protocol. OSs generally keep a directory of stored programs, indicating their names, beginning locations (track and sector number), and lengths (number of blocks). The directory has a specific format that includes how the alphanumeric file names are stored. In some systems this is in ASCII format; in others compression techniques are used—again, another protocol that must be met for interchange. In some systems the directory is organized in ways to reduce search time; in others entries are kept only in the order of creation.

Each volume has its own directory. In some systems these directories are isolated entities that may not be connected logically. In other systems, such as Unix, the various directories can be viewed as a hierarchy, with the directory of one volume serving as the node of another. The latter is a very powerful tool for the programmer.

Contiguous and Noncontiguous Storage. As programs and data are stored on disk, the operating system will use one of two basic methods to store the data. In one method all sectors associated with a file must be contiguous. That is, in a read operation, once the starting point of a file is reached subsequent physical sectors are read until the end is reached. This is done very quickly. However, when the file is written, enough contiguous space must be available to contain it. Otherwise the disk must be compressed, a time-consuming operation that rewrites the entire surface, shuffling all the unused space to the end of the volume by copying operations. This cannot be done when the system is functioning in real-time mode.

The second method writes a file into the first available space on disk. When that space is full it looks for the next available space and continues writing there. A link is kept that allows a read operation to logically follow the writing path. Although the storage mechanism looks logically contiguous to the user, the sectors involved are physically noncontiguous. As long as free sectors are available, this method allows storage in real time. However, as the files become more fragmented, many seeks are required to collect all the fragments. The system begins to "thresh" around with the heads in constant motion to and fro. At that point some collections algorithm must be run, or the real-time response of the system degrades.

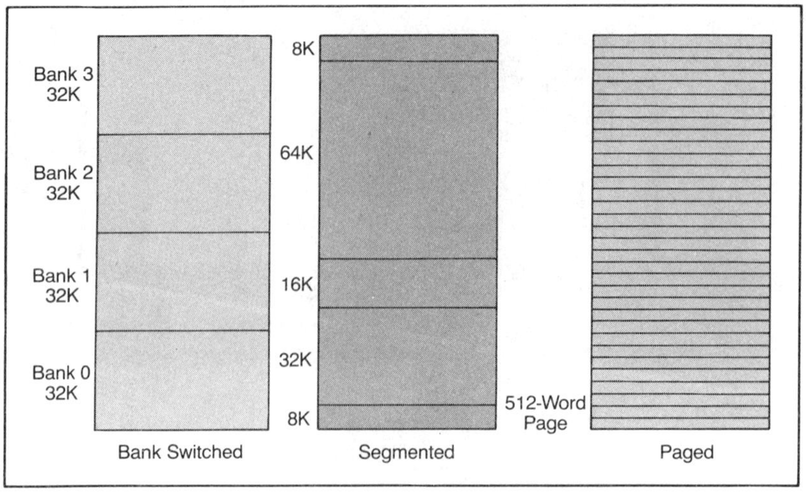

Figure 3. Memory can be divided in different ways. "8K," "16K," "32K," and "64K" refer to the number of computer "words" in each memory segment

In an ideal system both contiguous and noncontiguous storage exist concurrently.

Memory Management

A specific register in the CPU specifies the next address in memory to be used. This is called the program counter register. Processors with 16-bit program counter registers can address 64K addresses in memory. The required memory space can easily exceed the addressable space.

Bank Switching. Hardware solutions to the problem of addressable space on 16-bit program counter machines have been sought. One simple solution is to *bank switch* physical memory. This approach divides all memory into *n* banks, each with 64K addresses available. Simple hardware activates one memory bank at a time. The normal program counter is used to address relative locations in each bank.

Paging. Paging divides memory into small equal-sized units. *Page descriptor registers* are used, in conjunction with the program counter register and appropriate software/hardware, to "map" from logical memory space into a physical memory space.

Segmented Memory. Long programs require extensive space, short ones only minimal memory. A scheme can fence memory off into segments that are of variable length, usually some power of two. *Segment descriptors* contain information on where a segment begins and how long it is. This "modulo two" approach is a compromise between all segments having a size capable of handling the largest possible program that might be executed (very wasteful of space) and segments having an infinite number of lengths (a complex bookkeeping problem). Combinations of paging and segmentation are often employed (Figure 3).

The important questions to ask of such memory-mapped systems are: Can a program occupy more than 32K words, and is the memory mapping easily understood by the programmer? In many systems the answer to both these questions is no.

Virtual Memory. Look at an executing program as it attempts to run some procedure. If the required code is in a resident-memory page, it will be executed. If it is not, a "page-fault" exception will be raised, and the OS will do a seek to disk to bring it into memory. When it is available, execution will resume. Disk memory is now used as a *virtual memory.* Virtual array handling was discussed in the previous tutorial.

Contiguous and Noncontiguous Memory. In some OSs it is not necessary that the pages allocated to a given program be contiguous. Links are maintained that allow the physical memory to be mapped into what appears to be a logical unity by the user. This noncontiguous memory usage is similar to that described for disk storage. Hardware and software approaches are used to implement such features.

Although these memory management techniques have solved space allocation they may incur a considerable expenditure of time. Hardware designers implement storage of page and segment descriptors in high-speed *cache* memory. It is also possible to store entire blocks of program or data in cache memory. The rationale is that the next operation is related to the current, or just completed, operation. If the most recently used descriptors or program/data section are kept nearby in high-speed memory, they are readily available without paying the time price of a normal memory access

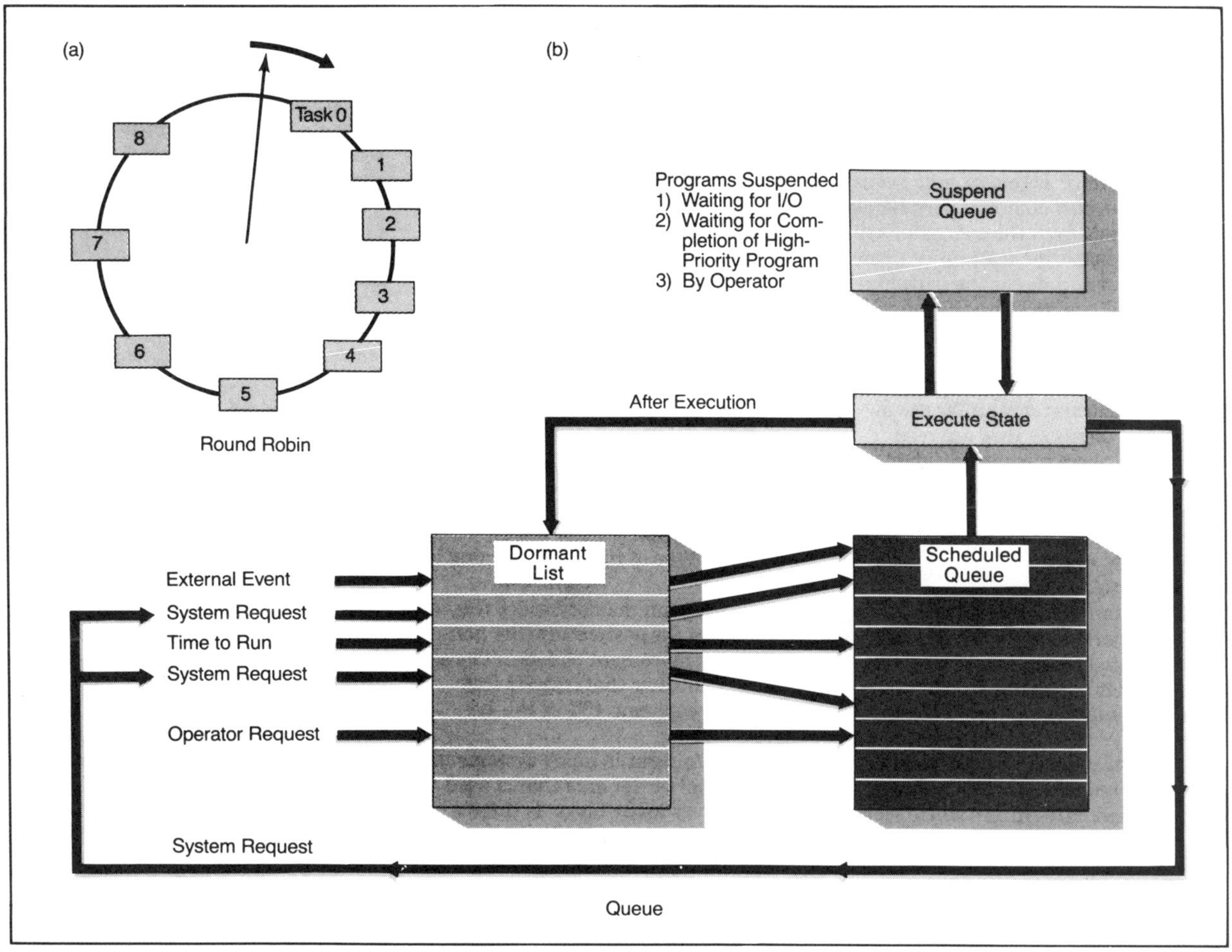

Figure 4. Multitask scheduling

or disk seek. Some systems claim to have 90% hit rates on desired items in cache.

Real-Time Lab Operating Systems

Why this obsession with time? Simply, if a laboratory system cannot respond in real time it might as well not be asked to do the task at all. We are looking at a world where the processor will be required to execute multiple tasks concurrently. Although the machine can only execute one task at a time, it can rapidly alternate between tasks in such a way that the various users see them as executing simultaneously. Allocation of CPU time to accomplish this can be done in a number of ways.

Round Robin. A round robin is a circular linked list of installed tasks. Each task in a round robin could control the processor until it voluntarily releases it to the next task. All tasks are interrupt-driven, and interrupts are serviced as requests are made. Tasks use round-robin time only between termination of one complete interrupt service sequence and the beginning of another. This requires careful programming, but avoids the high time overhead of a scheduler.

A round robin could allocate a time slice to each task. A system clock synchronizes the operation. Execution within a time slice allocated to a given task proceeds to the end of the slice. Tasks are removed from the circular queue only when they are completed, terminated, or blocked because of lack of access to a peripheral device or information from another executing task. This approach requires a scheduler in the kernel to coordinate the activity. As a system grows more complex, priority may be determined by such variables as entry time into the system, time to run, or declared importance. We have moved from real-time multitasking to the world of time sharing.

Queues. A system clock can be used to divide time into framing periods, separated by clock ticks. At the beginning of each period a scheduler creates a list of tasks to be executed based on interrupt requests, system requests, operator requests, etc. These are arranged in some priority sequence and then the queue is executed. All of the scheduled tasks must be serviced before the end of the framing period. If the system clock "ticks" too rapidly, most of the processor time is spent scheduling, not executing. This limits system clocks to a few hundred Hz (Figure 4).

It is difficult or impossible to predict the response time of many OSs. In some systems even interrupts are queued, that is, they are recognized, and the service request placed in a first-in first-out (FIFO) queue. How long it takes to finally service a request is determined by load. In many systems, data being collected from an ADC are stored temporarily in a buffer pool and flushed to disk when time and disk are available. Again, how soon that is done is determined by load. The user may wish to output data directly to a peripheral controlling his or her instrument. Can this be done? Usually not, since the operating system detects an attempt being made to control an input/output device. The OS queries the user to ascertain if this is an allowed operation and, if so, performs it. The system and the user are protected from incorrect acts by the user, but the protection costs time.

With these concepts in mind it is now possible to list some questions that a potential user might ask of an operating system intended for the laboratory. These questions have been generated by some idiosyncrasies of the popular OSs mentioned at the beginning of this article.

- Is it easy to install new drivers into the system? Is source code available for the most common peripherals and terminals from the computer vendor? Will the peripheral manufacturer provide source code for the peripheral? How will you read what is provided? (You must have a common disk protocol.)
- How is the disk directory accessed? Some systems buffer the directory segments as they are read. This keeps them memory-resident so they can be accessed quickly, as cache memory was used. For this reason, disk volumes must not be changed capriciously or old directory segments will be written onto the newly installed volume. Some systems read the directory periodically to see if you have changed disks! This can be very time-consuming. Is the directory structure suited to your real-time needs?
- Does the system support multiple-sector reads/writes? Some only support reading/writing a single sector at a time. Repeated seeks slow the system down.
- Does the system support absolute, relocatable, and position-independent code? Is it possible to interleave high-level and assembler code in the compilation process? If not, is the linking of high-level and assembler object modules well documented?
- Will the OS support transparent communication of binary information along serial lines? If not, all transfers will have to be done in ASCII, incurring a large penalty in time.
- Does the system support multiple users? Are reentrant versions of utilities available, or do all users have to have their own copy memory-resident?
- A *robust* system is protected from user error, makes it clumsy to program I/O, and is slow. A *fragile* system is very permissive about the actions it allows, is unforgiving, and executes rapidly. What are your requirements?
- Does the system support multitasking? How is execution priority established? Are real-time tasks locked into memory so that they cannot be swapped out to disk? Is it easy to install new tasks? Can tasks communicate with one another? Do the OS and its languages support coroutines or rendezvous? Can one task start another? Is it possible to terminate a process from the operator's console gracefully and without going through the kernel monitor?
- Are interrupts handled directly by their own service routines, or is a common interrupt handler used? Does the latter service requests immediately or queue them? How long does it take to execute?
- Is it possible for the user also to use the system clock to activate other events? Some systems only permit this if other critical functions, like the exception handler/abort processor, are given low priorities, making the system prone to fatal "hangs." The only solution may be the addition of another clock subsystem. If the system clock must be used to schedule all events, is the activation of tasks load-dependent?
- What is the disk architecture? Is it an industry standard in physical size, recording technique, and sectoring? Can reformatting of disks be done by the user?
- What type of memory mapping is supported?
- As program memory space fragments, does the system support noncontiguous placement of code? The alternative shuffler that collects free space may consume 40% of the system's time!
- Input/output in many systems involves a pool buffer area that is used commonly by many tasks. Is there a tendency for this area to become highly fragmented so that the system spends too much time "chasing" the linking pointers? Is it possible for a "pipeline" type of operation or excessive demand to nearly fill the pool buffer area? Many systems experience a fatal hang under these circumstances.

The Future

It has been traditional for computer vendors to develop and husband their own operating systems in a possessive manner. The personal computer market has demonstrated that portability in operating systems may be as possible as portability of languages. Some vendors have created an open environment with other systems houses and hardware manufacturers who wish to supply OS-compatible hardware and software products.

The popular MS/DOS, PC/DOS (Microsoft), CP/M (Digital Research), and the p-System (Softech) have demonstrated that industry standard systems can allow machine-to-machine portability in an *almost* user-transparent manner. The latter is a p-code-based system similar to the approach used for p-code Pascal. CP/M is a single-user OS; MP/M is multiuser, with some multitasking capability. These OSs are slowly evolving to include some of the attributes needed for a laboratory system.

This manuscript was prepared on a PDP-11 processor running RT-11, a DEC product that has become another type of industry standard. It is a single-user multitasking system. The disk hardware and driver came from General Robotics, the word processing software from Saturn Systems. All were quickly and fully compatible with the vendor's OS. DEC's RSX-11/M is another popular laboratory operating system suitable for multiuser multitasking. RSTS is DEC's multiuser time-sharing OS.

Both Unix and Idris are time-sharing systems. They do not normally support real-time multitasking.

There is a growing trend to put operating systems, operating system functions, and languages into silicon. This siliconization of software is typified by Intel's iAPX 86/30 and 88/30 Operating System Processors, which act as coprocessors to 8088 and 8086 CPU chips. The coprocessors arbitrate and handle interrupt requests. Many disk controllers are being conceived that handle file searching, hashing, indexing, and data compression in hardware. These steps will improve response time.

Magnetic recording techniques are improving rapidly. The next significant advances will result from the introduction of thin-film methods and perpendicular recording techniques. The dispersed particulate media now in common use, coupled with the ferrite core heads, limit writing density and resolution. Thin-film heads can be made by semiconductor technology, producing a head with coil and gap spacings suitable to high-density recording. Thin-film media minimize self magnetization and can be optimized for high writing densities. Finally, there is great interest in recording techniques where the magnetization induced in the media is perpendicular (PMR), not longitudinal (LMR). This removes the problems associated with high demagnetization values at transition boundaries. At the present, PMR densities are three times greater than LMR.

High-speed array processors and math coprocessors are providing the last element needed to allow the laboratory computer and its operating system to perform in real time. All of these tools will be needed as the next A/C INTERFACE tutorial addresses robots in the laboratory.

Acknowledgment

The author is pleased to acknowledge the helpful discussions with Michael Starling, Union Carbide Corporation, South Charleston, W.Va., and Jim Currie and Ian Chapple of the VPI&SU Laboratory Automation and Instrument Design Group that led to many of the philosophies presented here.

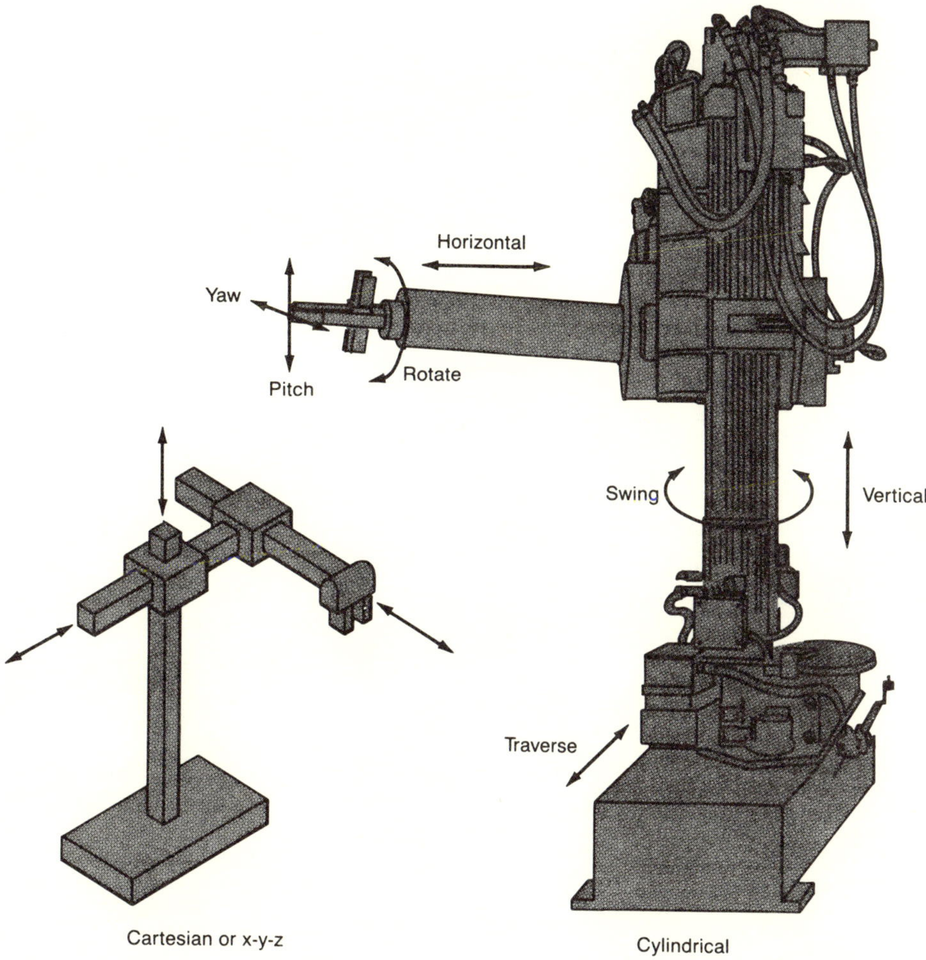

Figure 1. Common spatial configurations for robots

Originally published in
ANALYTICAL CHEMISTRY **1983** *55*(11) 1100A
and **1983** *55*(12) 1232A.

Robots in the Laboratory

Part I

Automation of a laboratory means far more than computerizing the lab. Previous tutorials in this series have focused on the capture, manipulation, and distribution of data in the electronic laboratory—all connected with computerization. Automation addresses the problems associated with the movement and manipulation of samples. These actions represent an ever-increasing cost factor and consume a major portion of every analyst's time. Industry has responded by using articulated computers that can see, hear, touch, and even talk: ROBOTS!

The industrial use of robots is driven by the simple fact that where human labor costs about $20/h, robots cost about $5/h. In production-line applications the quality of product is increased and the cost is decreased by installing robots capable of identification of objects, parts insertion, grinding, welding, painting, and palletizing. The analytical laboratory is a production environment in which it is necessary to identify samples, and then aliquot, dilute, concentrate, extract, crush or grind, filter, or move them prior to analysis. The analogy is obvious.

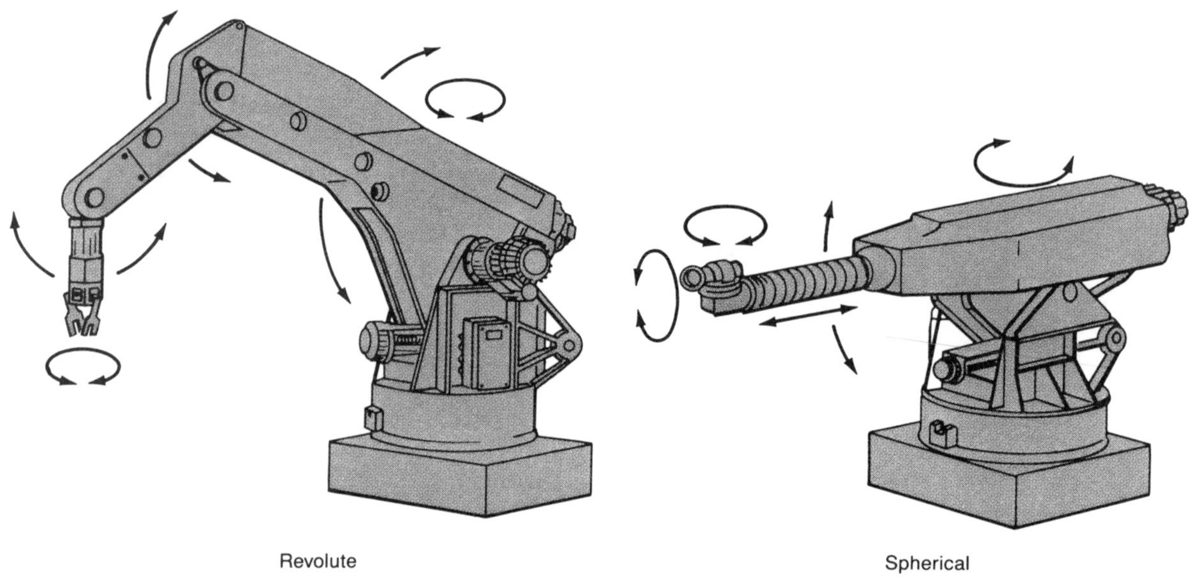

Figure 1. Continued

This tutorial will introduce the vocabulary of robotics, discussing the various geometries available, common drive systems, the exciting world of hands, how robots are programmed, and how they see.

Coordinate Systems

The performance of robots is intimately tied to their spatial geometry. Associated with each unique implementation is a body, an arm, and a gripper assembly. The body or arm may move so that the gripper assembly point sweeps out a space that can be viewed as Cartesian, cylindrical, or spherical. Joints are often added to the arm giving a jointed-spherical or revolute (movement downward and backward) configuration. At a minimum, the wrist assembly is capable of rotation, pitch, and yaw (Figure 1). The gripper assembly, in its most primitive form, consists of a pair of fingers that close to grasp an object. These degrees of spatial freedom allow the simple robot to perform many of the human arm and finger-pair operations.

Commercial robots are available in each of these spatial categories, and the prospective user should evaluate geometric needs carefully. Cartesian systems work well in planar-oriented applications. The remaining systems serve work station environments where the facilities to be serviced cluster around the central body. Cylindrically- and spherically-oriented systems reach toward things, while the revolute configurations reach over and around things.

The vendor should provide three-dimensional drawings of the accessible work space. For example, the cylinders and spheres mentioned are actually hollow, since the robot cannot manipulate space inside itself. Some work space restrictions are combinational, depending on the position of other degrees of freedom. For example, in some jointed arms the shoulder–elbow or rotate–pitch–yaw movements are coupled, and movement of one joint precludes complete freedom in the related joint.

Drive Systems

The speed, accuracy, and resolution of a robot movement is linked to the drive mechanism used. Both the motive force employed and its physical location are important.

- Hydraulic activation is used when heavy objects are to be manipulated. The power is easily generated, and the heavy hydraulic pumps and reservoirs required can be remote from the active hydraulic cylinder itself. This reduces the weight the arm must support.
- Where speed is required, pneumatic systems are employed; hands often are controlled pneumatically.
- For moderate load and speed requirements most commercial robots use electric servo motor systems. These work well in applications where large displacements are required.
- For lighter service it is possible to use digital stepping motors, which have been successfully applied where small load torques and displacements are involved.

Since electric motors are more suited to the chemical laboratory robot, it is essential to understand some of their strengths and weaknesses.

Servo motors. Servo motors are built around a feedback principle. A drive signal is applied to the system. The motor turns in response to this command. As the shaft of the motor moves, a tracking signal is generated that is related to the shaft position. This is fed back to the driving signal in a nulling operation. As long as the two signals are not equal the shaft continues to turn. When the two signals are equal, the shaft has assumed its assigned position and the motor stops.

Closed-loop hardware control. When dc servo motors are controlled by computers, the drive signal is generated by a digital-to-analog converter (DAC) that converts a binary number output from the computer into a voltage. Two common ways to generate the shaft position tracking signal use either a potentiometer attached to the shaft that functions as a voltage divider or a "synchro" device that outputs a signal proportional to its shaft position. Such position encoder feedback closes the loop in the motor hardware (Figure 2a). However, the computer control is open-loop.

Closed-loop computer control. For effective control of robot movement, the computer must be told what action has been taken as the result of the command it has given—true closed-loop computer control. A simple analog-to-digital converter (ADC) could monitor the potentiometer output, closing the loop (Figure 2b). Alternatively, a synchro/digital device could be used. A parallel input port on the computer could accept this binary information (Figure 2c). Digital/synchro and synchro/digital converters interconvert binary information and a signal proportional to the sine or cosine of angular revolution.

Another type of position encoder, called a digital shaft encoder, is often used. This encoder involves simple slots cut radially in a wheel, with a light-emitting diode (LED) on one side and a photodetector on the other. As the shaft and coupled wheel revolve, pulses are created that can be counted by digital integrated circuits or by the computer. Since this provides angular information relative to some starting point it is often called a *relative* shaft encoder. A more complex digital shaft position indicator encodes binary numbers along many radii of the wheel. These numbers are read by multiple LED/detector pairs placed on opposite sides of the wheel. This provides absolute position information and is called an *absolute* shaft encoder. Both devices use the parallel input port for closed-loop computer feedback (Figure 2c).

Speed of rotation is another element that should be monitored by the computer in a closed-loop system. A tachometer attached to the shaft can generate a voltage proportional to speed. The computer can sense that voltage with an ADC (Figure 2b). Digital tachometers include both of these functions in one package, and use a second parallel port for data feedback (Figure 2c). A shaft encoder device might be used to calculate speed. However, this places an extra computational burden on the computer.

Stepping motors. Stepping motors are of interest to robot designers because of their low cost and natural digital orientation. They are designed so that a simple current pulse on a winding causes them to rotate a small fraction of a turn, typically 0.9°.

The cost of a stepping motor can be kept low by having a computer generate the multiphase clock signals that are used to drive the motor, but this creates a tremendous software overhead for the processor. Some vendors use inexpensive integrated circuit controllers that allow the user to switch directions and step the motor by inputting simple digital signals.

The disadvantage of stepping motors is that their performance is often adversely affected by even moderate

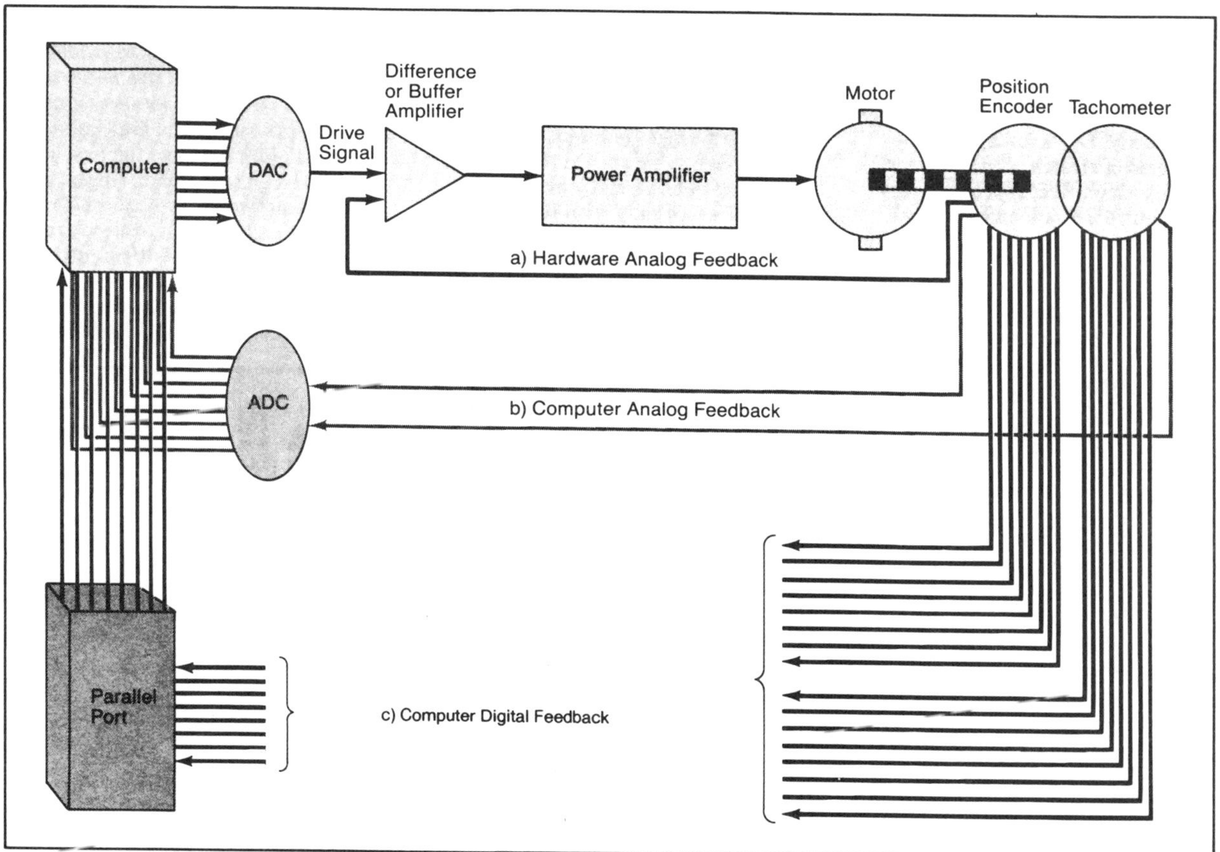

Figure 2. Stepping motor configurations

Hardware elements in yellow and signal paths in black across top of figure are elements common to all three configurations. The amplifier to the right of the DAC is a difference amplifier in configuration (a) and a buffer amplifier in configurations (b) and (c). **(a)** Closed-loop hardware control feedback path is shown in red. For this configuration, the tachometer is not used. **(b)** Closed-loop computer control with analog feedback is depicted in gray. **(c)** Closed-loop computer control with digital feedback is shown in blue increases in step rate due to resonance and torque loads. Under certain conditions, the motor will not respond to step commands and positional information on robot body or arm movements may be lost.

To relieve these problems, multistep motion should be started at a slow pulse rate and then accelerated. Since the motor is inherently underdamped, it tends to overshoot and oscillates for a time after it has reached its desired position; appropriate braking action should be used.

In well-implemented systems a closed-loop hardware system generates a pulse that signals the control system that a step has been taken. As higher step rates are requested each change in phase excitation must occur earlier to maximize torque. This is called *ignition advance*, analogous to the automated spark advance used in an automobile distributor. All these factors suggest that rather sophisticated stepping motor controllers and closed-loop operation are needed for well-controlled robot operation.

Regardless of which motor is used, the drive train is an important consideration. Gear trains, especially important in low-torque stepping motor installations, must have minimum backlash. Many systems place the motors at the joint to be powered to avoid the need for cable and belt drive linkages. This increases the load that must be carried by members and motors closer to the body. Drive linkages create problems of stretch, vibration, and slippage that affect reproducibility and require periodic maintenance.

Robots can be built to manipulate loads from a few ounces to hundreds of pounds, over distances of 1–2 m and at velocities near 1 m/s. But the laboratory scientist also is interested in the accuracy and reproducibility of that motion. Commercial systems quote figures that range from 0.1 to 0.001 in., depending on the load limit and construction. In most systems, reproducibility is much better than accuracy. Even in industrial robots it is common to find that the system's accuracy and reproducibility begin to degrade severely at long arm extensions, under heavy loads, and at high speeds. It is not always apparent what a quoted figure means. Worst case figures should be requested.

Hands

The human hand has some twenty degrees of freedom! It is capable of manipulating variously shaped objects in intricate ways using a sensory feedback system. Considerable work has gone into designing grippers and hands for robots. The standard robot hands have two or more rigid fingers that move toward one another. This motion may involve maintaining a parallel relationship between the contact surfaces or developing a closing angle (Figure 3).

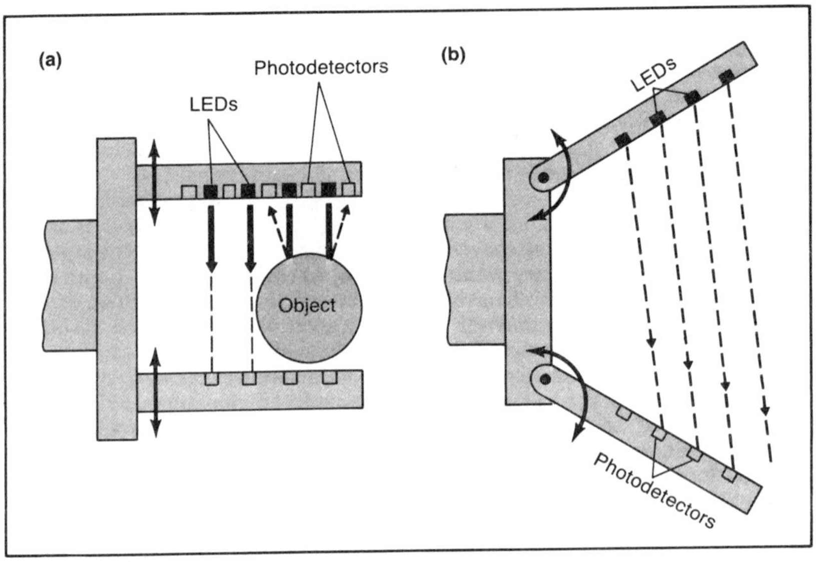

Figure 3. Two-fingered robot hands move to grasp an object. LED–photodetectors sense proper closure
(a) Parallel movement. **(b)** Angular movement

Syringes driven by stepping motors allow injection of reagents into various receivers; the ensemble can be made into a separate hand. If the robot is designed with replaceable hands, the arm can park its current hand and replace it with a hand designed to perform the syringe function.

Although some special purpose hands are commercially available, the user will eventually wish to design his own, and a survey of some interesting alternatives may help.

Both suction and magnetic hands have been developed. Pneumatic ballooning devices are designed to swell inside a hollow object, allowing the object to be picked up.

One imaginative attempt uses the cluster of interlocking fingers shown in Figure 4. The clutch pressure can be adjusted to accommodate the fragility of the material involved. No feedback is involved in the tactile contact.

Grippers that attempt to emulate the human hand also have been constructed. Figure 5 shows several different approaches. The multijointed fingers (Figure 5a) have the *first* joint retroflexive, like a human finger. The joints are activated by a motor-driven cable that causes them to close around an object. It is possible to monitor the drive current to the motor activating a given finger, providing force information that can be used in a force feed-

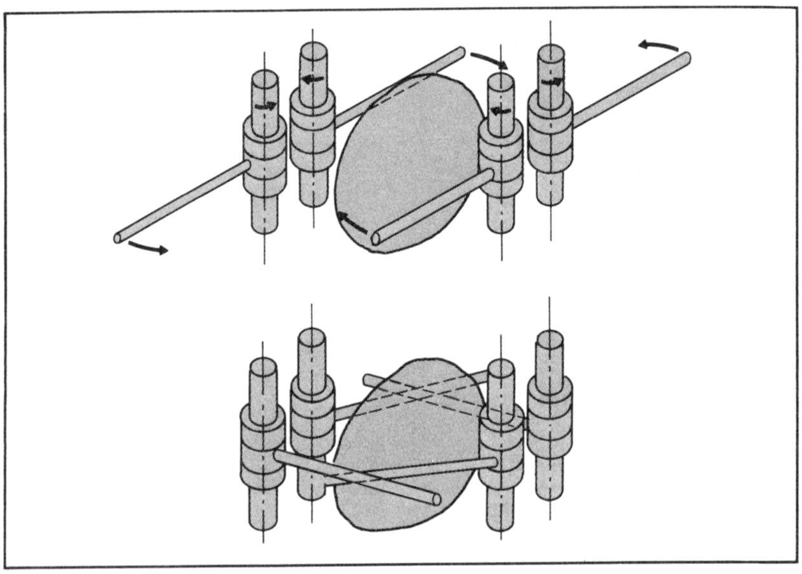

Figure 4. A cluster of interlocking fingers can be used to grasp an object

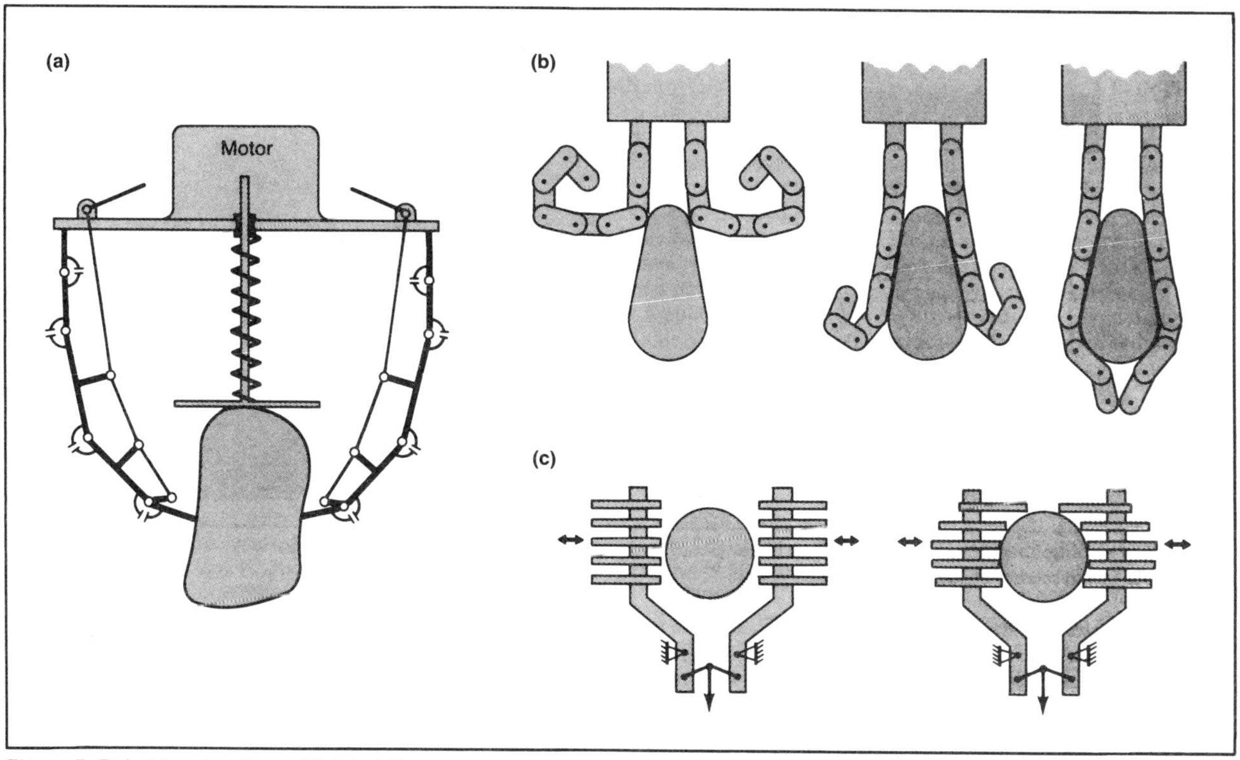

Figure 5. Robot hands with multijointed fingers
(a) Retroflexive joints like the human finger. **(b)** Joints close around an object. **(c)** Opposed set of parallel fingers close around an object

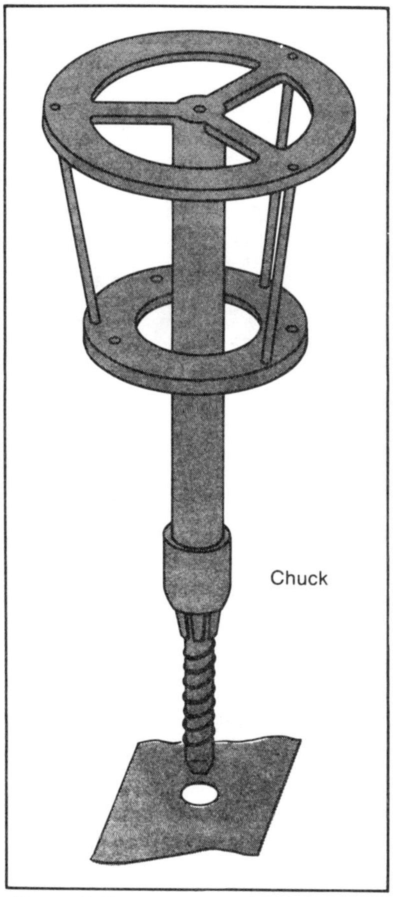

Figure 6. Remote center compliance gripper places an object in a hole
The lower ring is rigidly mounted, and the center of compliance is just below the chuck. The entire ensemble is mounted at the end of the robot arm

back arrangement to hold the object without crushing it. Imagine the problem of picking up an egg with two fingers, and then moving it in a pitch or yaw motion without breaking it. The new Salisbury hand has three triple-jointed fingers. *Each* joint is retroflexive. The drive motors used have force sensors installed. This hand is not anthropomorphic, and can do things we cannot, the ultimate goal of robotics. Figures 5b and 5c show other types of fingers.

Piezoelectric materials have the ability to alter their shape and position as a result of electrical fields applied to them. Several interesting micromanipulators have been developed using inorganic or polymeric organic piezoelectric materials. One small lobster claw gripper uses a dc signal to move a pincer toward a mating finger. An ac signal is used to sense force, providing an excellent force feedback gripper for handling very small objects. A micromanipulator with five degrees of freedom has been demonstrated using such technology. Polymeric plastic film also has been used to create small hollow cylinders that contract radially as potential is applied. This microinjector has interesting possibilities.

Placing shaped objects into similarly shaped holes is easy for the human. A complex process that senses force in multiple dimensions and then corrects the insertion process is involved. Although it is possible to build force sensors into complex finger arrangements and emulate this process, fabrication costs, software generation efforts, and program execution time are all high. An alternate approach uses a *passive* compliance system. A pin, free to move at its outer end, will eventually drop into a hole if the assembly is agitated slightly. The gripper shown in Figure 6 lets the remote center of the work piece move in compliance with the forces that the inserted end experiences, and the pin is guided into the hole. This approach is called remote center compliance (RCC).

Sight and Touch

Sight. The simplest optical detectors employ an LED light-source/light-pipe/photodetector assembly in the gripper to provide proximity information by transmission or reflectance. Multiple units in the gripper fingers have been used to permit coordination of wrist or finger motion as an object is picked up (Figure 3). Better visual feedback is being actively explored.

More sophisticated information on the shape and position of an object can be derived from linear photodiode arrays (Figure 7). These provide 256 to 1024 optically active elements on a single IC carrier. Each element can be viewed as a light-sensitive photodiode in parallel with a capacitor. Originally the capacitor is charged up to a fixed level. As light strikes the photodiode the capacitor is discharged to a level determined by the integral, over time,

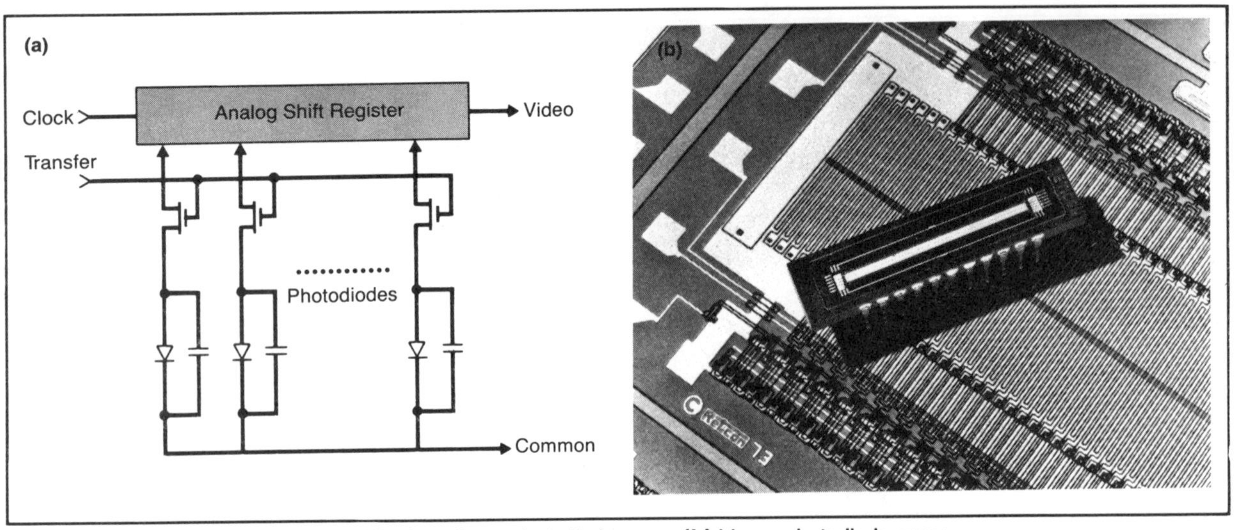

Figure 7. **(a)** Simplified schematic of Reticon linear photodiode array. **(b)** Linear photodiode array
Schematic and photo courtesy of EG&G Reticon, Sunnyvale, Calif.

of the light intensity striking the picture element, or pixel. Periodically each pixel is recharged, and the amount of current required generates an individual pixel signal that exits serially from the common video line of the IC. The serial string of pixel pulses can be integrated and digitally sampled, providing binary information on the light intensity reflected from an object. It is common to convert the information to a two-level grey scale (binary, with every pixel represented as a logic 1 or 0) in order to reduce the amount of information. Such linear scan information is used to detect edges of objects and orientations of simple objects.

For more complex objects and purposes a square photodiode array is used to create a solid-state camera. A 256 × 256 array is common. If a 16-level grey scale is used, 32 Kbyte of data are generated per frame. At 32 frames/s, 1 Mbyte of information must be processed! The magnitude of the information flow and the difficulties in emulating the pattern recognition capabilities of the human eye–mind combination make robot vision a formidable task.

A first step is to use variable and contrasting lighting to enhance image characteristics. It is often possible to produce an image that can be made binary. Edge analysis programs explore the stored image for points of transition. These points are stored as a logic 1 in an array where all other elements are logic 0 (Figure 8a). The resulting binary image is useful to the robot in determining the class of an object by template matching. In this technique reduced binary images of all anticipated objects, in every stable position, are stored. This may involve as many as 10 parameters, including the object area minus any holes, hole areas, number of holes, perimeter dimensions, aspect ratios, and so on. In complex shapes a concurve (*con*nected *curve*) description of the object is stored. This represents the line segments and arcs that make up the object, stored as curvature, length, and connectivity. A search and match routine identifies the unknown and its position.

As the objects grow more complicated, multiple-level grey scale images are equalized to normalize local and global contrast ratios in the captured image. This produces a more uniform image. The reduced image is then subjected to a cluster analysis program that searches for areas of uniform characteristics. Adjacent continuity allows the image of the part to emerge, where it can be represented by an even more reduced data set.

If a large number of different objects or foreign objects are involved, various pattern recognition techniques are employed. Radial lines can be drawn from the point shown at right in Figure 8b. The characteristic histogram plot of the intensity pattern found for each ray element can be

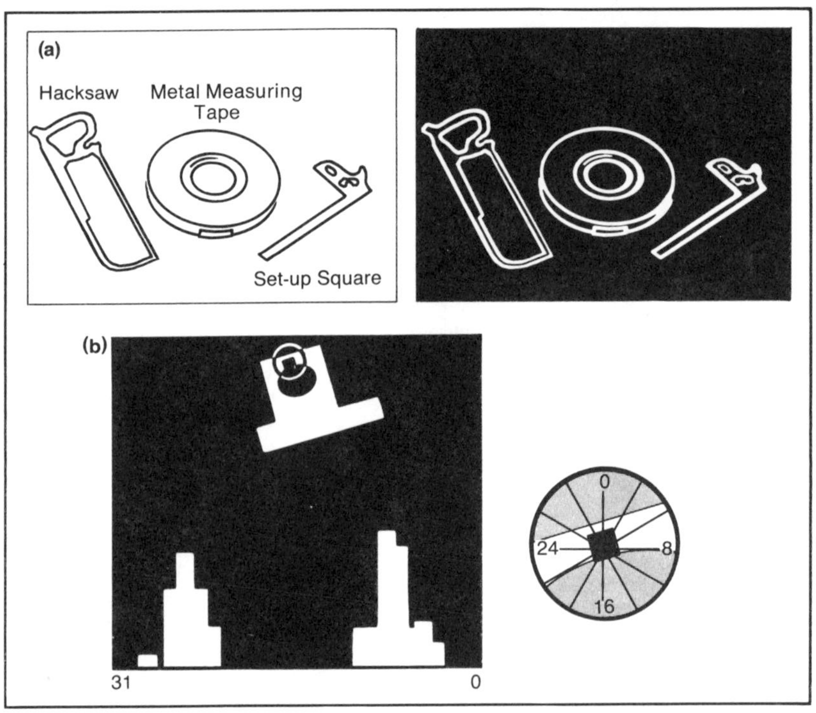

Figure 8. Edge analysis and cluster analysis
(a) Original objects (hacksaw, metal measuring tape, and set-up square) at left. Edge detection binary image at right gives outline of original objects. **(b)** Cluster analysis binary image is at top left. Binary histogram, bottom left, is obtained from radial profiles drawn from a point, right

used to distinguish between variously shaped holes and their relationship to edges, etc.

The texture of many objects can be enhanced by patterned lighting. For example, alternate patterns of dark and light bars create an identifiable pattern on the surface of a three-dimensional object. Laser scanning and ultrasound techniques also have been used to elicit shape information.

As the amount of data increases, array processors, coprocessors, and other hardwired devices must be used to abstract the visual information in relevant time. The frame of visual information is "grabbed" by a separate memory device. This frame grabber/buffer can then process the data and make the reduced form available to the robot's computer by direct memory access (DMA). This approach may take only fractions of a second, in contrast to the minute or so that the control computer would take. Correlation and Fourier transform compression operations have been implemented in this way.

Touch. The least expensive hand implementations use microswitches to detect closure on an object. This provides a yes or no response to contact with the item. Conductive and capacitive effects in thin films have been used to provide a quantitative response. Wire grids and conductive–compressive polymer films have been used to create macromatrices that can sense the quality and quantity of depression as an object is grasped. The possibility exists of fabricating 100 × 100 micromatrices, using IC technology, to produce sensing pads for fingers that would provide thousands of *tact*ile *el*ements, or *tactels*, that could be used like pixels for sensing.

Programming

In relationship to their motion in space, robots may be divided into point-to-point and continuous path types. In the former the spatial positions of the points the gripper should transit are stored in sequence together with the dwell times. As each move is initiated, the new coordinate values are "jammed" into the motor control and the movement is executed. This usually involves simultaneous movement in many degrees of freedom. The most direct, efficient path is invoked. Although this appears to be a simple task, it does involve the concurrent execution of multiple tasks—viewing the control of each motor as a separate operation. If three degrees of freedom are involved, each with different displacements, it will be necessary for the multitasking operating system to execute the three movements at different rates so that the end motion will appear uniform to the observer, and all movements will start and stop synchronously.

In continuous path control, time (or occasionally, space) is divided into fixed increments, and the coordinate commands for each motor are issued at every clock tick. This allows the gripper or tool to follow a predefined, complex path. This may be taught to the robot, or equations may be used to calculate the predefined path, or sensor techniques can be employed to make the robot track a desired path. All three have been used in welding robots that create complex bead tracks. In a point-to-point system, as the point spacing decreases, the path tends to become continuous.

The schooling of robots can involve explicit programming, lead-through, and walk-through approaches. In explicit programming a series of statements expressed in a suitable robot control language is written and then executed. The best languages are interactive and immediate during development, and compiled for run-time application. Macro- or subroutine libraries are provided to make program generation simple. Typical commands such as MOVE_TO, PICK_UP, PUT_DOWN, ROTATE, and ELEVATE are available. Positions in space are defined and given names. These may then be used as arguments for the above primitives. A FORTH-like language is ideal for this environment since parameters on the stack can be used as arguments for the words defined in the FORTH dictionary. Another very viable approach uses dictionary entries in which parameters are stored along with a function name. Each dictionary entry belongs to a predefined type that uses the parameters in a prescribed way to cause execution of the movement. Regardless of the language used, certain attributes are essential:

- Since robot motions are best implemented using table-driven software, a pointer variable language is useful.
- Since external sensor input is commonly used to select alternative pathways for robot operation, an extended set of conditional and repetitive macros is essential.
- Many operations in robotics are repeated at different points in space. These operations should be able to execute on a relative position basis as far as the programmer is concerned, and not involve absolute coordinate posi-

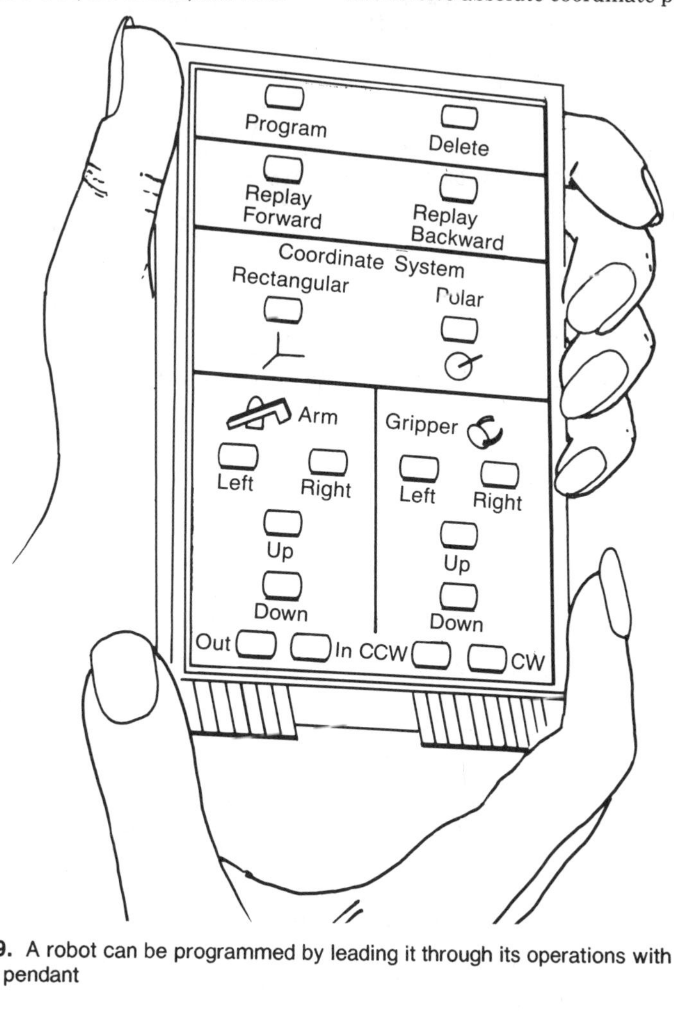

Figure 9. A robot can be programmed by leading it through its operations with a training pendant

tions. The language should support reentrant and recursive code.

• The robot is envisioned by the user as alternately working in Cartesian or polar coordinate space, or often mixtures of the two simultaneously. Coordinate transform packages must be available. Where revolute architecture is involved the user should not be concerned with the multiple motions in joint-space needed to implement what the user sees as a simple sliding motion. High-speed math operations are needed, requiring a broad spectrum integer math package. Excellent segmented and interpolated trig function algorithms are needed.

• The language should be extendable to provide for the definition of new operations.

• Concurrent multitasking must be supported, and well-developed task coordination commands should exist. This will permit multiple motors to move in synchrony, and allow coordinated use of multiple arms.

In lead-through programming, a training pendant (Figure 9) attached to the control computer has keys for each basic movement of the robot. As these keys are struck the robot executes the requested operation and the computer stores the command. In this way a program is transparently created for the user. The program may be named, stored, retrieved, and then executed at a future date.

Some systems use special function keys on the computer keyboard in lieu of a pendant. Others have soft keys associated with a CRT display. The keys' functions can be changed by software and are indicated by alphanumeric information displayed on the screen. In many systems, as the joints are iteratively moved into a target position by the operator of the pendant, the executing program strips out extraneous moves and only stores the most efficient point-to-point coordinates. If this is not what is required, then the points must be placed closer together.

In walk-through programming the robot activators have the position feedback mechanisms activated, and the operator moves the components manually through the operation. The motion information is sampled periodically (e.g., 60 Hz) and the information stored. When the program is replayed, the motion replicates the training cycle's continuous path motion. Well-designed systems permit editing of the final sequence to correct minor errors, or to add small changes.

Time

Previous tutorials have focused on the effect that language and operating system choice have on the speed of executing programs and the ability to coordinate tasks. Robot control stresses both of these areas. Total software control, with clock-paced interrupt-driven service, may only support simultaneous activation of *three* stepping motors. Hardware support is needed for relevant time execution. In continuous path operations, the feedback normally includes corrections that are *proportional* to the error signal, as well elements that are a function of the *integral* of the tracking error (PI control). Additional feed-forward, or *preview* control, has been used successfully to prevent overshoot (PIP control). This can easily take 10% of the CPU time. True control of a multidimensional arm involves a complex set of nonlinear differential equations; joint interaction, friction, stiction, resonance, and momentum will cause loss of control, accuracy, and precision if not included in the control algorithm. Solutions to the kinematic equations are time consuming. It has thus become common in industrial robots to use a microprocessor to control each drive motor, with a fast central processor performing more global functions. Separate vision processors are used to collect and process such data, permitting visual coordination in real-time. As multiple robots are used in the truly electronic laboratory, networking will be essential to transfer data and commands from one station to another.

Money and People

The fiscal analysis of robotics is both easy and promising. General Electric Corporation estimates that the total cost of installing a $3000 robot is three times its purchase price. A $20,000 system involves about a factor of two. Return on investment experience from Unimation suggests that the pay-back period for a $35,000 system is a little over two years for a one-shift operation. The two-shift lifetime of a quality robot is demonstrated to be 8 y.

Implementation is another matter. The successful installations have educated and accustomed technicians to the concept at an early stage. 3M has run contests to solicit the best ideas for using computers and robots in its lab. The reward? A robot of your own! Pick simple applications for the first effort. Permit in-house programmers and maintenance personnel to develop. Provide adequate spare parts. Do not push the system to its limit in speed, load, or reach. Select a vendor carefully, with strong emphasis on service and technical support in software and work station components.

Next month the capsule reports, Part II in this series, will look at some typical industrial laboratory applications.

Part 2

The following capsule summaries discuss the application of three commercial robot systems to laboratory problems. The Microbot Alpha (Microbot, Inc.) is a revolute design driven by stepping motors in an open-loop configuration. The Zymate robot (Zymark, Inc.) involves a cylindrical geometry, with dc servo motors and potentiometers as position encoders for feedback. The Rhino (Rhino Robot, Inc.) is a revolute spatial design, driven by dc servo motors with digital shaft encoders in the feedback loop. The Microbot Alpha is programmed via a teaching pendant; the Zymate is commanded by a set of soft keys associated with the CRT attached to the microcomputer programming console. The Rhino is programmed via an RS-232 port to its intelligent controller card; any small microcomputer can be used to provide the string of control characters. Other laboratory robot vendors include Perkin-Elmer and GCA. The three systems range in cost from $2000 to $20,000.

Procter & Gamble Company
Corporate Technology Division
Advanced Instrumentation Group
P.O. Box 39175 MVL
Cincinnati, Ohio 45247

Contributors: Grover Owens, Rod Eckstein, and Tom Franz

A company engaged in the development and manufacture of many consumer products requires the routine chemical analysis of a large number of samples. Once the analytical method is fully validated and established, its execution is usually a boring, monotonous task. About two years ago, our Advanced Instrumentation Group was in the process of developing and applying powerful microcomputer systems to laboratory control and automation problems. We realized at that time that the full power of the microcomputer in the laboratory environment would not be felt without a flexible, articulate mechanical interface. Using the intelligence supplied by the microcomputer, this mechanical interface could solve the problem of moving samples from place to place. Depending on its complexity and level of articulation, the mechanical interface might even be able to mimic the human arm and hand in performing sample preparation tasks such as weighing and pipetting. With these considerations in mind, we turned to the rapidly developing field of robotics in search of this mechanical interface.

Two years ago the field of industrial robotics was coming of age. A number of robots, or articulated arms, were available commercially. However, the majority of applications at that time were in the heavy manufacturing industries requiring >50-lb payload capacities from large hydraulic robots. Somehow, the thought of a several-ton behemoth bolted to the laboratory floor and leaking hydraulic fluid profusely did not seem appropriate for the analytical laboratory. A few smaller, electrically driven units, notably the Unimate PUMA robots, were available, but their cost exceeded $30,000. Hence, we turned to a small, inexpensive hobbyist–educational robot, the Minimover-5 from Microbot, for use as a development tool. Coupling this unit with an Apple II microcomputer, a pH meter, an electronic balance, a pipettor, and a pressurized solvent dispenser made it possible to totally automate a pH titration (*1*). This was an extremely valuable learning experience that resulted in a generalized concept for a laboratory robotic work station.

The three obvious components for such a work station are the robot, the microcomputer and its interface hardware, and the specialized instruments that the particular application might require. Two less obvious components are software and special mechanical interfaces. These critical components are the glue that holds the entire system together. The system software supervises the communication and cooperation of the work station elements. The software also schedules the sequence of events and performs whatever computations are required. Special mechanical interfaces encompass any modification of equipment that is required to provide the robot with precise, reliable mechanical coupling with the system elements, including special finger attachments for gripping test tubes, holders for pH electrodes and pipet tips, and racks or magazines for presenting empty test tubes for fresh samples.

Desirable attributes for a laboratory robot include:

- positional feedback from each joint,
- five or more arm motion axes plus gripper actuation,
- three-pound payload capacity,
- dedicated microcomputer controller providing a standard interface protocol for straightforward connection to any laboratory computer,
- low cost (<$10,000),
- rugged construction for durability in the industrial environment,
- position repeatability to ±0.02 in.,
- flexible or interchangeable grippers,
- programmable speed of motion from 0 to 15 in./s,
- large work envelope, accessing 4–6 ft of bench top, and
- hand-held teach control, or pendant.

We have not yet found a robot that meets all of these requirements. However, we have been able to develop and apply a robotic work station using a Microbot Alpha robot that does have a number of desirable features.

The Microbot Alpha robot (Figure 1) is a rugged, articulated arm priced under $10,000. This robot is capable of handling loads up to 1.5 lb at speeds programmable from 0 to 20 in./s. This five-axis arm is mounted on an aluminum base casting that houses the stepper-motor drive circuitry. A separate control enclosure contains the required power supplies and a dedicated 6502 microprocessor, which can communicate with virtually any lab computer over an asynchronous RS-232C serial communications interface. This robot also has a hand-held teach control that greatly simplifies programming. The Alpha uses timing belts and metal pulleys to transmit the torque of stepper motors mounted in the body to stainless steel aircraft cables that control arm motion. This robot implements a limited form of closed-loop control using optical sensors on each axis. These sensors uniquely define a "home" position that the robot can be commanded to seek at any time before or during a sequence of moves. "Homing" is the only means by which this robot can correct for motor slippage caused by a collision or trying to lift an object that is too heavy. This minimal amount of positional feedback often suffices; however, actual joint encoding is a superior, but more costly, feedback method. The Alpha can repeatedly position a two-fingered cable-driven gripper to ±0.02 in. of any desired position within its 18-in. hemispherical work envelope.

We have developed a 4-ft-long translation base. The base is energized by a stepper-motor-driven rack-and-pinion assembly and rides on two linear ball bearing slides. The translation base stepper motor is controlled by the Alpha's dedicated microprocessor. This greatly increases the robot's work envelope, giving it access to about 6 ft of lab bench space.

This translation base lets the robot use a work station that includes a pH meter, electronic balance, centrifuge, and solution-handling system, all of which are interfaced to an Apple II microcomputer. Electrode and dispenser tip holders have been fabricated that allow the electrode and tip to be positioned reproducibly by the robot. The centrifuge has been modified by the addition of an optoelectric detector for indexing purposes and a pneumatic piston, which holds the centrifuge rotor in place during tube loading and unloading.

The software that supervises the activities of the robotic work station is written exclusively in BASIC. Utility programs make use of the robot's hand-held teach control to record desired movements of the arm and translation base. Coordinate transforms are performed by these programs, which convert the arm orientation to X, Y, Z coordinates of the gripper (in inches) and pitch and yaw (in degrees) of the wrist. The positions are saved on disk files for later use by the applications program.

This robot and an 18-in. linear translation base have been installed on a Praxis-900 pulsed NMR instrument (*2*). In this case, the robot is responsible for positioning test tubes containing fat samples into and out of 146 holes on the face of the NMR. The standard cable-driven gripper on the Alpha robot was replaced by a commercially available pneumatic gripper, allowing precise test tube positioning during the 4–6 h solid-fat-content determination. This particular installation is approximately three months old and required six man-months to implement. Nearly half of this time was spent developing utility software that will greatly facilitate future ap-

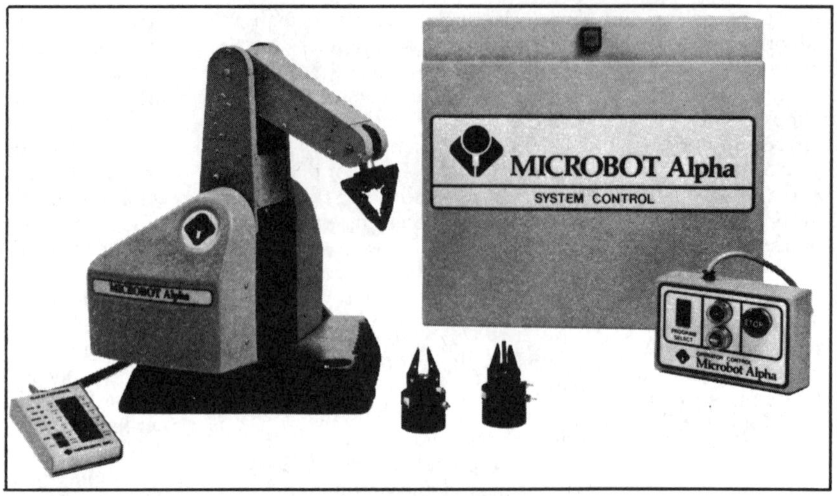

Figure 1. The Microbot Alpha robot
Photo courtesy of Microbot, Inc., Mountain View, Calif.

plications. The system is maintained and operated by a single technician who was trained during installation and start-up. We estimate that this installation will result in a yearly time savings of 600 man-hours. The cost of a second pulsed NMR instrument will also be saved because the productivity of the existing unit has been increased by the robot, allowing it to handle a heavier sample load.

Several other applications using similar work stations are currently under development. These experiences are teaching us that the design and construction of specialized mechanical devices such as translation bases, turbulent mixers, electrode wash stations, and centrifuge interfaces account for a large portion of the total development time. Each application has different requirements for these devices, and the success of the entire system rests on their reliability.

We look forward to the development of more intelligent low-cost robots equipped with complete position feedback and error recovery capabilities. Improvements in the speed and capacity of the work station microcomputer will come with the use of the rapidly emerging 16-bit personal computers.

Merck & Company, Inc.
Automation & Control
Rahway, N.J. 07065

Contributors: Nancy H. Woo and Arthur S. Rosenberg

Over the past few years, a number of commercial robot systems have become available in the marketplace. Most of these have been used in manufacturing operations, automating repetitive and labor-intensive procedures in materials handling, fabrication, and packaging. Many scientific organizations are searching for ways to improve the productivity of their scientists; therefore, it is not surprising that this technology is now finding a niche in the laboratory.

Although there are parallels between research and manufacturing activities, there are also some fundamental differences. Most notably, manufacturing tends to be a static environment; once operational procedures are established they tend to remain in place for relatively long periods. On the other hand, laboratories, particularly those engaged in basic research, require flexibility. The types of experiments done, and therefore the operations that a robot will be required to perform, change from day to day, depending both on the type of study being conducted and the results of prior experiments. Change is the name of the game.

The first general-purpose laboratory robot installed at Merck was in the Enzyme Kinetics Laboratory of the Biochemistry Department. This laboratory conducts kinetic studies on various enzyme systems to characterize them and to determine the mechanism of enzyme action under a variety of reaction conditions. A typical experiment entails taking spectroscopic measurements on a series of reaction mixtures, each differing from the other only in the concentration of one reagent. The mixtures may contain as many as five or six different reagents, in volumes ranging from $<5\ \mu L$ to $>100\ \mu L$. The preparation of these mixtures and the determination of their spectra is time-consuming and tedious.

A second type of experiment involves thin-layer chromatography (TLC) of a labeled substance that can be obtained only in dilute solutions. In these cases a large sample volume must be applied to the TLC plates in the smallest possible physical area. This is generally accomplished by spotting the plates dropwise, that is, applying a small droplet of sample, letting it dry completely, then applying a second droplet on top of the first, letting it dry, and so on until all of the sample has been applied. It is clear that this procedure is tedious and difficult if done manually. Because many of the activities in a laboratory are of this type, we decided that a significant increase in throughput as well as standardization of experimental procedures would be achieved if robotic systems were installed.

Our major requirement for a robotic system is that it be capable of performing multiple tasks, including the ability to operate micro-syringes, pick up test tubes or small flasks, pour out container contents, and pick up other small objects. Our need for a flexible system is far greater than our need for extraordinary precision or speed. An equally important consideration is the ease with which a user can "teach" the robot to do new tasks, and hence make full use of its flexibility.

In November 1982, a Zymark Zymate laboratory sample preparation station was installed (Figure 2). It consists of a microprocessor controller, a robotic arm with detachable hands, and a dispensing station. This

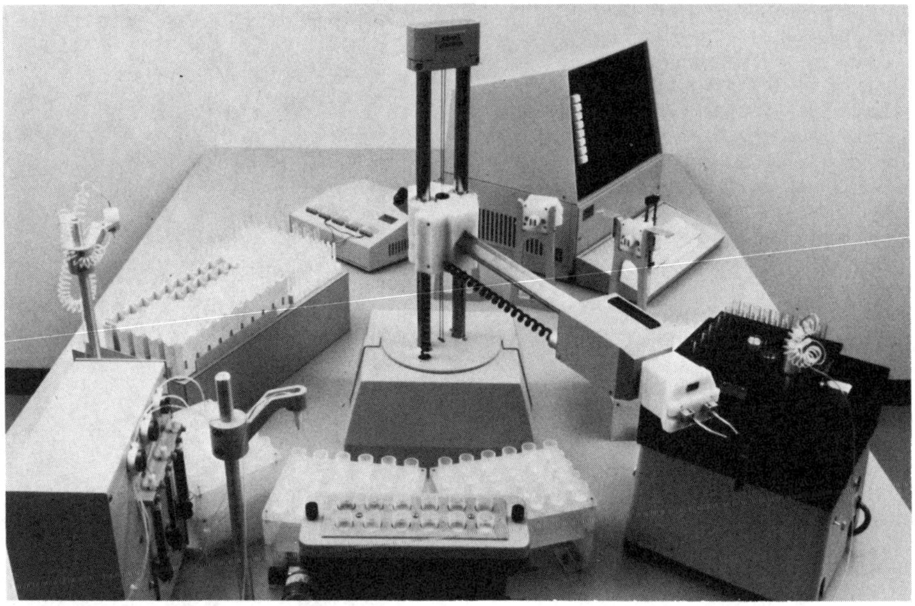

Figure 2. The Zymark robot
Photo courtesy of Zymark Corp., Hopkinton, Mass.

robot can be outfitted with a number of different specialized "hands" that can be changed *automatically* under program control. While the robot as initially offered did not have all the specialized capabilities we required, Zymark agreed to help develop the hands Merck wanted.

For a robot to make up enzyme reaction mixtures, it must be able to measure reagent quantities varying from 5 μL to over 100 μL. The simplest way to accomplish this is to use syringes; thus hands were developed to manipulate commercially available syringes. These hands are designed to fit Hamilton syringes in the size range of 50 μL to 1 mL such that the syringes can be manually removed or replaced. When the system is used to make up reaction mixtures, several syringe hands are used, one for each reagent. A typical sequence of actions for the robot in making up a mixture is the following: The robot uses a syringe hand to measure out a specified volume of one reagent, delivers that reagent to a cuvette in a spectrophotometer, changes hands so that a different syringe is used for the second reagent, measures out the required volume of the second reagent, and repeats these operations until the mixture is complete. The reaction mixture is stirred in the cuvette, and absorbance measurements are made. These data are acquired and stored on a small laboratory computer for later processing. When data acquisition is complete, the laboratory computer signals the robot controller to empty and wash the cuvette before preparing the next reaction mixture.

Changing hands (and, therefore, syringes) permits the use of a dedicated syringe for each reagent, minimizing cross-contamination, and allows the use of the optimal size of syringe for each reagent. Linking the laboratory computer to the robot controller facilitates automation and coordination of each experimental step: sample preparation, data acquisition, and cleanup. As a result, a series of experiments can now be conducted without human intervention and with greater precision. We are in the process of upgrading the laboratory computer to allow real-time data processing and analysis. Our ultimate goal is to create an enzyme kinetics facility in which the design, implementation, and analysis of experiments are fully automated.

The robot has been in routine use since January 1983. Much of the initial period was spent exploring applications beyond those originally conceived and in learning optimal operating procedures. Among the new applications found was the spotting of TLC plates described earlier. The syringe hands designed for the enzyme reaction mixture task are also ideal for this task. Use of the robot for TLC plate spotting has significantly increased the percentage of successful TLC runs. The vast majority of these efforts have been made by laboratory personnel without assistance or intervention by computer or robotics experts. The robot has been very well received. Most of the initial problems with the system arose from deficiencies in the original version of the user interface software. This software has been extensively modified, at least partially in response to our comments.

The success of this installation has led us to examine other laboratories, both production and basic research, as potential sites for future robots. Among the applications currently being explored are dissolution testing and radioimmunoassays using commercial kits.

Imaging Systems Division/3M
225-3S 3M Center
St. Paul, Minn. 55144

Contributor: Peter Bellus

The era of the personal computer is upon us, and 3M is actively implementing personal computer solutions to a variety of laboratory and manufacturing problems. A necessary prelude, however, is generating awareness of the power of personal computers as solutions to these problems. 3M has been active here also. Recently 3M's process engineering forum sponsored a contest for the use of personal computers in solving engineering problems. The constraints on the range of allowable uses were purposely left broad to encourage creative responses. Five finalists were selected, one of which is the concept described here.

Small, inexpensive, but powerful (in the software sense) personal robots are demonstrating applicability to laboratory problems much as personal computers have. They are particularly suited to product testing, where a large number of small samples must be manipulated in a reproducible manner. One can cite the following advantages to using a robot in a product testing application:

- frees highly trained technical personnel from boring, repetitive tasks,
- ensures reproducibility from sample to sample,
- allows testing to occur in a "hostile" environment without danger or inconvenience to personnel,
- reduces labor costs by obviating the need to hire technicians for occasional large testing programs, and
- reduces cost associated with improperly performed tests by poorly trained or inattentive personnel.

3M's Imaging Systems Division is a manufacturer of heat-processed photographic imaging films and papers, and as such has an extensive need for finished product and raw material testing. A program is under way in the 3M/Imaging Systems Laboratory to adapt a small robot to two different product testing applications.

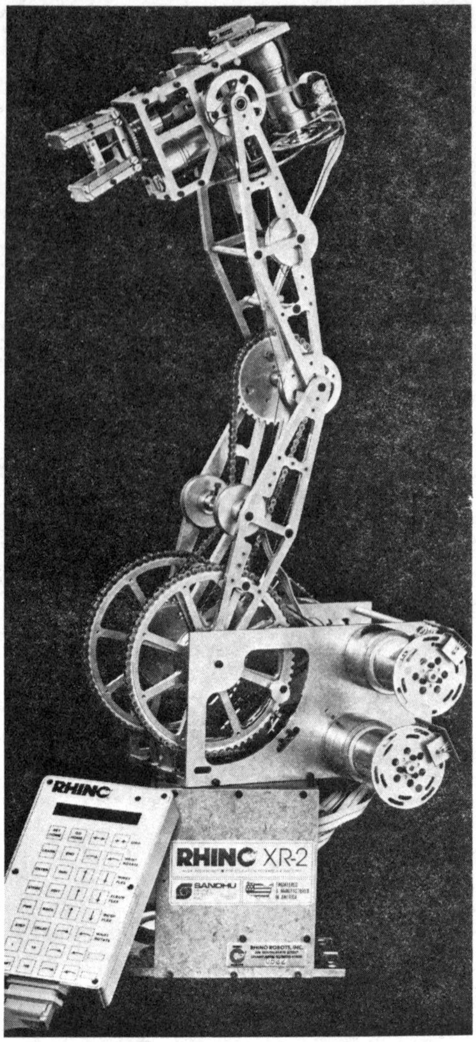

Figure 3. The Rhino XR Mark II robot with teach pendant at left can be operated without use of a general-purpose computer
Photo courtesy of Rhino Robot, Champaign, Ill.

The first application now running successfully is a procedure to determine the basis weight, or weight per unit area, of paper stock. The equipment employed is a Rhino-XR-1 robot, with the hand modified to accept a rubber vacuum gripper, an Apple II+ computer with disk and printer, two RS-232 serial interface boards, and a Setra 1000 balance with the RS-232 serial interface option.

The Rhino robot (Figure 3) is controlled via commands sent by the Apple over a three-wire RS-232 serial link. Positional information can be returned to the Apple over the same link. A BASIC program written to control the movement of the robot provides several features lacking in the software provided by Rhino:

- detection of positional errors,
- ability to reset the robot to a "hard" home position before continuing the sequence (the detection and correction of positional errors is essential in any robotic equipment that must perform for long periods unattended), and
- ability to read an instrument (i.e., the Setra balance) at a programmed point in the sequence and record the data on the disk.

The system functions as follows: The robot positions the vacuum gripper onto a stack of paper samples, precut to 4 × 6 in. with a precision die. A solenoid valve is actuated through an unused joint control port on the robot controller, and vacuum, applied to the gripper, allows one sheet of paper to be lifted. The robot positions the sheet over the pan of the balance and releases the sheet by closing the solenoid valve. The robot then returns to the home position. The balance is set to continuously transmit its weight reading over the RS-232 serial port. The Apple reads the balance at this point, waiting for a pair of identical weights before accepting a weight as valid. Before repeating the cycle, two positional checks are made. The first determines whether any joints are out of the home position and reindexes them. The second checks to see if the weight last recorded is zero, in which case all joints are reindexed and a second attempt is made to retrieve a sample.

The system is capable of processing nearly 100 samples without intervention and promises to be of great benefit when large numbers of basis weights must be measured, as in a pilot coating run, where raw base and coated product basis weights are measured, and coating weight is calculated by difference.

A second, more ambitious, application is being planned for the robot system. In this application, a product must be exposed to light through one of several filters for varying lengths of time, passed through a heat-processing step, and the reflectivity (or image density) recorded. Because the samples involved are unexposed photographic materials, the operation must be carried out in darkness.

Equipment is being designed to handle the exposure, filter-changing, and heat-processing portions of the operation. The Rhino has demonstrated that it can present samples to a densitometer for image density measurements. A MacBeth RD-514 densitometer is interfaced to the Apple via a John Bell Engineering parallel I/O card. The necessary software to drive the interface is incorporated as a subroutine in the robot control program. The reading head of the densitometer is mounted with urethane foam underpadding, so that the head has some "give" when a sample is pressed into contact with it. A porous glass filter funnel, ground flat, is used to hold the sample via applied vacuum.

This part of the system has been tested successfully, but, with a $\frac{1}{4}$-in.-diameter measuring spot and a 3 × 3-in. sample, positional accuracy has not been critical. In an actual testing application the location of the spot will need to be known within $\frac{1}{4}$ in. Although this is a factor of eight greater than the claimed resolution of the encoders, mechanical play and cumulative positional errors make such accuracy impossible without modification to the system. Positive feedback position sensors on the densitometer head as well as a means to detect positional errors and reestablish the robot in a "hard home" position will be necessary.

Personal robots, such as the Rhino-XR-1, have application in product testing. The ability to easily program the robot for new tasks makes it cost-effective, even when more classical automated test fixtures would not be.

Dow Chemical U.S.A.
Michigan Division, Analytical Laboratory
Midland, Mich. 48640

Contributors: Mark L. Dittenhafer and James D. McLean

A commercially available system using a robotic arm to automate sample preparation and analysis has been successfully applied to electroanalytical procedures. Use of the robotic system has enabled a large series of newly synthesized compounds to be screened for electrochemical activity, with a minimum of time and attention required from an analyst. This frees the analyst from the tedium of hundreds of routine operations and makes him or her available for more imaginative tasks.

Flexibility of the robotic system readily permits all of the typical modes of polarography (dc, sampled dc, pulse, and differential pulse) to be automated. Analysis of impurities in virgin and used acid samples required only slight modifications to the computer program used for the screening analysis. Precision of the robotically performed electrochemistry procedures was excellent. The RSD was approximately 2% for sampled dc and differential pulse techniques, compared to the 5% normally quoted for these analyses when performed manually.

Flexibility and unattended operation are the major advantages of robotics in analytical chemistry. But other benefits include increased productivity, excellent precision, and reduced analytical cost.

The centerpiece of the system is a Zymate robot (Figure 2). The arm can turn 360 degrees on its base, has a maximum reach of 24 in. from the center of the work area, and has a maximum working height of 22 in. above the bench top. Motions of the robotic arm are used to mimic those of a human arm during sample preparation and analysis. The arm is equipped with specialized hands that can be automatically interchanged. The hands can grasp objects, operate syringes or pipettors, and control the switches of an analytical instrument.

In performing the electrochemical techniques, the robotic system cleans, fills, and deaerates the electrochemical cell, adds samples, performs standard additions, starts and stops the polarographic analyzer, and records the data. Unlike many automation systems that require highly customized instrument modifications, the robotic arm does not require that the polarographic analyzer be altered. The polarographic analyzer is not hardwired into the system and can easily be removed for manual operation.

To maintain maximum flexibility, a unique bench top was designed. The 6 × 6-ft main bench-top was constructed in-house from a $\frac{5}{16}$-in.-thick aluminum sheet and has a grid of threaded holes spaced at 6-in. intervals. With the threaded holes and brackets, objects can be conveniently secured to or removed from the bench top. The robotic arm and its appurtenances are affixed to a smaller bench top that rests on a lab cart. This assembly allows the robot to be moved to other instrument locations. The arm and controller are mounted separately, so that the arm can be controlled from 12 ft away. When the arm is working with hazardous or noxious materials, it can be placed in a fume hood.

Although the robotic movements are slower than human movements, this has not been a rate-limiting step. The time-consuming segment of analyses studied thus far has been the cycle time of the analytical instrument. For example, the polarographic analysis cell requires a 2–5-min purge between each 2-min scan. During these periods, the robotic arm has ample time to ready itself for the next step. The robot requires approximately the same analysis time per sample as the manual procedure, but, with a large number of samples, the process becomes highly efficient. The robot moves from one sample to the next, 24 h per day if necessary, without taking breaks, answering the phone, or attending meetings. The robot can work in the dark with light-sensitive materials and in inert or hazardous atmospheres. Because it can work continuously, it has the ability to produce massive amounts of data, for which the analyst had best be prepared.

Programming ease is the key to effective applications of robots. When a manual procedure is automated, it is difficult to foresee all of the necessary details. Many movements that are taken for granted need to be programmed as individual steps. For example, the electrochemical screening experiment requires over 340 steps per analysis cycle!

The most efficient planning approach is to list all of the steps required for performing the analysis manually. A written procedure helps define what programming and hardware accessories will be needed. It will also help identify sequences that will be frequently repeated, such as attaching or parking the hands. Programming becomes more efficient if these frequent operations are written as individual subprograms, which can be called upon when needed within the body of the main analysis program. Even with thorough planning, some debugging of the program will be required.

The robotic arm has also been used to automate sample injections for a commercial nitrogen analyzer. Planned applications include the automation of: analyses with ion-selective electrodes and fiber-optic probes; analyses that require timed reagent addition; scintillation counting; synthesis reactions; and titrations.

The ultimate goal is to develop a system with multiple analysis capability, which persons without training in analysis or computer programming could use to analyze samples automatically. Sample preparation, analysis, and documentation should all be performed. Self-diagnostics should be included to allow evaluation of the quality of results. A system with these features would be useful in R & D, process control, and quality assurance laboratories.

References

(1) Owens, G. D.; Eckstein, R. J. *Anal. Chem.* **1982**, *54*, 2347.
(2) Owens, G. D.; Eckstein, R. J.; Gage, D. R.; Madison, B. L.; and DePalma, R. A. "Applications of Microcomputer-Controlled Robots in Automated Analysis," 34th Pittsburgh Conference on Analytical Chemistry and Applied Spectroscopy, Atlantic City, N.J., 1983.

Originally published in
ANALYTICAL CHEMISTRY **1984** *56*(1) 68A
and **1984** *56*(2) 303A.

A Laboratory *Son et Lumière*

Part I

Sound and light are the sources of our two most important senses. In the human–machine dialogue so essential to effective use of computers in the analytical environment the *son et lumière* approach expands our capabilities and increases our awareness. This tutorial will examine some of the techniques being used to provide computers with voice output, voice input, image generation, and image analysis. These capabilities merge the scientist with his instrument, and immerse him, with increased awareness, in the data produced.

Imagine controlling instruments with both our hands *and* voices, receiving prompts and information back audibly while our eyes are occupied with other tasks, then seeing the data in graphic form, with color highlighting and color animation. Or imagine capturing images in the lab and eliminating spurious noise, increasing contrast, detecting edges, converting to pseudo-3-D images, and measuring areas. This world already exists. To use these tools effectively analytical chemists must know how they work, so that they can have models on which to build new techniques.

Son

The computer-generated voice is often monotonous. Male voices may predominate. Voice input commands to computers have a limited vocabulary, and more errors occur with female speakers. Why? To answer these questions let us explore the source of human speech.

The human voice is created by a mechanism akin to the whistle created by air escaping from the gently pinched throat of a balloon. The rapidly escaping air creates a low-pressure area that causes the throat to constrict. Internal pressure forces the throttled opening to expand. This rapidly alternating set of events creates a whistle. Muscle-controlled vocal cords provide a similar reed that produces various sounds. However, a tubular resonator (the throat) and a variety of tunable resonant cavities and muscle-controlled "stops" are essential to create the harmonic structures present in the human voice. The glottis, oral cavity, tongue, teeth, and lips all function interactively to produce recognizable speech.

Languages are cleverly constructed with respect to vocabulary and grammar so that a complete thought can be expressed by voicing speech units in "puffs" of vocal energy, with soft or silent intervals embedded. This mechanism conserves the air supply. Our speech apparatus generates about 100 basic sounds, or phonemes (like the *p* in pat). These are interconnected to form words. Depending on the surroundings, each phoneme can take on one of several variant forms, called allophones (like the *p* in pin or spin).

Equipment that produces a spectral analysis of speech as a function of time provides important information. From such studies, the complex, burst nature of speech is obvious, and speech scientists can abstract the several basic frequencies used in generating speech elements, called formant frequencies (Figure 1). These are the resonant frequencies of the vocal tract. The quality of speech is carried in the harmonic frequencies. Utterances can be broken down into short time segments and each subjected to an analysis that reflects all the frequency components involved during the period. This can simply involve subjecting the speech segment to a series of frequency-sensitive filters. The output of each filter represents the energy content in each passband during that segment. Alternatively, the data can be digitized and subjected to Fourier transform to provide similar data. Another approach views the speech process as a noise source that is subjected to a series of filters. This speech analysis process entails calculating the filter parameters.

Voice Output. Speech can be created via computers by several techniques. The simplest method requires that the sound stream be digitized at a rate twice that of the highest frequency components required for intelligibility or acceptability. The digital data are then sent to a digital-to-analog converter (DAC) to recreate the sound (Figure 2a). Given a 4-kHz upper limit and a dynamic intensity range of 256, this generates 64 kbits of information per second. Storage of

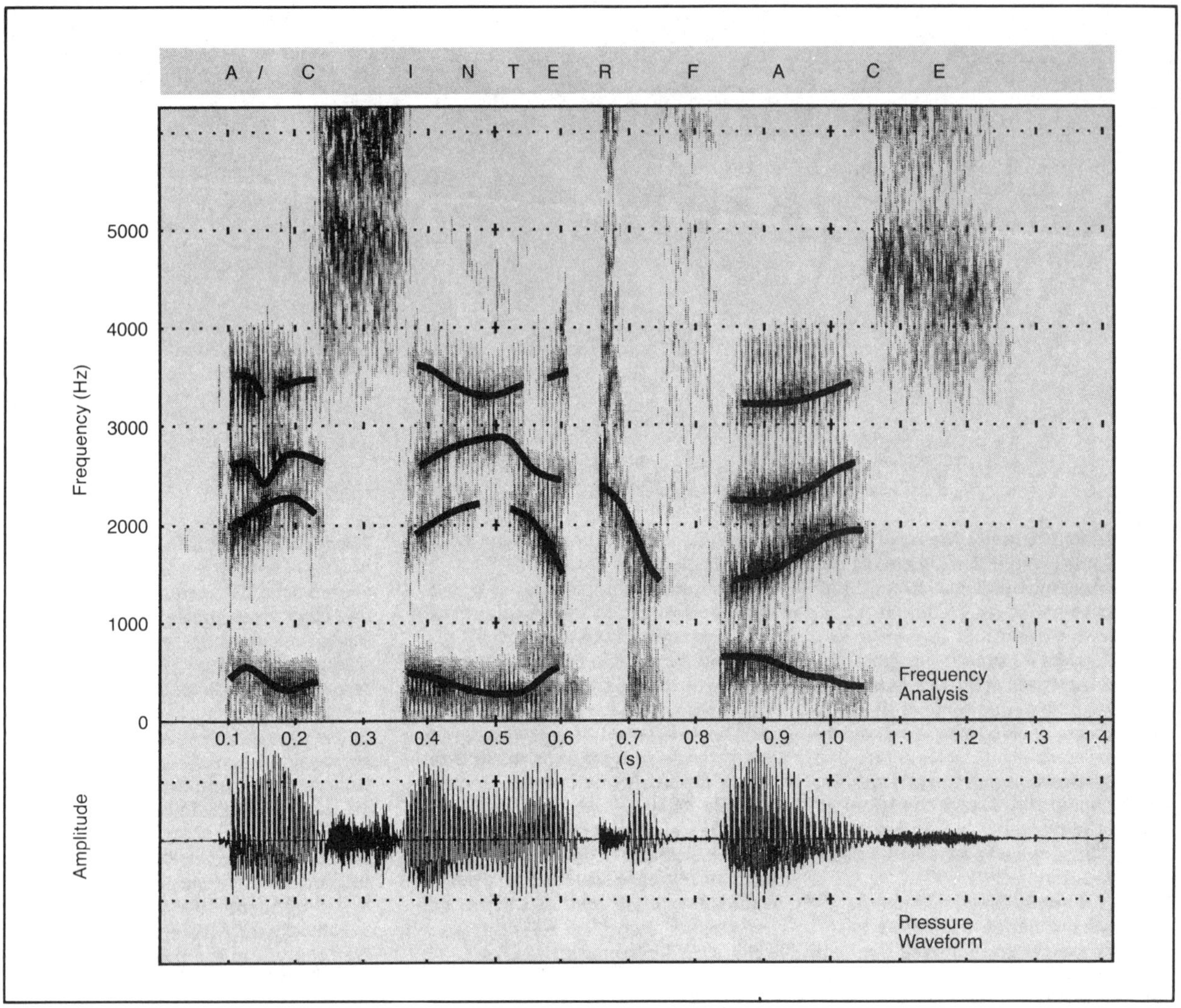

Figure 1. A frequency analysis of the words "A/C Interface"

The formant frequencies are centered around 500, 2000, 2500, and 3500 Hz. The burst nature of speech is very evident, as is the large amount of information speech contains. Courtesy of George R. Doddington, Texas Instruments, Inc., Dallas, Tex.

such a large amount of data is currently impractical, and some very imaginative techniques are used to compress the data.

Moser encoding, popular in some hand-held products, combines a number of techniques, all of which reduce the amount of necessary data. First, the speech is differentiated, lowering the dynamic range. Then, the burst pattern of speech is used to eliminate the silent areas, which carry no information. Subunits that recur require a single storage of the repeating unit. The bursts are then aligned so that the wave pattern is symmetrical about zero. Since the burst has little amplitude at the beginning and end these are "chopped off." The burst can be considered symmetrical, and only half of the remaining information need be stored. Finally, instead of storing each amplitude as an 8-bit digital number, only the differences between each succeeding value are stored. With proper techniques this *delta* encoding can quarter the storage requirements. All of these techniques are multiplicative, and the result is speech that can be stored at a cost of about 1 kbit/s. The speech produced from the digital data is rather flat, but acceptable in restricted environments.

Frequency analysis encoding (Figure 2b) is also used to generate speech. A digital table of energies associated with each passband in the sound segments can be created by filter or Fourier transform techniques. Since the human ear is relatively insensitive to phase relationships between the various required frequencies, the regeneration process involves summing the individual components.

Linear predictive encoding is the most viable approach. The parameters describing the filter coefficients of the modeled vocal tract, with a suitable noise source, are used. Because of the linear nature of the time-dependent recursive filters the technique is termed linear predictive encoding (Figures 2c and 3).

These latter two techniques require about 2 to 3 kbits of information/s to produce acceptable voice sounds via DACs (Figure 2b, 2c). The voice quality is excellent.

Text-to-voice stores phonemes and/or allophone sounds in read-only memory (ROM), using one of the techniques described above. Supported vocabulary is also stored in ROM, along with pointers to the allophones that would be used to generate the spoken word. ASCII character strings (where seven bits of information are used to encode each letter or number in the text) are sent to the speech device from a host computer (Figure 2d). These character strings are interpreted, and if the word is found in the vo-

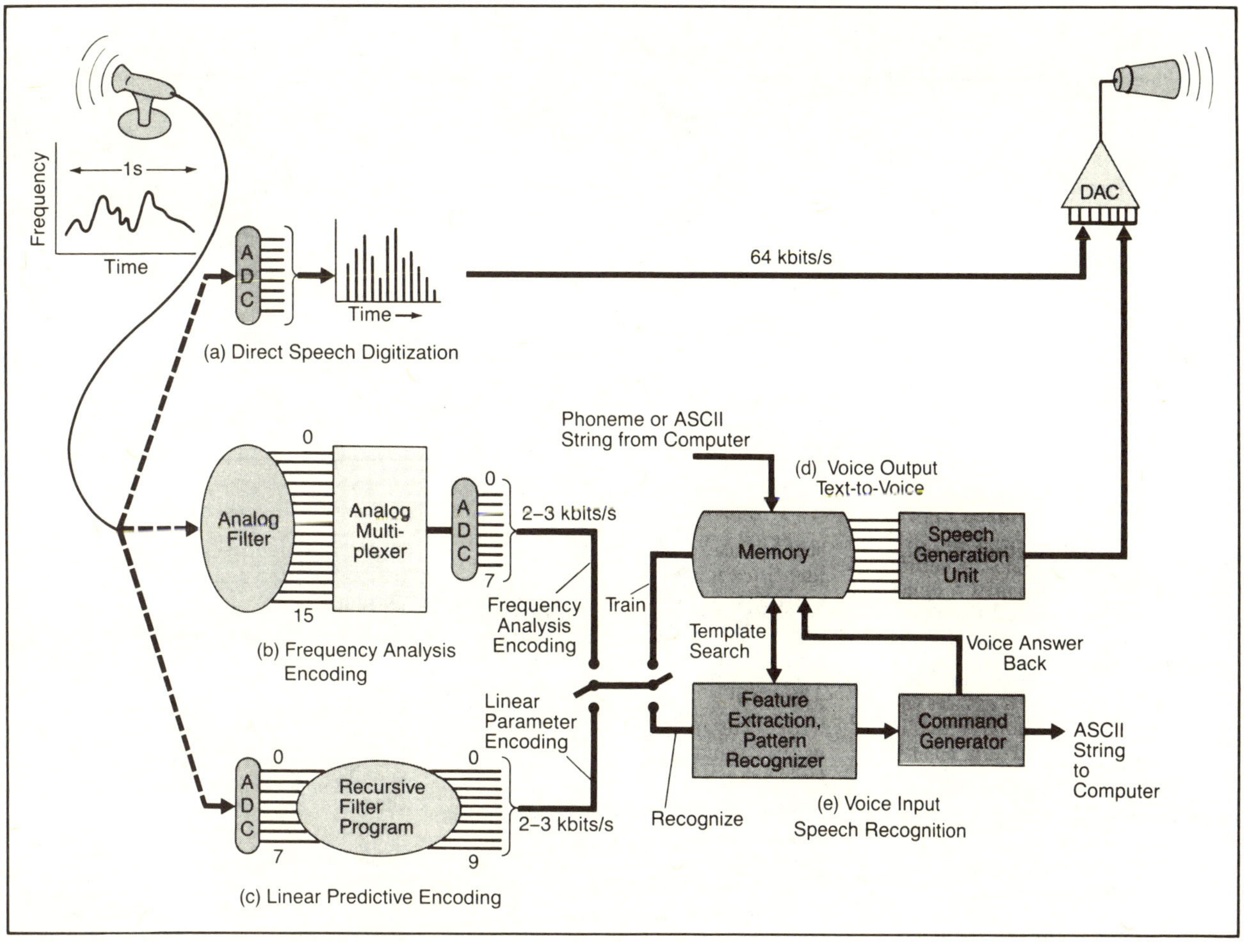

Figure 2. Various ways to implement voice input and voice output

cabulary ROM the corresponding sound is heard. Rules are available in some implementations that allow unlisted words to be pronounced. A program chooses the correct phoneme from the letter sequence and selects the correct allophone depending upon the surroundings. If the rules fail, the word is spelled out. Languages that do not exhibit numerous anomalies, such as Spanish, are served well by this approach.

Voice Input. Peering through a microscope you identify a new cell type in a biological sample. How can you enter the datum into a computer without breaking away from the eyepiece? Running an electron microprobe occupies both hands with sample movement. How can you change a gain setting? Simple, just talk to the device.

Speech input is more difficult to implement than voice output. In principle, direct digitization, frequency analysis, or linear predictive techniques could all be used to characterize a speech waveform. Search–match techniques would then be used to scan a previously stored library for equality (Figure 2e). If a "hit" occurred, the speech identification device would transmit a predetermined ASCII sequence to the host computer. In practice there are difficulties.

Direct digitization of the speech waveform envelope by analog-to-digital converters (ADCs) does not work well, since the envelope characteristics are very phase sensitive. Two sounds will "hear" the same, but have quite different envelopes due to phase differences in the sources.

Frequency analysis and linear predictive techniques are often limited to applications involving discrete utterances. These consist of speech of a few seconds in length with embedded pauses that do not exceed about 100 ms. The utterances may be words or short phrases. The utterance is divided into 20–50-ms windows and analyzed. The first step produces a series of descriptors for each window using the techniques already described.

A set of frequency-sensitive analog filters can produce a set of 16 numbers that represents the integral of the energy content of each passband (Figure 2b). A related technique digitizes the entire utterance and then sends it, via DACs, through a single filter repetitively, at faster and faster rates. The single filter produces the same spectral energy information. Fourier transform techniques can produce a similar set of numbers that represents the power spectrum magnitude in certain passband regions.

Alternatively, relatively inexpensive microchips can analyze the utterance by linear predictive techniques, which produce a set of 10 filter parameters (Figures 2c and 3). Such devices can even compress continuous speech in real time! These digital data must be compared with a library of stored utterances.

Templates are created in a training process in which the user is asked by the computer to say the various words in the vocabulary randomly and repeatedly (Figure 2e). The parameters are "averaged" in a process that also normalizes both the amplitude and time variables. These library parameters are then tested. The testing utterance cannot be compared to the templates in a direct search–match manner. The human voice is too variable; it changes the duration and stress of each phoneme according to the human

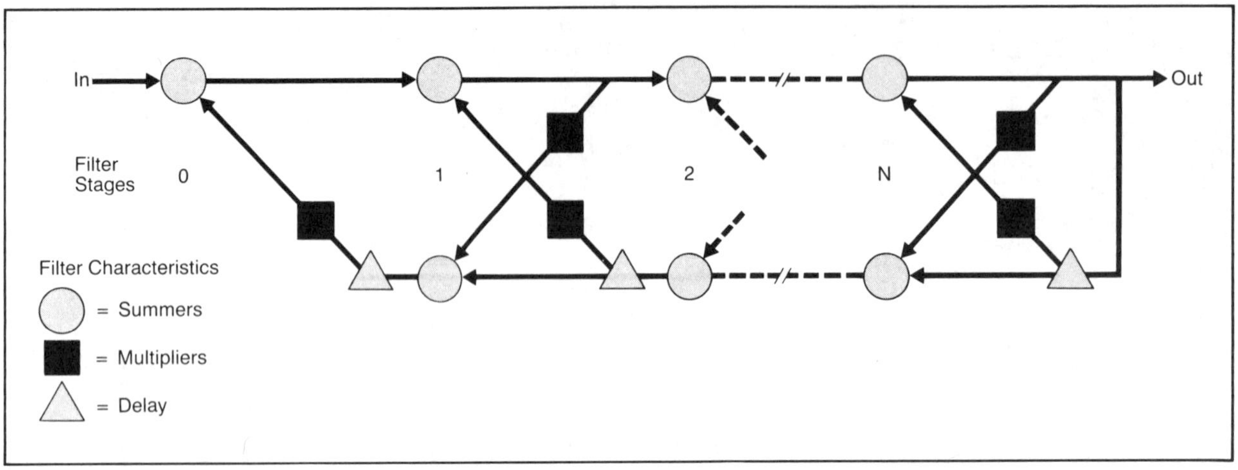

Figure 3. Recursive lattice filter used in linear predictive encoding voice synthesis

condition. The most successful approach locates pronounced areas of similarity between the test utterance and a template and then compresses and expands the unknown accordingly. A summation is made of the differences that exist between the two. All vocabulary candidates are treated this way, and a list is made of the best "hits," along with their scores. The goal is to make the uttered word and the identified word agree, and have a score much better (lower) than any other candidate. If this fails, other words must be chosen for the voice control function. At the end of this iterative process the system is trained for a single person and may be switched to recognize mode (Figure 2e). Speaker independence is not commonly found in the lab because of speed and cost factors.

The discrete-utterance speaker-dependent systems currently available for the lab can handle vocabularies of a few hundred words. The user may still be faced both with failures (no identification) and errors (incorrect substitutions). The better systems recognize that at any moment in the human–machine dialogue only a few words from the entire vocabulary need to be distinguished from one another. Other utterances included in the user's vocabulary would be nonsense. Only acceptable library templates are scanned in the search–match process, so that the comparison process involves only about 10 words at any one time. Such systems work well even in noisy environments.

We now have a tool that allows users to train systems to respond to their own voices, employing any words they choose. Each user command results in the transmission of an ASCII string to a host computer to activate a required function (Figure 2e). The voice utterance and the ASCII command string need not be the same. In systems that can store speech, using the techniques described above, the instrument can respond vocally in prechosen words with a "personality" that can be determined by the user. Such personalized systems are not a luxury. They can provide security, since speaker-dependent systems are oriented to one person. They provide technicians with a mode of control that does not require a knowledge of the arcane ASCII command strings required by most operating systems. They provide a control mode that can augment and assist busy hands. Finally, they can provide an aural output that permits the eyes to continue performing other tasks.

Lumière

Traditional computer output is a list of numbers. Graphic presentation of data is more effective. Color graphics are capable of communicating even more information (Figure 4). Multidi-

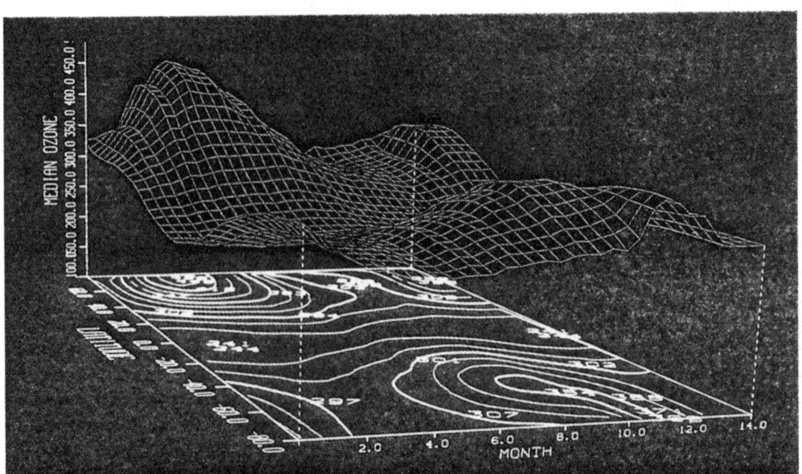

Figure 4. A three-dimensional view of chemical data

One frame of a stereo pair. Courtesy of Stanley Grotch, Lawrence Livermore National Laboratory, Livermore, Calif.

mensional plotting and 3-D images of objects are often the only way a human can comprehend the data message. How do computers create such images?

The gamut of color sensations can be created by adding together varying amounts of three-colored light sources, each covering a selected passband in the short, medium, and long wavelength region between 380 and 780 mμ. Classically, the primary additive colors are red, blue, and green.

CRT Terminals. Black-and-white cathode ray tubes (CRTs) have only a single electron gun and a phosphor-coated screen that emits visible light when struck by electrons. White or green phosphors are common. Color CRTs use three electron guns, one for each of the primary colors. The screen is coated with phosphor triads composed of a red-, blue-, and green-emitting phosphor dot at the corners of an equilateral triangle. A shadow mask located in front of the phosphor screen has a small hole located over each triad to help direct the electron beams. The CRT display is a rectangular array of visual points created by selectively exciting phosphor areas. Although it is possible to build displays in which individual points are randomly addressable, or displays in which images are created by vector strokes oriented at any angle, the most common displays are raster scan devices.

Driven by scan electronics, the electron beam moves (rasters) across the tube face through one horizontal line at a time, retracing rapidly to the left at end-of-line. At end-of-frame the beam vertically retraces to the top left.

The minimum picture element is called a pixel. (Typically there are about 2.5 triads/pixel in a high-resolution display.) Each pixel area is excited by electron guns controlled by DACs, which in turn receive their digital input from computer memory. A row of pixels is called a scan line. A collection of scan lines is a raster frame. If a frame is displayed at a rate of 60 Hz the human eye fuses separate frames into a continuum. This generates the need to handle a very large bandwidth (60 MHz). A 30-Hz rate reduces bandwidth, but leads to flicker. This flicker can be reduced by dividing the raster frame into odd- and even-numbered lines and alternately displaying each set (a field). Such displays are termed *interlaced.* There are trade-offs in cost, brightness, and flicker between interlaced and noninterlaced displays.

Bit-Mapped Displays. The raster frame is driven from a high-speed frame buffer memory that can be viewed as a matrix whose planar array dimensions, in bits, are determined by the number of pixels on the screen; e.g., 512 × 512 or 1024 × 1024 bits. If the picture is "sampled" at 512 pixels/line the time to generate and display one pixel is 100 ns! If the horizontal and vertical resolution is increased by two the display time/pixel would be just 25 ns. But resolution is the key to acceptable displays. Two hundred fifty-six elements of resolution lead to displays in which "smooth" curves appear with very jagged edges, referred to as granularity or aliasing. This can be not only visually distracting, but the data may be misrepresented or distorted by the lack of resolution. Low-resolution displays (256 × 256) are suitable for peer distribution. Presentation graphics requires 515 × 512 to 1024 × 1024 displays. Picture quality is delivered by 2048 × 2048 high-resolution CRTs. Commonly a rectangular frame is involved, rather than a square one.

Depth gives a third dimension, z, to the matrix. This depth, expressed in bit-planes, is determined by the color characteristics required. If this is one bit deep (one bit-plane) only monochrome black-and-white information can be displayed; the single electron gun is turned off or on by the binary bit pattern stored in this bit-plane (Figure 5a). A monochrome display having three bit-planes could support eight levels of gray scale, since the DAC driving the electron gun could be set at eight beam current levels. The simplest eight-color display would also have three bit-planes: one bit for each color gun. A logic 1 would turn that color beam on; a logic 0 would leave it off. Such a display mixes the primary colors in an off–on manner giving very saturated colors (Figure 5b). A more versatile approach allocates two bits to each color gun allowing four energy levels for each (six bit-planes). This system can create 64 different colors by properly mixing the primary additive colors. Allocating eight bits to each color gun gives a palette of 16 million colors, but requires 24 bit-planes and a total of 3 Mbyte of memory.

The intent of these palettes is not to encourage concurrent use of such a panorama. Rather the scientist can use color texture to describe surfaces, where shadows and three-dimensional character must be expressed or use related hues to express chemical connec-

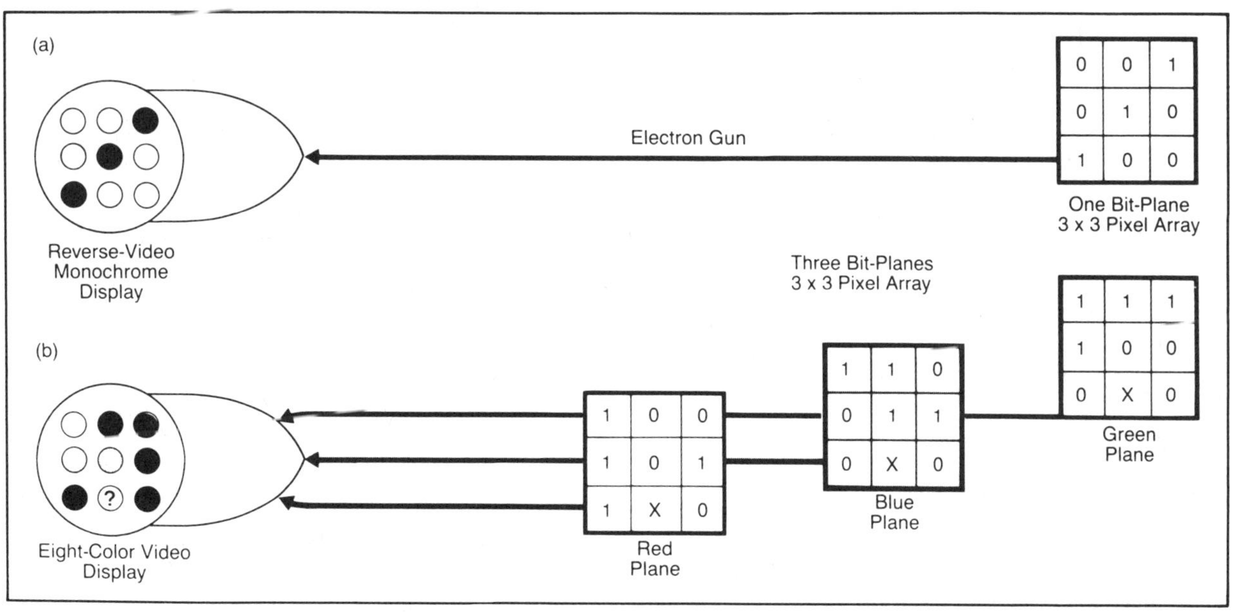

Figure 5. The number of bit-planes in the matrix determines the color variations that can be output to the screen

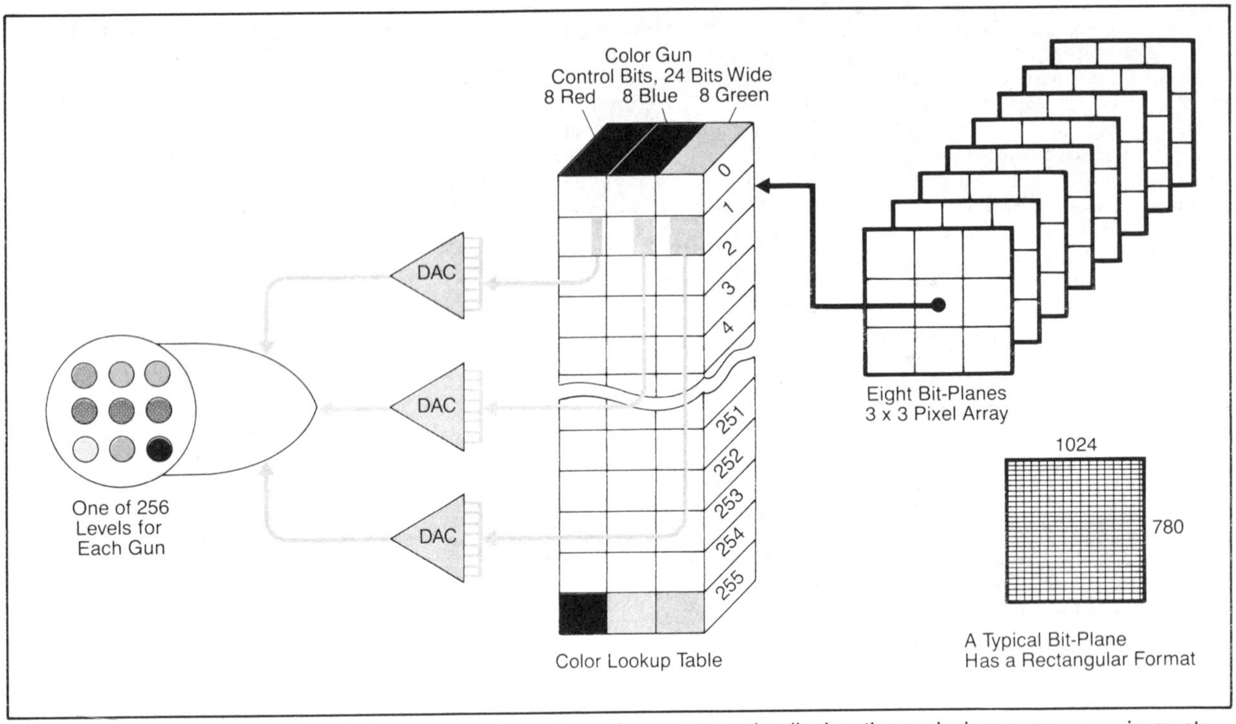

Figure 6. Bit-planes map through a lookup table to determine colors output to the display, thus reducing memory requirements

tions. Since even the shallowest matrix involves megabits of memory, and all palette colors will never be used concurrently, most systems "map" from this matrix, through a color lookup table, to the CRT screen. The binary word describing each pixel in the matrix points to an address in the color map lookup table. The contents of this memory cell determine the values output to the DAC associated with each color gun. An 8-bit-deep matrix can address one of 256 different locations in the lookup table. Each word in this table might contain 24 bits, eight for each primary color. We can now display 256 colors chosen from a palette of 16 million (Figure 6)!

It is possible to alter the contents of the lookup table in a very short time. Changing certain memory cells during vertical retrace will then change all pixels referenced through that cell during the next scan. (Imagine rewriting every affected cell in the memory matrix!) The array maps to the tube surface geometrically; the bit-planes map through the lookup table to color. This allows the chemist to change colors rapidly, illustrating connections within a data set. Simple animation can be done by changing the values in the lookup table.

Character vs. Graphics CRTs. ASCII characters occupy seven bits (128 possible characters). Their display requires many more pixels, e.g., a 5 × 7 or 9 × 15 matrix. It is possible to map from ASCII code through a character generator, conserving considerable memory space (Figure 7). Alpha-

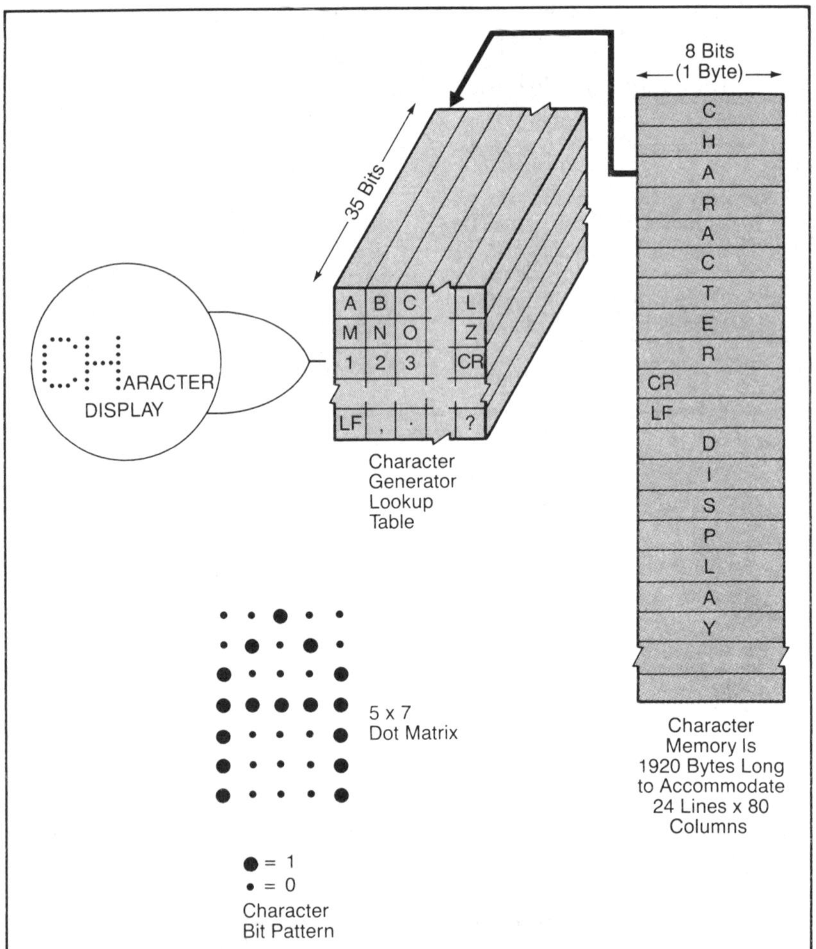

Figure 7. Letters and numbers are bit-mapped through a character generator to conserve memory

CR = carriage return, LF = linefeed

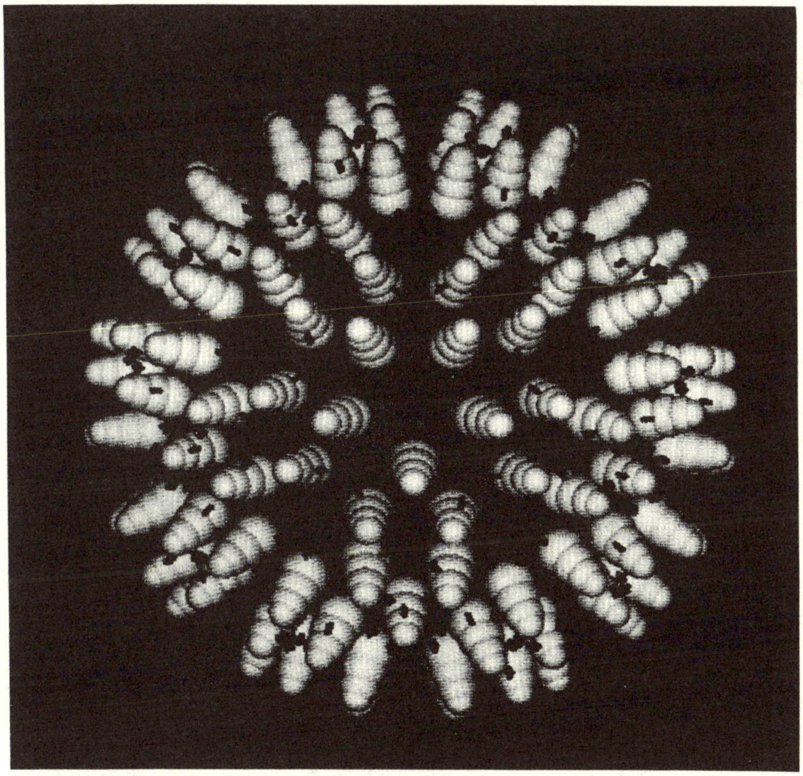

Figure 8. The use of related colors, light reflection, and perspective to provide a sense of depth
Courtesy of Richard Feldmann, Molecular Graphics Lab, National Institutes of Health, Bethesda, Md.

numeric terminals possess *only* this capability. They can draw graphs only by using symbols in the 24-line × 80-character/line positions. Some of these terminals can have options added that allow bit-mapped addressing. Good bit-mapped displays can superimpose alphanumeric information over graphs and drawings. Often one or more bit-planes are reserved for such overlay functions. Overlays also allow simple animation over a fixed background.

3-D Plotting. With an extensive color palette it is possible to produce pseudo three-dimensional representations of data and objects (Figure 8). The perspective problem in Cartesian space is easily solved. The outlines of solid objects can be generated by sweeping out in "space" volumes defined by rotating polygons. Clever algorithms solve the hidden line and surface problem, so that acceptable visual surfaces remain. The fundamental laws of reflectance of light are then used to generate color and shading information that draws on the color palette for implementation. Both diffuse reflectance (cosine-dependent, Lambert's law) and specular reflection (mirror reflection) models have been developed. The eye then sees depth by perspective, shadow, and color differences. The result is very lifelike images.

Stereo views can easily be created by changing the angle of view between two frames of the same subject, photographing the two images, and using the appropriate stereo viewers (Figure 4). Even more startling, 3-D images can be created by interlacing stereo images on the screen or by using side-by-side presentation. These may be viewed in relevant time.

Graphics Standards. Unfortunately, software is the most difficult problem. The user would like to use powerful graphic commands without having to know the steps taken in high-level languages to accomplish his/her goal. Standard graphic functions include drawing vectors, generation of conics and wire frame polygons, and filling these structures with accent and color. Advanced functions include zoom, rotate, translate, and scale operations. Application programs deal with the geometric and topographical description of objects by graphic primitives and Cartesian transformations. A viewing package translates these descriptions into image data suitable for the bit-mapped matrix of a display device. Then, a device driver, written in machine code, outputs the data to a particular device. Incompatibilities in coordinate systems, scaling, graphic primitives, and display protocols make transportation of graphics from one system to another a nontrivial task.

The International Standards Organization (ISO) has adopted a Graphics Kernel Standard (GKS) that addresses the application and viewing levels. The American National Standards Institute (ANSI) has endorsed GKS. Although GKS supports only 2-D displays, the Core Standard, developed by the American Association for Computing Machinery Special Interest Group on Graphics (SIGGRAPH), does address 3-D graphics.

The North American Presentation Level Protocol Syntax (NAPLPS) has been developed because of the intense interest in videotex, which involves text–graphics exchange between central data bases and terminals via commercial communication channels. AT&T has adopted and enhanced this standard.

Vendors provide a wide range of offerings, from enhanced graphic boards that plug into standard buses of microcomputers to turnkey systems. In addition to resolution and interlacing specifications, the buyer should examine the speeds at which oblique vectors can be drawn (100 000 to 4 000 000 pixels/s) and polygons filled (4 000 000 to 40 000 000 pixels/s). Some of these terminals will emulate industry standard terminals (Tektronix 4010, 4014, the DEC VT family, or IBM 3270), which provides flexibility. The speeds supported by serial ports range from 9600 to 39.2 kbaud. Some devices have standard printer ports that allow direct attachment to color ink jet printers that provide color hard copy.

Image Processing and Analysis

Technology is moving so rapidly that the rigid distinctions between graphics and image manipulation are blurring. Peter Jansson of Du Pont has defined computer graphics as a "numbers-in/picture-out" technology. Image processing is "picture-in/picture-out," and image analysis involves "picture-in/numbers-out" (Figure 9). A chemical example? 2-D electrophoresis plates, in which molecules are separated by molecular weight and polarity, often produce fuzzy, hard-to-decompose images. Boolean/algebraic manipulation of the image by techniques called erosion and dilation, with background subtraction, can create pseudo-3-D views that are umbral (moon crater shadow) in nature. Important features are easily seen. Then quantitative numbers can be abstracted from the display.

Image processing usually implies capture of the image of an object by a 256 × 256, or larger, solid-state camera device. As each pixel in the captured image is scanned out it is digitized (eight-bit resolution) and stored in memory. Imagine a very large

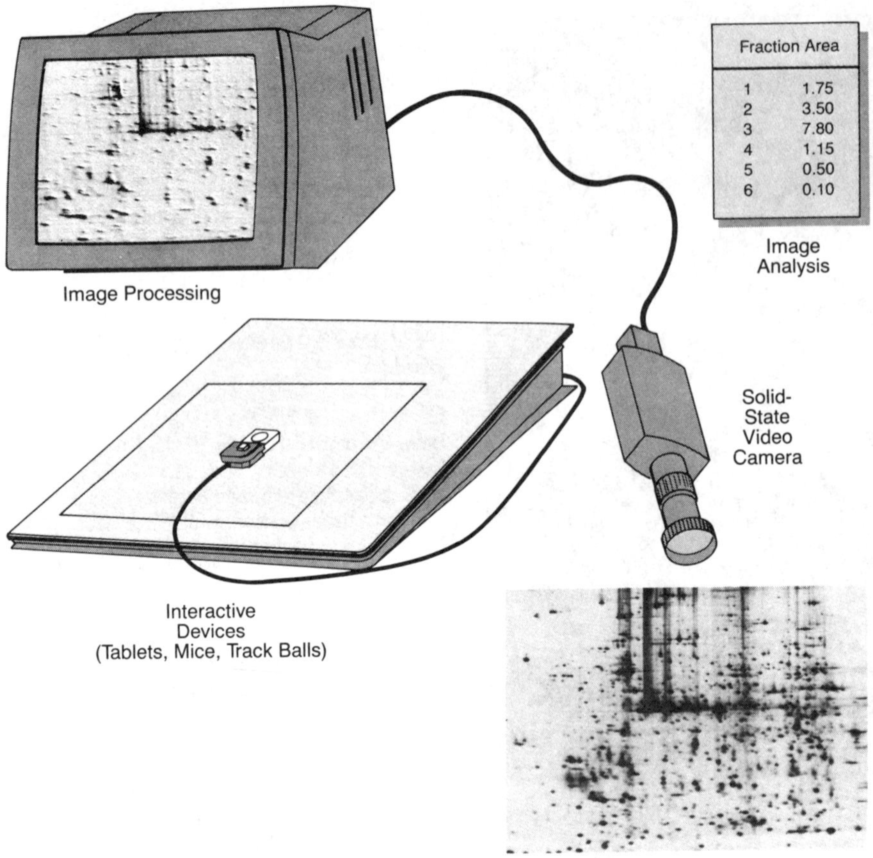

Figure 9. Image processing and analysis of a 2-D electrophoresis gel

Photographs courtesy of N. Leigh Anderson, Molecular Anatomy Program, Argonne National Laboratory, Argonne, Ill.

graphics memory capable of storing many such frames. Imagine that many such frames can be stored in a virtual manner so that a prospective user need only specify a size, coordinate system, origin, and name. A small window of a frame can be viewed by specifying name, size of window, and origin. Several windows may be viewed simultaneously by "splitting" the screen, or frame windows can overlay one another. Screens can be manipulated algebraically or in a Boolean fashion.

Successive images of the same object can be added together, and the result normalized, yielding a less noisy image. This is similar to ensemble averaging. A region of interest can be brought into our display and viewed. By panning left/right or up/down the desired area can be found. To zoom in on this area, take a section and move the adjacent pixel values into even-numbered positions, and then replicate each value into the adjacent odd position. Unfortunately, the result is a jagged image. To correct this, take the entire filled frame and replicate it; then move it horizontally one pixel position; replicate the original and move it vertically one pixel position; then again diagonally. Add them together, and divide by four (Figure 10a). The

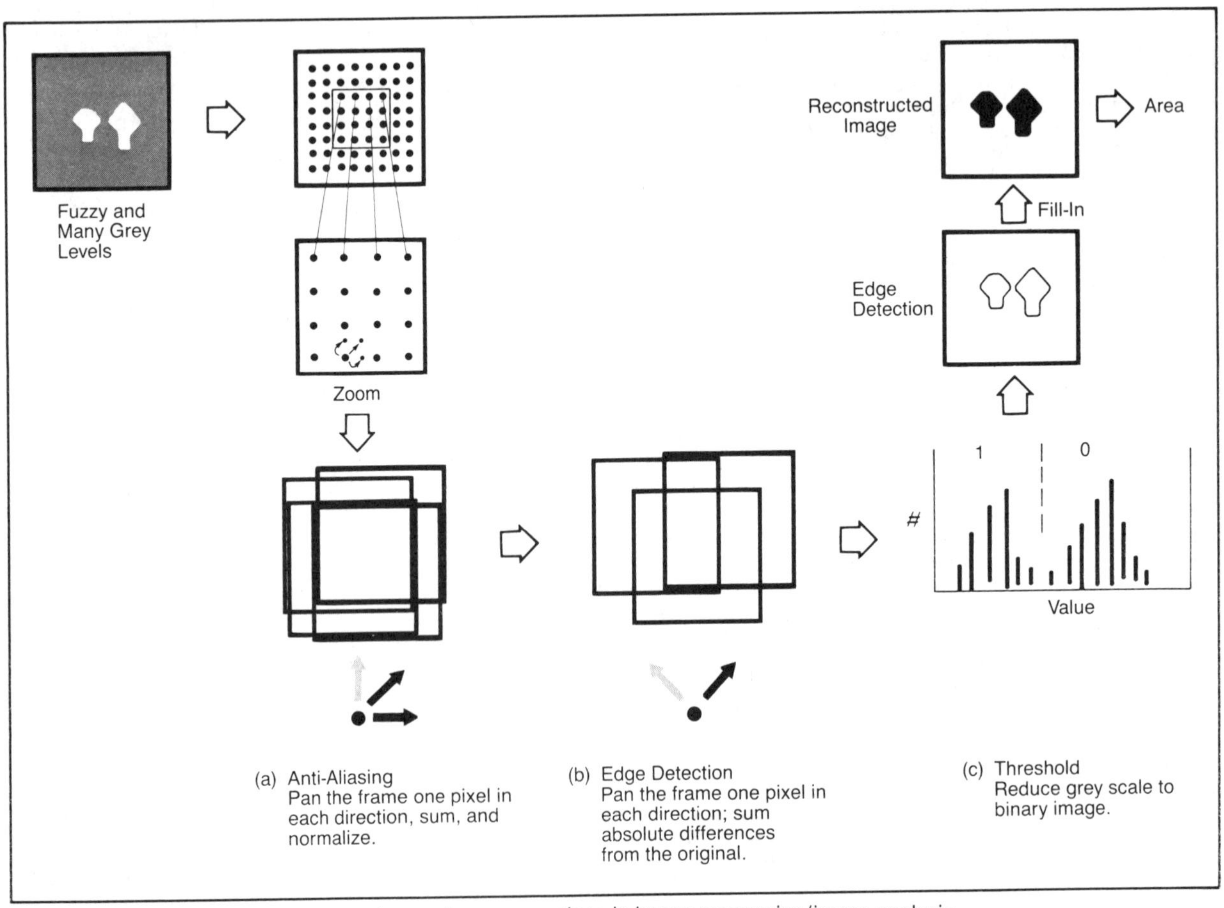

Figure 10. Zoom, smooth, edge detection, and area operations in image processing/image analysis

result is a perfect zoom. The process is sometimes called anti-aliasing because it tends to remove the granularity of digital video displays. Some displays have anti-aliasing hardware.

How is the area of an enclosed space determined? The original must be replicated and the origin shifted one pixel position in both the x and y direction into the first quadrant; subtract the two images. Repeat this by a motion into the fourth quadrant. Sum the absolute values of the difference arrays. This Robert's cross gives information as to the location of edges in the image (Figure 10b). Then a histogram plot of pixel values is made. This plots the number of pixels having a common value. Often this is bimodal and allows the image to be "thresholded" into a black–white binary image (Figure 10c). Now duplicate this binary image and move it in four directions (left, diagonally left, up, diagonally right) applying Boolean operations (ANDing and ORing) between the intermediate images. This directional filter yields a final image that is a clean representation of the edges of the object. The area within the edge can now be easily established.

The pixel values within an image can be multiplied or divided by a constant to change the contrast of the image. Addition or subtraction of a constant can change the brightness (Figure 11). In cases of nonuniform lighting a "blank" can be taken and this image used to correct a subsequent image so that it has a uniform background. This involves multiplying images. Much of this can be performed in real time with inexpensive hardware. Plug-in cards for several standard buses are available, providing image-processing capabilities in microcomputer packages. The price of fully implemented work stations is also dropping.

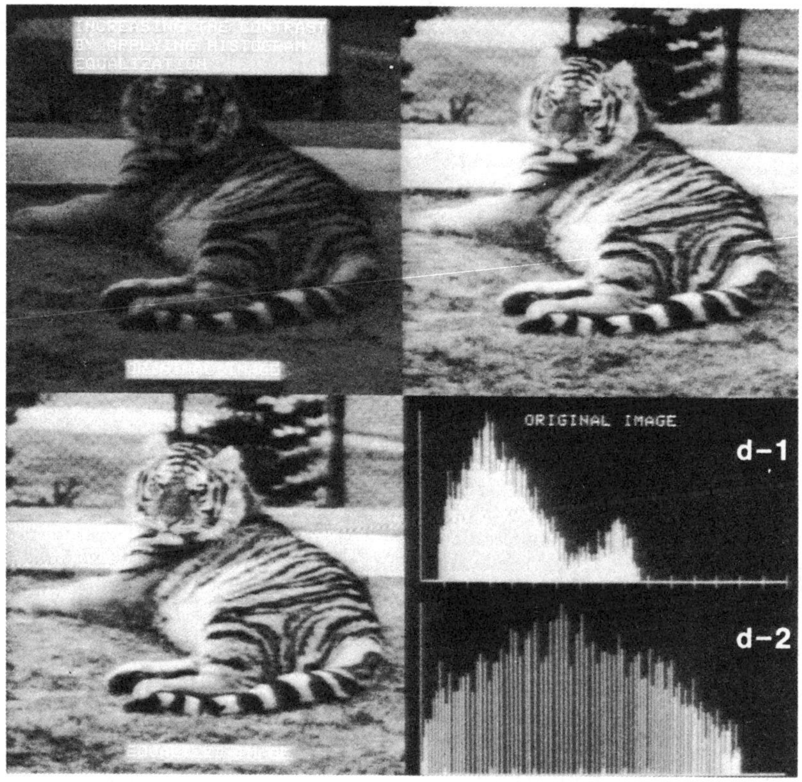

Figure 11. Image processing
(a) Original image; (b,c) improved contrast images derived by expanding the pixel histogram as shown in d; (d-1) pixel histogram of a; (d-2) pixel histogram of b,c. Courtesy of Lexidata, Billerica, Mass.

Interactive Graphics

And now the action of the *son et lumière* begins. Simple graphic tools allow the user to move the cursor about the screen with function keys (up, down, right, and left). More human-oriented interactive techniques are available. These allow rapid manipulation and choice of objects on the screen.

Light Pens. As the light beam rasters across the surface of the CRT a photocell held in front of the tube experiences a change of light intensity as the beam passes in front of it. This can be used to activate circuitry that records the x,y position of the beam at that moment. What does this allow us to do? The cursor can be moved to this new position. A vector can be drawn from a previously selected point to the new position. A segment of a line can be picked up and stretched or bent. Or the computer can be instructed to select the operation displayed on the CRT screen in the area of the beam. This is interactive graphics with a light pen.

Touch Screens. With a matrix of light-emitting diodes (LEDs) and photodetectors arranged around the screen periphery, a finger can block one horizontal and one vertical beam, establishing the x,y position. This touch screen can be used like the light pen. The same effect can be created with a transparent electrical conducting sheet placed in front of the CRT screen. This sheet is a resistive element that can serve as a voltage divider for an imposed potential. A second transparent conducting screen is supported one mil in front of the first. When a finger depresses this it "taps" the voltage divider like a potentiometer wiper. The potential gradient is rapidly applied, first in the x, then in the y direction. ADCs are used to produce x,y coordinates.

Tablets. Off-screen devices can be used for digitizing spectra or drawing pictures and letters, as well as manipulating the cursor and choosing screen images. A tablet consists of a grid system of horizontal and vertical wires embedded in a plastic body. A detector coil is placed above the grid. All of the columns are successively pulsed with an ac signal source. Then all of the rows are sequentially pulsed. The detector coil will produce pulses that can be time correlated with particular transmitting wires, providing x,y coordinate data.

Mice and Track Balls. Imagine a reflective sheet with a grid of finely spaced horizontal and vertical nonreflective lines. The detector assembly consists of an LED and a four-quadrant photodetector. The light from the LED is bounced off the surface of the grid and strikes the four-quadrant assembly. As the detector is moved across the surface the photodetector pairs pulse on and off, providing a pulse for every line crossed and directional information. The x,y pulse strings are used to interact with the display. This is an optical mouse.

An electrical mouse has two orthogonal shafts and two idler shafts that form a nest for a ball. As the ball is pressed against a friction surface it turns in the nest, causing the active shafts to rotate. As these revolve they generate pulse trains proportional to the ball's rotational component along the Cartesian coordinates, as well as directional information. If such a device is turned over, so that the ball is seen nestling in the shaft cradle, we have a track ball. These are manipulated with the palm of the hand.

Joysticks. These devices have a vertical shaft gimble connected to two orthogonally mounted potentiometers.

The wipers of the potentiometer indicate an x,y displacement. ADCs are used to convert the analog potentials to digital numbers for coordinate purposes. Alternatively, binary up–down counters and DACs may be used to null the potential, providing a type of digital servo. Coordinate position is read directly from the counters.

All of these devices have one or more push buttons or capacitively coupled switches that signal the computer to read the x,y position and/or take other actions.

Selection of an interactive device should involve careful consideration of the accuracy, reproducibility, and resolution of the system, particularly if digitization from hard copy or drawing is the goal. A resolution of 1 to 5 mil is quoted for tablets, with accuracies of 10 to 25 mil. Reproducibility is seldom specified. For screen-located selection devices, the distances are much larger.

Next month's A/C INTERFACE will describe typical laboratory examples of these interactive techniques.

Part 2

E. I. du Pont de Nemours & Co., Inc.
Experimental Station
Wilmington, Del. 19898

Contributor: Peter A. Jansson

Image Processing

Our laboratory formally began digital image-processing research in 1969. The early commercial image analysis instruments, intended primarily for particle size analysis, were severely restricted in their ability to deal with problems more complex than determining simple size distributions in micrographs of particles. Yet a host of potential applications existed including image texture analysis, precise geometrical measurements, and pictorial pattern recognition. These applications all require the power of a true general-purpose digital computer. The minicomputers available in 1969 could not deal with the masses of data that even relatively low-resolution digital images produce, typically ¼ megabyte per image. Analysis was thus relegated to large batch-mode mainframe computers.

The first major breakthrough came with the introduction of inexpensive but complete 16-bit computer systems. Coupled to a video camera and digitizer, such computers made it possible to develop convenient self-contained special-purpose systems with custom-designed front-end optics and application-specific software at modest cost. Successful developments proliferated within the laboratory.

Our first units relied on shading correction schemes to compensate for uneven illumination and photometric nonlinearity. In response to application demands for greater accuracy and precision, our image analysis systems now measure true sample photometric properties, essentially devoid of corruption by sensor and system characteristics (*1*). They are insensitive to light level variations and video sensor changes. We have addressed video camera geometrical distortion with rapid image-warping algorithms and spatially dependent correction techniques.

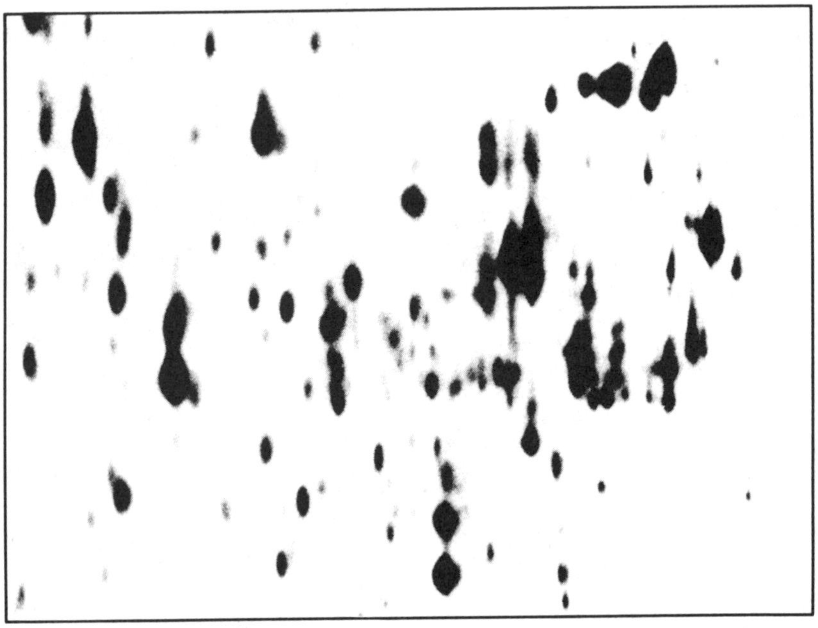

Figure 1. 2-D electrophoresis gel separation of *E. coli* proteins

Even the best video cameras yield signals with too much noise for many applications. We now apply pipeline image processors to average video images in real time, at a rate of 30 full frames/s (*1*). Pipeline processors can perform arithmetic or logical operations on entire ¼-megabyte video frames in 1/30 s. These operations may be cascaded to perform such useful functions as edge enhancement, smoothing, and restoration by inverse filtering. Such pipeline processor subsystems are now available for many microcomputers at steadily decreasing cost.

The memory address space limitations of 16-bit computers have been overcome by our own multimegabyte RAM memories with memory-mapping features. We have constructed special high-speed image-processing logic devices. When called for by the particular requirements of a given application, we have designed scanners based on laser and photodiode array technologies.

The image-processing effort at Du Pont has grown to include a core group of technical personnel and technicians representing spectroscopy, physical biochemistry, electrical engineering, mechanical engineering, and physics. The techniques of applied mathematics, especially numerical analysis and statistics, often enter our work. The instruments we develop frequently incorporate "front-ends" such as Schlieren systems and microscope interferometers which require expertise in optical engineering. Our concern for accurate photometry had its roots in our prior quantitative spectroscopy experience. This experience has also provided a valuable basis for future work in image restoration (*2*).

We are thus able to confront the entire development cycle for image analysis instrumentation from the physics and chemistry of the sample to the optics, electronics, and computer technology required in a stand-alone unit that includes both hardware and software. The following paragraphs describe a recent application that has successfully drawn upon the resources described.

The recently developed technique of high-resolution 2-D gel electrophoresis (*3*) has enabled researchers in the life sciences to separate essentially all components of a complex mixture of thousands of proteins into a pattern of spots (Figure 1). Although the absence or presence of a particular protein can be observed visually, a thorough examination of a typical pattern is painstaking and laborious. It is not possible to accurately determine the quantity of protein in a spot by these means. Furthermore, identifying and cataloging large numbers of spots is impractical without the aid of computers.

Commercial scanners are ill-suited to this application. Video-camera and photodiode-array-based units suffer from light scattering that severely limits the range over which they can accurately measure optical density. The more accurate drum scanners are expensive, slow, and incapable of scanning wet gels. Protein quantitation software developed by pioneers in the field has been potentially efficient *or* capable of resolving overlapping spots, but not *both* (*1*). These limitations motivated us to develop a high-resolution laser-based scanning system specifically for stained wet gels and autoradiograms. This system digitizes optical densities with up to 12-bit precision over formats as large as 2048 × 2048 pixels. High-capacity RAM-based image memories and digital processing electronics that work in concert with a new, uniquely efficient, unpublished spot-resolving algorithm complete the package.

Figure 2. Laser-based 2-D gel scanner–analyzer

The unit is packaged as a desk-size console with full-featured color display facilities and an accompanying scanner module (Figure 2). By using the display's pan and scroll plus 16 levels of zoom, the operator can monitor the analysis and interact if he or she wishes. Figure 3 shows a color-contoured display of a portion of a gel image. In this display, selected optical density ranges have been encoded as pseudo-color bands.

Future advances in solid-state electronics cannot fail to propagate image-processing technology into new areas presently inaccessible because of the remaining limitations of cost, speed, and memory capacity.

Acknowledgment

The author is pleased to acknowledge the contributions of J. G. Elias, his collaborator in the development of the 2-D gel instrument, and of K. K.

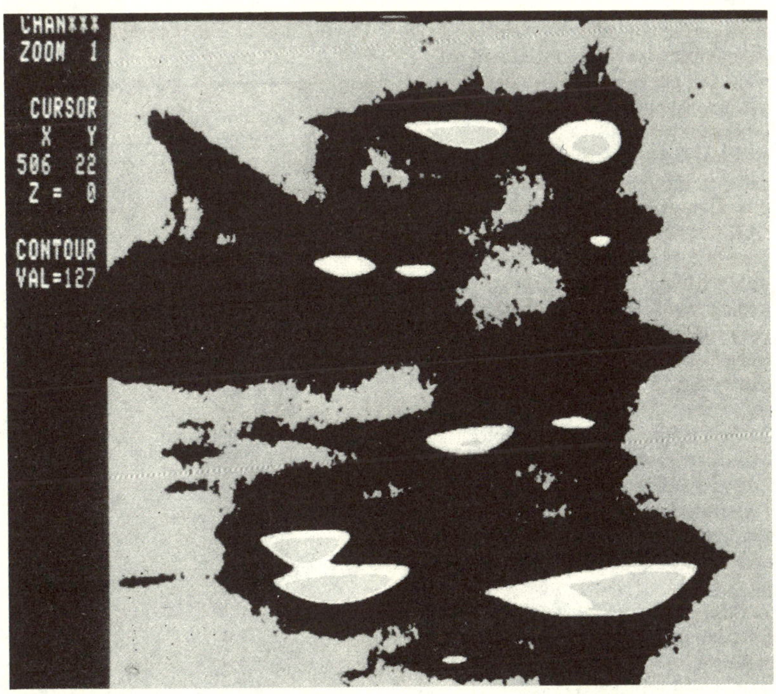

Figure 3. A 2-D gel image

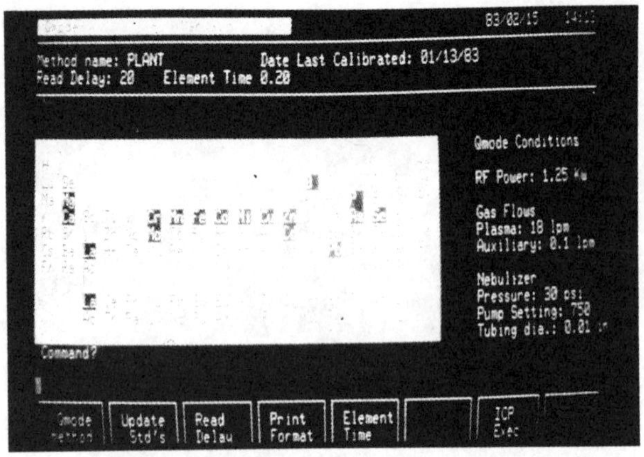

Figure 4. A screen from Perkin-Elmer ICP/6000 software

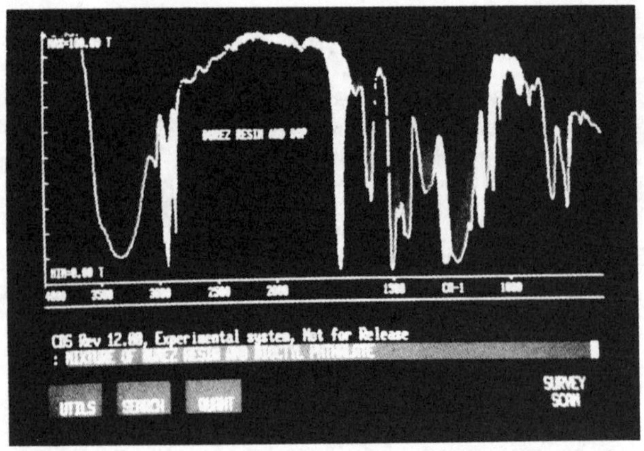

Figure 5. A screen from Perkin-Elmer CDS-3 IR software

Lonberg-Holm and E. M. Chait, who lent their expertise and invaluable support to this project.

References

(1) Jansson, P. A.; Grim, L. B.; Elias, J. G.; Bagley, E. A.; Lonberg-Holm, K. K. *Electrophoresis* **1983**, *4*, 82–91.
(2) "Deconvolution with Applications in Spectroscopy"; Jansson, P. A., Ed.; Academic Press: New York, N.Y., in press.
(3) O'Farrell, P. H. *J. Biol. Chem.* **1975**, *250*, 4007–21.

Perkin-Elmer
Computer-Aided Chemistry Group
Norwalk, Conn. 06856

Contributors: Peter V. Passalacqua and J. Fenton Williams

Color Graphics

The advent of color graphics has provided the analytical chemist with a powerful tool. This tool, although relatively new, is already assisting the chemist in performing a variety of tasks.

An effective and simple use of color is to highlight text on the screen to draw attention to important information. For example, in Figure 4 the periodic chart contains selected elements printed on a red background. These are the elements to be analyzed in this particular test sequence. Not only does the display highlight the particular elements of interest, it is also used as a status flag. As the analysis proceeds, those elements which have been determined revert to a standard color. Thus, the operator can easily monitor the progress of the experiment.

Figure 5 shows an array of soft key *labels* located across the bottom of the display. Soft keys are touch-sensitive devices located on the bezel directly below the video monitor. This arrangement permits easy, single-key interaction with user software. Each key can be "relabeled" under computer control as needed. On the screen shown, auxiliary modules can be selected via the blue soft keys. An optimized spectral scan is initiated by pressing the red soft key, and then the key label turns green, indicating run status. Within the hierarchy of a dynamic menu selection, individual levels are color coded and the default choice is highlighted in a unique shade of that color. For example, the default soft key is light blue for a blue menu. The patterns of various options in a colored menu become very familiar, making learning easier and reducing operator error.

Another common application of color is to distinguish between different items presented in the same two-dimensional display. This is particularly important with multiple data curves and intertwined spectra. Color can serve to distinguish individual curves in a plot. For example, a spectrum of a mixture might always appear in red, the interferent in green, and the difference (analyte) in yellow. In Figure 5, the difference areas between the intertwined IR curves are flooded with color. The specific color of the difference area indicates which curve has greater absorbance in that region. With this type of display it is easy to see the overall strength of the absorbance and the contribution of individual species in each spectral region. There is also a need in other forms of spectroscopy to recognize the areas of band overlap. In Figure 6, the dark blue represents the overlap area, a region where the presence of one component will interfere with the determination of the other by ICP techniques. The operator can easily deter-

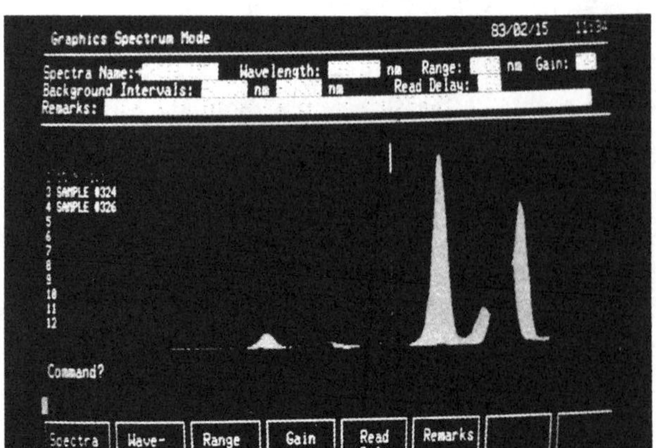

Figure 6. Overlapping spectral regions are lighter

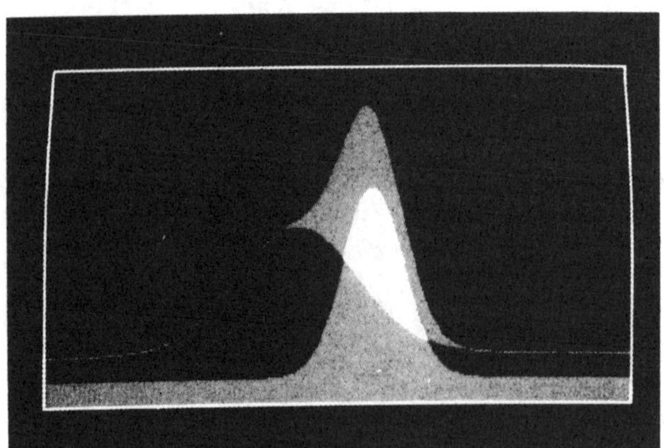

Figure 7. Notch filter absorption spectra

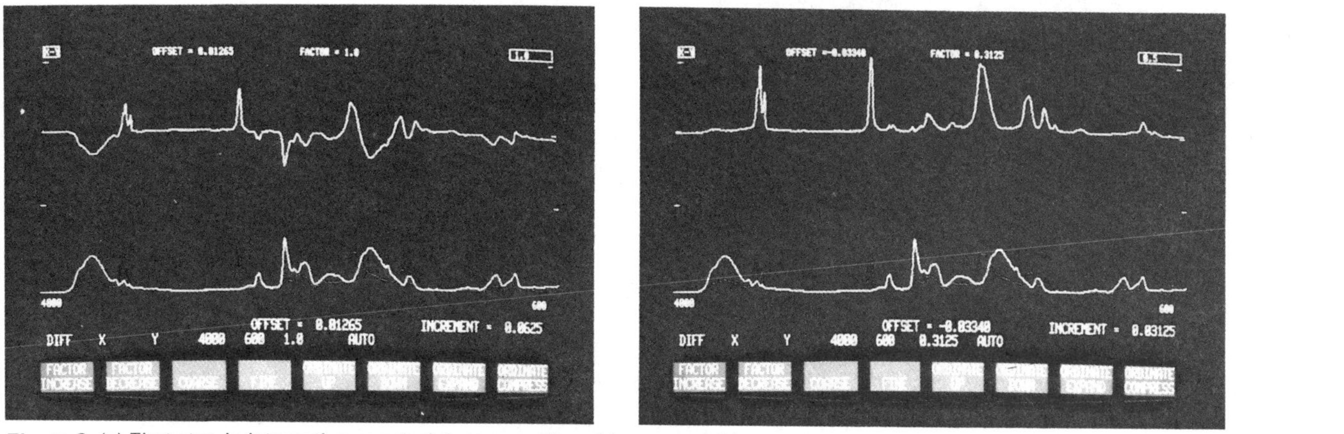

Figure 8. (a) First step in interactive spectral decomposition; (b) successful interactive spectral decomposition

mine the best wavelength at which to perform an analysis.

Many laboratory computers today have bit-mapped color displays which allow for virtually instantaneous color changes through the entire color palette. The example in Figure 7 shows four colored regions typical of analysis problems encountered in optical notch filter examination or spectral decomposition. Bit-mapped color graphics allow an instantaneous display of the total integrated area in a single color to alternate with this display. Such animation is also useful in interactive decomposition of composite spectra.

Difference algorithms using interactive color graphics are especially useful to the analyst for decomposition of curves made up of two components, i.e., a mixture consisting of an analyte and an interferent. Figure 8 shows a two-component absorbance spectrum (X) in red, the pure interferent absorbance spectrum (Y) in green, and the trial difference spectrum (Z) in yellow.

$$Z = X - (N * Y), \quad \text{where } N = 1$$

In an interactive routine with fast graphics screen updates, the chemist can vary the factor N until the interferent is quantitatively stripped from the mixture. Figure 8a represents the display early in the interaction; Figure 8b is the final frame in the sequence, with the yellow spectrum difference values all zero or positive. Such interactive graphics is often a more satisfactory solution to these problems, in comparison to complex spectral decomposition software running without human contact.

Spectral resolution enhancement is another technique which benefits from interactive graphics. For example, Figure 9 shows the results of a two-parameter optimization process which required operator judgment based solely on an interactive graphics display. The original spectrum (top panel) is smoothed with a tunable window and scaled by a variable factor before being subtracted from itself. The bottom panel shows the resolution-enhanced spectrum obtained via the interactive optimization. Notice that band splitting in the carbonyl region of the dioctyl phthalate is now evident.

Color, which adds a third dimension to chemical displays, greatly improves readability and is commonly used in contour maps, density plots, phase diagrams, chromaticity charts, etc. The colors in Figure 10 indicate to the analyst the effectiveness of each possible combination of parameters for a chromatographic separation. The most cost-effective alternative can be easily selected from the display.

We all need to become more imaginative in the application of color and more aware of how it can be used effectively. This includes an appreciation of the associated physiological and psychological factors (*1*). The use of well-designed color displays in analytical instrumentation will lead to increases in both productivity and accuracy.

Reference

(1) Greenberg, D; Marcus, A.; Schmidt, A. H.; Gorter, V. "The Computer Image: Applications of Computer Graphics"; Addison-Wesley: Reading, Mass., 1982.

Contributor: Jack Frazer

Multidimensional Graphics

It is time for scientists to escape from the traditional methods used to present data and to adopt and adapt the graphics tools now available. Multidimensional plotting is particularly

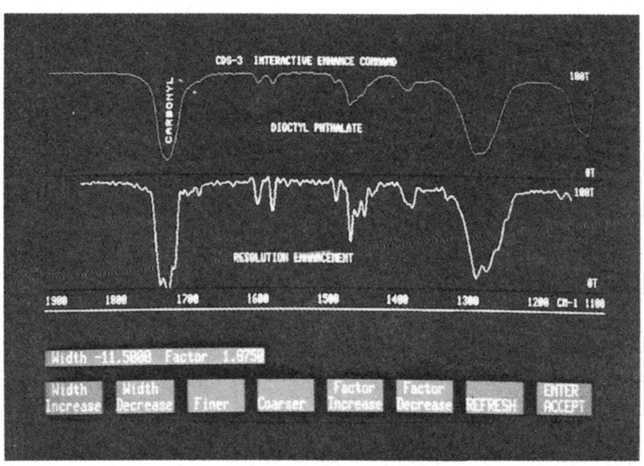

Figure 9. Interactive resolution enhancement

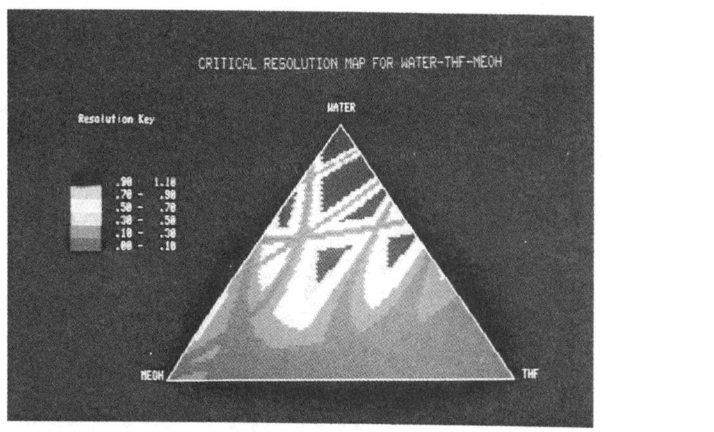

Figure 10. LC solvent phase diagram

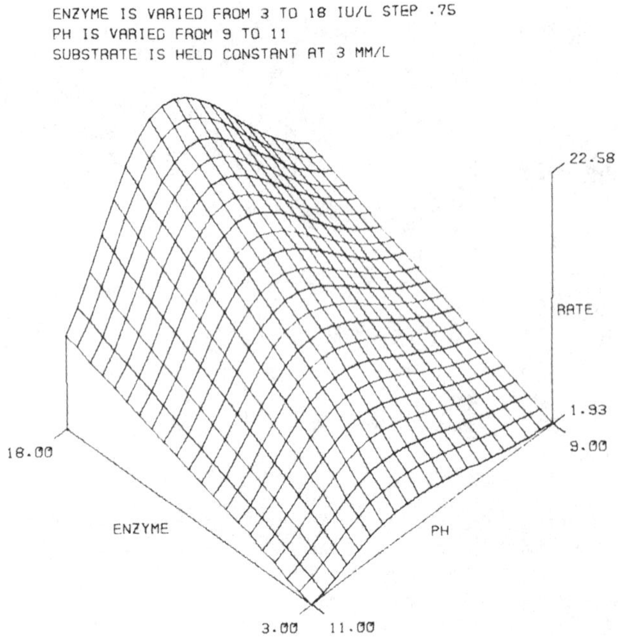

Figure 11. Enzyme reaction response surface

essential to proper analysis and reporting. This capsule summary describes several approaches that hopefully will stimulate applications to your area.

For the analytical chemist one of the most important uses of interactive graphics is in support of automated experimentation. Graphics supports self-adaptive control of experimental equipment, data acquisition, and modeling of data concurrent with the experimentation.

Two fully automated, graphics-supported systems have been constructed for the study of enzyme-catalyzed reactions. One is a continuous-flow unit designed for initial velocity measurements employing standard UV–VIS spectrophotometry. The other incorporates a UV–VIS array spectrophotometer and is designed principally for long-term stop-flow measurements (*1, 2*).

Graphic representation can be used both to help develop improved designs requiring less experimentation time and materials and to display meaningful results during the course of the experiment (*2*). In a typical overnight experiment one obtains several hundred initial velocity measures, each for a different set of reaction conditions. The data can conveniently be displayed as response surfaces, as shown in Figure 11, where every intersection represents a measurement. Data of this type are required if one is attempting to identify the most probable mechanism among several rather complex candidates (*3*).

Most papers analyzing such results contain several plots or graphic representations. To improve understanding and reduce mathematical calculations the tendency has been to search out transformations that result in linear plots. These have been helpful; however, given the capabilities of new inexpensive computers one can avoid these transformations and obtain different and very meaningful graphic representations of data and analyzed data sets.

For the purpose of illustration we will use data obtained from a study of an alkaline-phosphatase-catalyzed reaction. In the past the interpretation of such data was conducted by interpretation of Lineweaver–Burk and other plots, obtained as a result of transformations (*4*). Such a series of single- and multiple-linear line plots has been used to evaluate five possible mechanisms for this phosphatase reaction by the traditional method. After considerable work, it was concluded that the *hyperbolic-mixed system* provided the best representation of the data (*5*).

These traditional data analysis approaches can be replaced by nonlinear analysis techniques that also provide improved estimates of the parameters being studied. In addition, these numerical techniques can be supported by compact graphic representations which make it easy to communicate a complex data set to the scientist. One such graphical method involves the generation of a residual surface showing the degree of agreement between experimental data and a kinetic model.

Only a casual look at the residual surfaces for *uncompetitive* and *noncompetitive* kinetic mechanisms is required for the scientist to know that these two models are not representative of the reaction (Figure 12a). The residual surfaces are not flat. A careful look at the residual surfaces for *competitive, linear-mixed system*, and *hyperbolic-mixed system* mechanisms quickly leads to the selection of the hyperbolic-mixed system as the most likely mechanism of the five considered (Figure 12b). The residual surface deviations for the latter are equally distributed above and below the plane. A relative novice could interpret the data. Unlike the more traditional techniques, all data were used in obtaining each set of parameter estimates. To analyze the same data set by the traditional technique would have required the development and examination of hundreds of linear plots by a skilled kineticist.

Often it is desirable to obtain visual representations of the effects of four variables. This can be accomplished on a color terminal using transparencies. On a monochromic terminal one variable, e.g., the dependent variable, can be placed on the time axis, with the result that one sees a dynamic display where each frame shows the relationship of three independent variables at a fixed level of the dependent variable. For the kinetics research described above, such a four-dimensional representation was developed with a dynamic display of a reaction-rate surface as a function of three independent variables. Figure 13 shows three sequential frames from such a presentation, interactively displayed to the user in real time.

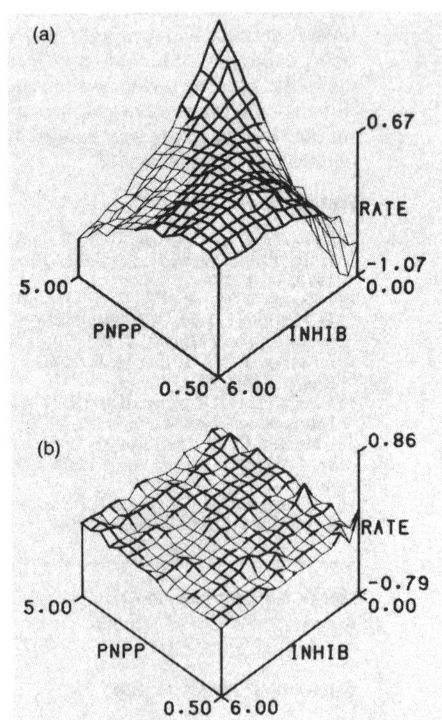

Figure 12. (a) Poor residual surface (uncompetitive kinetic model); (b) good residual surface (hyperbolic-mixed model)

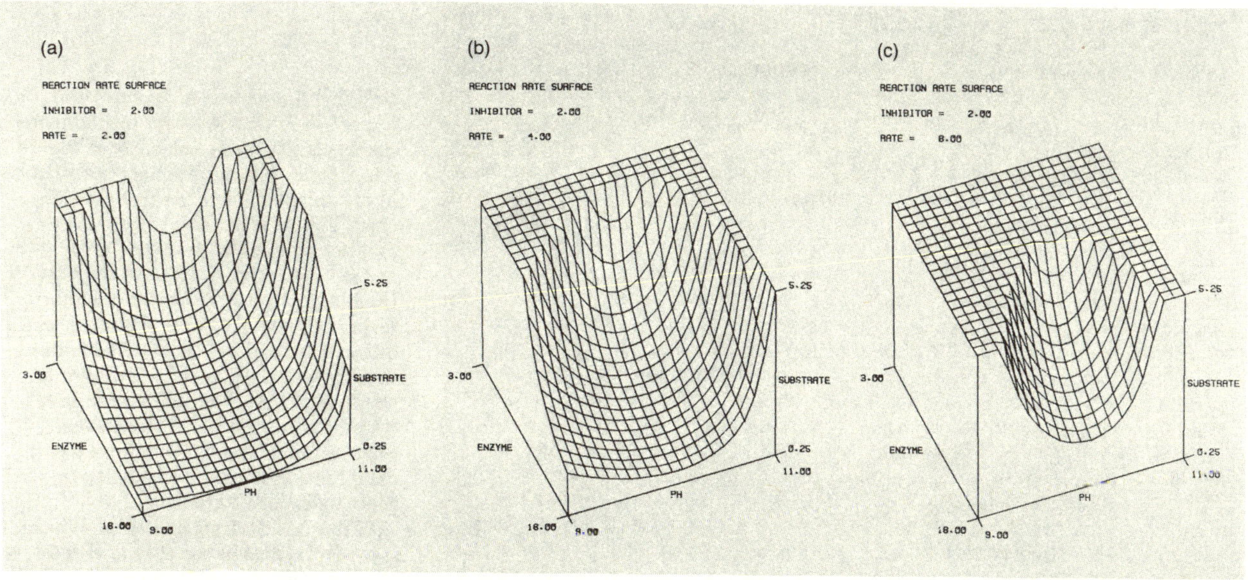

Figure 13. Four-dimensional plotting example
The time domain is used to represent variation in the remaining dependent variable (reaction rate)

There are also useful graphics-oriented programs for analyzing experimental data for which the form of the mathematical model is not known. As one example, a program called D Model (*6*) can be used to explore the form of proposed models and test their fit to data sets. Mathematical functions are defined simply by entering them through the keyboard. Complex functions can be built up from simpler functions previously defined. The scientist can interactively change the model parameters to produce a desired fit to data sets. The program does linear least-squares curves for any linear model including discontinuous functions. The program is now being expanded to include nonlinear curve fitting. The power of such multidimensional interactive graphics will markedly change the way analytical chemistry is performed.

References

(1) Frazer, J. W.; Rigdon, L. P.; Brand, H. R.; Pomernacki, C. L. *Anal. Chem.* **1979,** *51*, 1739.
(2) Frazer, J. W.; Rigdon, L. P.; Brand, H. R.; Pomernacki, C. L.; Brubaker, T. A. *Anal. Chem.* **1979,** *51*, 1747.
(3) Frazer, J. W.; Brand, H. R. *Anal. Chem.* **1980,** *52*, 1730.
(4) Segel, I. H. "Enzyme Kinetics"; Wiley-Interscience: New York, N.Y., 1975.
(5) Herget, C. J.; Pomernacki, C. L.; Frazer, J. W. *Anal. Chim. Acta* **1980,** *122*, 403.
(6) Balaban, D. J.; Wang, J. L.; Frazer, J. W. *Anal. Chem.* **1983,** *55*, 900.

Merck & Company, Inc.
Automation & Control Group
Rahway, N.J. 07065

Contributor: Nancy H. Woo

Voice Input

The introduction of microcomputers has resulted in a large expansion in the computer user community and in increased attention to the operator–computer interface, the method by which one communicates with a computer. Computers have become commonplace; they are no longer the exclusive domain of data-processing professionals and they are being used in nontraditional environments. Today's users include many who are not skilled typists as well as those who need to communicate with a computer while performing other tasks, some that may occupy their hands and/or eyes.

Alternatives or adjuncts to keyboards are required in many applications to optimize the use of the computer. These devices include touch screens, digitizers, and voice interactive systems, including both voice recognition systems and voice output systems. Voice systems are interesting because they allow the operator the maximum freedom from a keyboard and because they simulate (albeit not perfectly) a normal mode of human communication—speech. The most common voice recognition devices (VRDs) today are speaker-dependent discrete-utterance recognition units.

The ability of a VRD to replace or supplement a keyboard comes from the fact that most VRDs can be programmed to transmit a prespecified string to a computer upon recognition of a given utterance. A VRD can be programmed to transmit a carriage-return/line-feed ASCII string (CR/LF) to a computer upon recognition of the utterance "carriage return." However, there is no requirement that there be any semantic relationship between the string transmitted and the utterance. "Go," "c'est fini," "adios," "drop dead," and "brillig" are equally acceptable as the verbal cue to transmit a CR/LF. Users can "personalize" their vocabularies *and still enter exactly the same information into the computer.* VRDs offer an atypical degree of flexibility in human–computer interactions.

The first installation of a VRD at Merck was in the Hematology Laboratory, Safety Assessment Department, Merck Sharp & Dohme Research Laboratories. Among the assays performed in this laboratory is microurinalysis of rat urine samples, a task which involves examining urine slides under a microscope. The data from these examinations form part of large drug safety studies. The standard procedure in this laboratory had been to manually record observations on paper and to key them into the computer later. This procedure is inefficient and prone to transcription errors. Because of the stringent requirement for accuracy, all data must be verified and any errors found must be corrected. The laboratory personnel asked if they could try using a VRD attached to their terminal to determine if it could solve some of the problems of data entry.

The microurinalysis programs are part of a large set of integrated programs. We did not want to invest a significant manpower effort on voice data entry, so it was extremely important that the voice recognition system be capable of functioning without modification of existing applications programs. It was also important that it be accurate so that the time required for data verification could be minimized. In addition, the system had to be easy to use, so that laboratory personnel could modify vocabulary without extensive programming knowledge.

In April 1981, we installed a Threshold Technology 600 voice recognition unit in the hematology laboratory for evaluation. This unit has sufficient memory for a vocabulary of up to 200 utterances, it has a tape drive for permanent storage of each user's vocabulary, and it can be programmed to transmit ASCII strings upon recognition of an utterance. The object of the evaluation was to obtain data that would allow us to measure differences in efficiency resulting from voice data entry and to determine if use of these devices led to any changes in the error rate. The comparison in efficiency was made by collecting "time expended" data for both manual and voice modes of data entry. Existing standard procedures were followed for verifying data and correcting errors. Time spent in error correction (data verification) was included in the evaluation of efficiency.

The errors which occur in manual entry of data are standard typographical errors, such as typing a "2" instead of a "3." To detect these, the biologists check the data immediately after typing it in and later proofread the printed laboratory report, character by character—a painstaking task. When data are entered using a VRD, an error will occur (the wrong string will be transmitted to the computer) if the VRD "mis-recognizes" the input utterance (associates it with the wrong reference pattern). For example, an error occurs if one says "three" and the VRD recognizes the input as "Lee" and transmits "Lee" to the computer. The biologists were asked to check voice-entered data in the same way that they checked keyboard entries.

The results of this evaluation are rather interesting. The biologists found that their efficiency increased when they used voice data entry. The actual gain in throughput using voice was at least 80%. This was in spite of a significantly *higher* detected error rate for voice data entry, reaching as high as 5% compared with a manual error rate of 1%. However, the time and attention required to verify data entered by voice was less than for data entered using the keyboard. The biologists pointed out that this was a consequence of the fact that mis-recognition errors occur because of *phonetic* similarity and often result in substitution of *orthographically* distinct strings. Hence, one knows immediately if the VRD has mis-recognized "three" as "Lee" if it transmits "Lee" instead of "three." The typical typographical error, hitting "2" instead of "3" is not so blatant and requires more attention to detect. The biologists are convinced that *verified* data entered using voice recognition are more accurate because they can easily find any substitutions. It should be noted here that the biologists have full access to the keyboard at all times. Data can always be entered manually, even while using the VRD.

Voice data entry has been accepted enthusiastically by all of the laboratory personnel. An Interstate Electronics VRT300 voice recognition board installed in a DEC VT100 terminal has now been installed as a permanent part of their facility.

One area in which we are planning to install voice recognition terminals is in a tissue culture laboratory where data are to be entered by personnel working in sterile laminar-flow hoods. For this application we are considering voice response verification of the data entries as well as a display on the CRT. Another promising application is the use of VRDs as sophisticated, flexible keypads to facilitate management access to computer functions such as data base systems. VRDs can be programmed to transmit any string the user specifies upon recognition of given input; in particular, they can be programmed to transmit sequences of commands in response to a single spoken command. They can, therefore, fulfill the same functions as special function keys or a keypad with the critical difference that they can be locally reprogrammed and hence have enormous flexibility.

Originally published in
Analytical Chemistry **1984** *56*(6) 725A
and **1984** *56*(7) 855A.

Managing the Electronic Laboratory

Part I

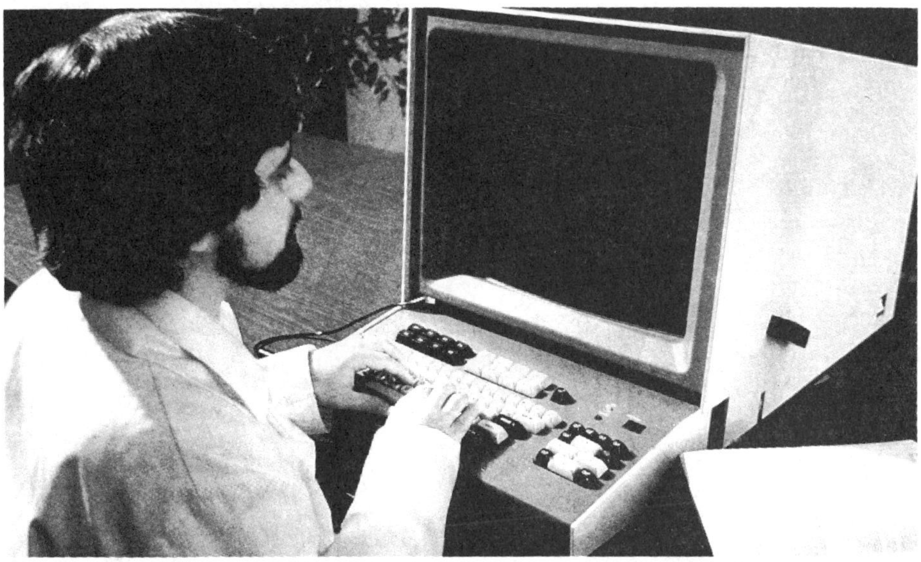

Figure 1. Designing and searching molecular structures, a typical service in the electronic laboratory
Photo courtesy Molecular Design Ltd., Hayward, Calif.

Imagine a laboratory built upon the technological principles of the electronic office. On Tuesday morning a memo from your supervisor appears in your electronic mail. Scanning the memo on the CRT screen of your work station, you see that a major purchaser of your company's untreated fiber is experiencing difficulties achieving uniform dyeing. The customer has asked for assistance from your analytical laboratory, and a "fix" is needed by the following Monday! The unautomated laboratory would have little hope of addressing this problem, but you have help at hand, the *electronic laboratory*.

The solution to this problem might proceed according to the following scenario:

- First, keystroke a reply at your work station to the lab manager indicating receipt of the request.
- Then, call your personal CALENDAR from the lab computer and cancel all nonurgent appointments. Memos of explanation are created and routed to the affected individuals. Accessing the CALENDARS of your lab personnel allows you to delegate attendance at important meetings to individuals who are free; again, appropriate electronic memos are transmitted.
- A telephone call to the customer service contact—who has further technical information available from the user—finds her away from the office. Your verbal message is digitized and stored, and the system begins to track her down. In the meantime you face the task of searching for background data.
- Your lab system has two important data bases, SIGFIBER and SIGDYES (special interest group on fibers and SIG on dyes). These allow relational access to internal company reports that contain data pertinent to your problem. More than 50 SIGs exist within the company, allowing access to information that would normally be lost across political and divisional boundaries. Copies of the source documents begin to appear at the regional high-speed printer.
- Next, accessing the corporate data base RESOURCE, located on a large central computer, allows a search for personnel who have worked on such problems in the past. Your company realizes that skill-awareness tends to get lost in career shuffling, and data tend to disappear into "cellulose black holes." Since these are your most important assets, the computer group has provided a convenient way to retrieve this vital information.
- Finally, a free slot is found through CALENDAR for scheduling a task force meeting late Tuesday involving all the personnel needed to help on the project.
- The customer service representative has returned your call via electronic phone (she's "dialed" your terminal). Additional technical information is dumped to your office printer, and samples are on their way.
- At the meeting, your lab personnel work with the experts discovered by the RESOURCE search, and a schedule is set up for tests to begin on

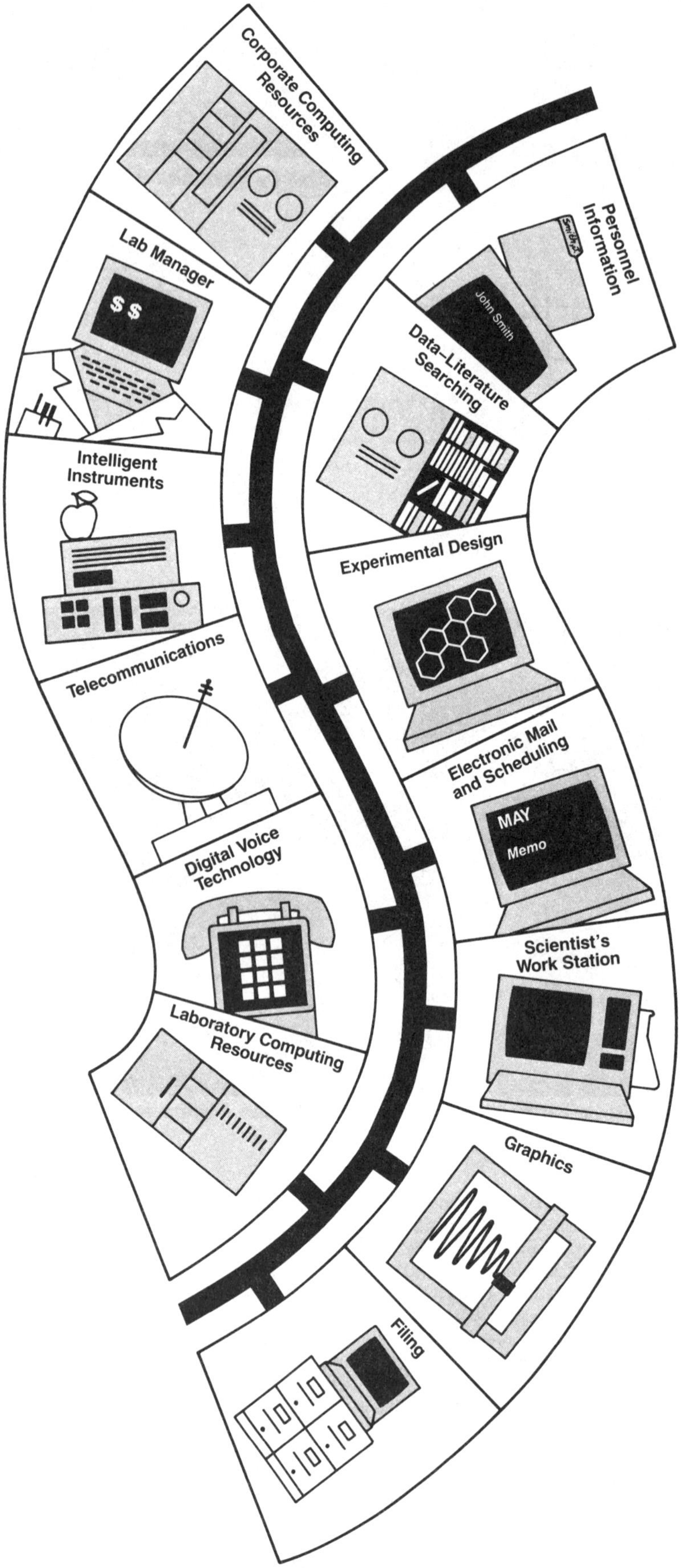

Figure 2. The rational electronic laboratory

Wednesday. Priority samples are submitted through the laboratory information management system (LIMS).

- On Wednesday, as data are entered into this system by analysts at their own work stations, you have immediate access to it via your CRT. Your work station allows you to capture items such as numeric data, spectra, and chemical structures and permits you to enter this information into your own electronic notebook, which supports both tabular and textual entry. Good plotting and statistics packages are available, as well as modeling programs.
- By late Wednesday it is obvious that some trace contaminants are showing up in the fibers with nonuniform dyeing characteristics. These organic molecules need to be identified.
- An electronic memo is dispatched to key personnel advising them of the progress to date.
- On Thursday morning, submitting spectra of the organic contaminants to remote spectral identification libraries leads to an unusual collection of substructures.
- Access to data bases that allow you to draw these chemical substructures and search on them generates a list of candidate molecules and further vital literature information (Figure 1). These services are located on the lab's VAX 11/780, but you access them through your work station computer over the local area network (LAN), which connects all such services!
- An *expert system* is available on the corporate mainframe. It has been developed by capturing the technical expertise of a dozen highly qualified scientists within the company who have demonstrated skills in this area. By noon the system has returned a number of possible sources and solutions to the problem.
- Further analyses are priority scheduled for late Thursday through the LIMS/LAN system. By late evening clear evidence is returned to your work station that points to a specific source of the trouble. The fibers have been stored in an area where adsorption of organic fumes has occurred on exposed surfaces.
- After drawing the structure of the organic molecule concerned on your CRT screen, a search is requested of the stockroom file maintained by your organic synthesis lab. This returns the information that a sample of the organic compound is available, and you send a request for the sample to be delivered to the applications lab.
- A quick test in the applications lab is scheduled for Friday. This demonstrates that a simple pretreatment will cure the problem, and their report is on your CRT by noon.
- Final reports are drafted using

scientific word processing tools Friday afternoon. Various reports, differing materially in content and depth, are prepared for the customer, customer service, and the lab manager. These are submitted for approval over the LAN. Sign-offs are obtained by late afternoon.

• Submitting these reports to distribution, you begin thinking about how your electronic laboratory can help you catch up Saturday morning with the backlog this hectic week has created, yet still leave time to take the kids to the matinee of "StarWars IV."

Speculation? No, many of the tools described are already available in the chemical laboratories of such companies as Procter & Gamble (Cincinnati, Ohio) and Burroughs-Wellcome (Research Triangle Park, N.C.).

Coping with the design and implementation of the electronic laboratory is a tangled, complicated problem. Perhaps half the attempts to automate laboratories fail, if one defines failure as an inability to use the technological resources currently available to their reasonable limit. Implementations fail not because of technological weaknesses, but because of the human condition; inadequate analysis and misunderstanding of the problems, political and ego tensions, insensitivity to user needs and motivations, poor diplomatic skills, or failure to establish proper educational programs. This is just plain poor management! Satisfactory installation requires the cooperative overlap of three areas: corporate computing, scientific lab automation, and word and speech processing. The traditional immiscibility of these areas needs to be overcome. Otherwise, separate, duplicate, incompatible, proliferating facilities will be developed. This tutorial discusses the vocabulary, problems, and alternative solutions associated with the management of the electronic laboratory (see Figures 2 and 3). Key issues such as system planning, the role of personal computers, training, education, software support, system acceptance and growth will be explored. The capsule reports that follow next month will describe some industrial experiences.

Needs, Both Real and Imagined

The goal of automation is to provide services beyond what is currently available, not merely to provide old tools in a different way. This often requires educating the participants in how computers function so they can develop a vocabulary with which to express their desires. Increased productivity is a desirable target, but increased creativity is a better one. The system should help people rise above their data-saturated environments so they have new perspectives on the

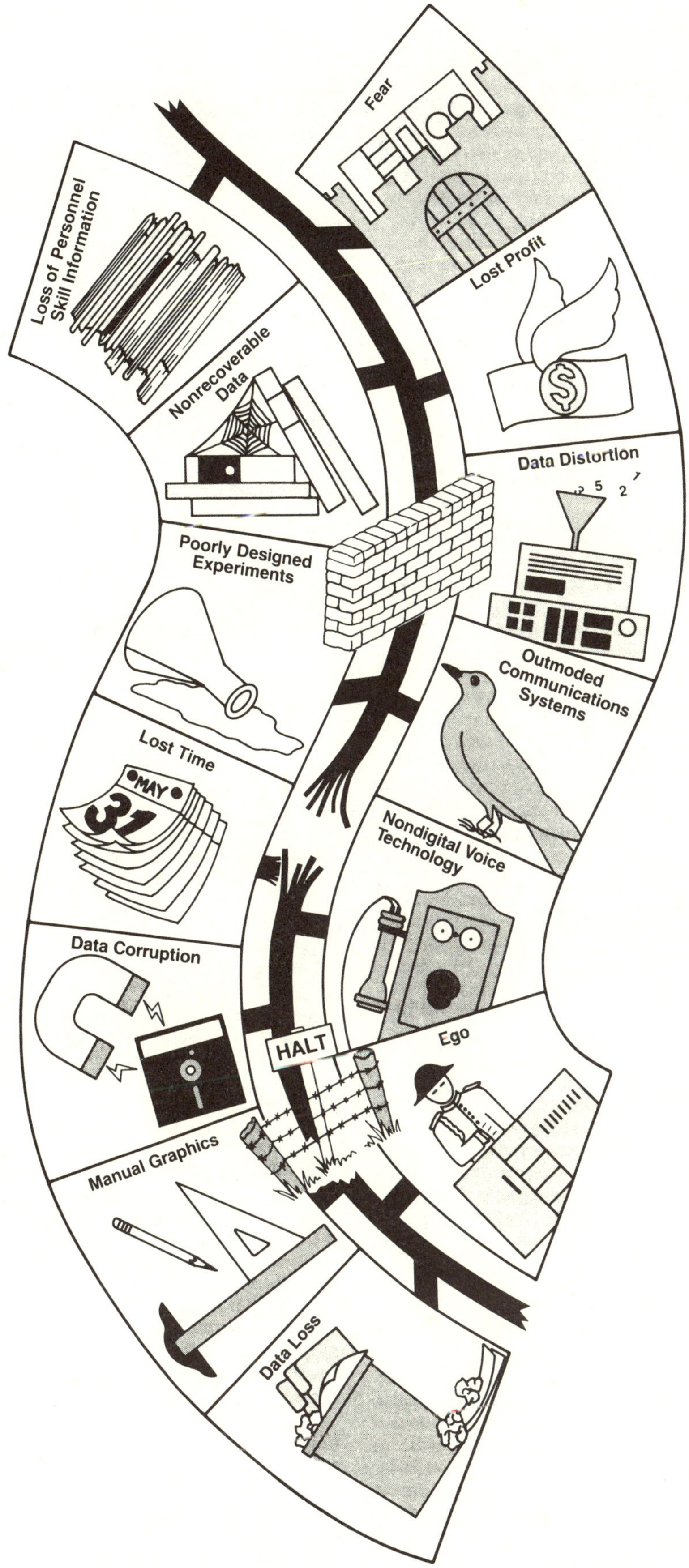

Figure 3. Problems that can arise in the electronic laboratory

problems they face. It is essential that the lab manager and his people become involved in system planning, since the needs of an analytical service or research lab are often quite different from those of a business office, and require skills and understanding often lacking in computer group personnel. It also is essential that the scientist retain control of the environment; chemistry should control the way computers are used, rather than the reverse.

Understanding and communicating your laboratory's needs are not simple. Successful endeavors are characterized by task teams, carefully designed questionnaires, and personal interviews that draw out scientists' and technicians' needs.

A good questionnaire will contain many cross-checks. For example, a request for a sophisticated statistics package in one part of the questionnaire does not correlate with another response from the same person indicating little use of repetitive sampling and standards. Groups of individuals will often develop a sense of need for very sophisticated items. Yet they may be unwilling to involve themselves in the education necessary to use these packages when delivered, or the system may prove too complicated for them to use on an intermittent basis. Tragedy abounds when users get what they ask for and it is not what they really want (or need).

Different groups of users have differing needs and perceptions. Managers are interested in an overview of operations; scientists and technicians are concerned with concept and detail. Groups at all levels must be involved in the analysis, a step that also helps diminish fear.

Most successful installations have asked prospective users to quantitatively rank system facilities and needs in some way, and have used an Oracle approach in order to arrive at a consensus. This involves sharing first go-round results among group members, and between groups, so that an iterative feedback mechanism leads to final system specifications.

Conflict Resolution

Regardless of the technique employed, the "noise" in such a study is very high in today's environment. Everyone has become an instant expert in computers and their applications.

On the one hand, the traditional computer centers, or automation and control groups, are often characterized by critics as being inflexible, slow to respond, and expensive. Caught off-guard by the personal computer wave, many are enthusiastically playing technological catch-up, while others are responding out of fear or with grudging acceptance.

On the one hand, critics of the computer neophyte claim that, in the personal computer arena, he is experiencing the false starts, mistakes, and learning experiences undergone by minicomputer and microcomputer advocates during the 1970s and by mainframe personnel in the 1960s. The only difference is the lower capital cost of the personal computer. There are a large number of these lay computer-scientists. They usually have a well-equipped vocabulary, but a poor grasp of how things work, and an even poorer understanding of systems design.

However, cooperation of the two groups described is necessary for success. Personal computers with well-designed hardware and software will be selected and integrated into a system where long-term support services are ensured.

If, instead, the two groups collide, then a situation develops that Hoffmann-La Roche calls a ticking time bomb. Five years from now the ill-chosen personal computer software tools will not be compatible with developing technology and hardware support will be nonexistent.

The only rational approach to the electronic laboratory requires a coherent modeling attempt that focuses on *who creates data, who uses those data*, and *how those data are manipulated, interpreted, and reported.* Both the fiscal and legal imperatives of today's analytical laboratory and production environments must be considered.

Most of the successful planners have used a "troika" system, with computer personnel, laboratory management, and users involved in a representative democracy. If this sensible approach fails at implementation time, it is usually because the real power continues to lie in the computing center's hands. Where such vested institutions exist, some laboratory managements have set up separate groups that may subliminally or overtly threaten the computer establishment. This Draconian measure may focus the latter's attention, but it is a strategy that requires great administrative care and attention. What may result are separate developments that have trouble coordinating and communicating with each other.

What has not worked very successfully is the lone *ombudsman* or *amicus curiae* set up to interface unofficially between small lab groups and the computer center or automation and control groups. In many instances the ombudsman is asked to walk a narrow path between two extremes: on the one hand, providing advice and consultation only, which tends to leave lab personnel floundering; and on the other hand, becoming a technical center charged with total installation and maintenance, which spreads the available energy and talent too thin. Without power, money, and a secure organizational position, the ombudsman seldom survives pressures from colleagues, power struggles, the usurping of credit, and the assignment of blame, even if he manages to survive an environment where he is always "putting out fires." There must be time to keep current in an area whose half-life of technology is a few years. Otherwise, these personnel eventually become less competent and stay, or become frustrated and leave.

The most successful automated labs have created a user-responsive steering committee or board of directors to establish policy for the entire facility. No Potemkin village, such a group wields the weapons of budget disposition and personnel control. This is often difficult to inaugurate because of corporate boundary lines, egos, and tradition.

Data and Software Assurance

The typical laboratory is a mixture of instruments from a variety of vendors. Previous tutorials have discussed the real-time control and data acquisition problems involved. Other articles in this series have addressed the protocol problems associated with operating systems, disks, languages, LANs, and LIMs. Communications is the most essential aspect of the instruments of the 1980s. Instrument vendors are beginning to realize their obligations, and the necessity for them to adapt to, and adopt, standards. This will make it easier to connect the multiple levels of intelligent instruments and computer facilities required to construct the rational electronic laboratory.

The LAN provides the envelope and postman necessary to deliver data from one point to another. The LIMS packages provide the secretary to store and retrieve the information. When and if the instrument vendors concur on a relatively neutral data format that standardizes the messages from their instruments that are stuffed into these envelopes, the open system concept will be a reality. It will be "open" for easy inclusion of new nodes. Users will be free to manipulate the data collected in exciting new ways in order to extract more and different information. The current dichotomy existing in labs with dual-track record systems (paper and electronic) will disappear, as the two are merged by the electronic laboratory. But what about data and software assurance?

Assurance at the Instrument. At the point of data capture, programs written by instrument vendors interact with that data. Users have a right or an obligation to assure that the programs involved are certified; that is, showing that the results are valid within the ranges and conditions of the sample universe. Who certifies? It is either user or vendor. Some laboratories are insisting that vendors provide algorithms or software listings for all packages that may be used in regulated areas. The user must be willing to sign releases and pay license fees for this information.

Assurance at the Computer. But what happens after the data leave the intelligent instrument? The traditional approach has been to store data safely away in inviolable computer rooms. Does this step also deny easy human access? Arcane commands are often necessary to invoke the deity when it deigns to communicate with you at all. Personal computers, on the other hand, are at your command by the rubbing of the genie-lamp keyboard. Some require that you merely point at what you want done—no typing necessary. Everything is under your control. And therein lies the problem.

The Large Computer. In some cases, the regulatory, legal, and fiscal imperatives of a situation demand that capture and manipulation of certain data be performed in a rigid and controlled environment. Toxicological, pathological, and personnel data are all sensitive. They comprise data that are expensive to replicate, that are used over long periods of time, and that require security. Professional computer groups can assure the quality of the software used to manipulate these data, and they can provide human-cordial access methods.

The Personal Computer. On the other hand many research findings are possible to repeat (albeit at some expense), often transient, and commonly less sensitive. Therefore some companies permit research activities to employ personal computers.

Data are more vulnerable and data loss and poor documentation are more prevalent in personal computer systems. More troublesome is user-generated software that is incorrect. Quality assurance of data is an important concern. Poor-quality software will certainly generate poor-quality data. Who is responsible for software testing and assurance in your laboratory, and how is personal computer software controlled?

Some labs reason that local software problems are no better or worse than those created in a pencil-and-paper or calculator environment. Other labs are convinced that the problem is more serious, since software is an amplifier of human ability, both good and bad.

Therefore, many labs have implemented acceptance programs to screen local software before it is made available for general use. Few labs have begun to address this problem at the individual user level, but increased attention and action by the regulatory agencies will force such concern. Significantly, the Pharmaceutical Manufacturing Association held a three-day symposium in January of this year on software assurance problems.

Even good software manipulates and massages data in different ways as the user modifies it. Who changed what programs? How? When? What data were treated with each program version? Since most scientists do not document software well, these questions become extremely important in a lab subject to regulatory examination.

The laboratory manager should not be seduced by the ready acceptance of personal computers in the office environment. The business milieu involves simple, well-defined manipulation of information that is collected by well-established procedures. Good general-purpose programs are available because of the large, and therefore profitable, user base. Any potential for distortion occurs *after* the storage process, as users apply spreadsheet, statistics, and "what-if?" packages to relatively small data bases. In contrast, the laboratory involves a large amount of data subject to potential distortion *before* their storage, during data capture and reduction, often by complex algorithms. Personal computers also can easily lead to multiple data sets derived from the same source that differ, destroying the concept of a single electronic data record.

The Micro–Mainframe Connection. Unless some control is initiated over personal computer activities, data will be corrupted. One extreme is to forbid the use of personal computers and rely on user-cordial equivalents in larger computers. It is certainly possible to provide such services. But if the response time of the computer services group (or the response time of the actual computer system) is excessive, then people will choose to use personal computers. Even when response time is excellent it may be useful to encourage personal computers and larger local computers. Studies have shown that computer use can increase 100% per year in the middle of the S-shaped curve relating load and user sophistication. It is more cost-efficient to perform some jobs on personal work stations, so that trivial loads do not slow the response time of big systems.

The users can then run nonsensitive computer programs at their own volition. But what of sensitive areas? The solution is a good micro–mainframe connection.

Some laboratories have standard forms that they require employees to use. Likewise, many laboratories insist that all handling of critical data at the personal computer level use standard programs, which are downline-loaded from a central host. These programs are maintained by a central group.

Some companies require that research notebooks be photocopied periodically to permit recovery from loss, theft, flood, and fire. For similar reasons, some companies periodically upline-copy critical updated files containing data or programs from personal computer disks. This prevents system crashes, operator errors, and magnetic screwdrivers from causing problems due to lost information.

Standards. Personal computers and microcomputers do need to be controlled and standardized. A reasonable list of specifications includes a 16- or 32-bit CPU running a standard operating system on a standard bus with plenty of empty slots for add-ons. An adequate power supply and ventilation to handle new peripherals is mandatory. Finally, choose a vendor who will exist five years from now. Such standardization assures that both hardware and software maintenance will be affordable. In-house personnel can board-swap from a central repository, and software support centers and exchanges can be developed. Such standardization also makes it feasible to easily incorporate the personal computers into a net. This provides the support services capable of addressing the problems of lost data and altered programs.

Some Comments on Education

Involving users in the decisions that relate to computer systems is common sense. Educating them prior to the decision will generate more intelligent input and better acceptance. Continued education after the installation is also essential. It saves both time and money. However, since it is common to get little credit for didactic skill in a university environment, it is not surprising that companies also usually give little credit for this activity.

It is important that management make an overt commitment to education. Methods of evaluating and rewarding the personnel involved are sorely lacking in many lab environments. Job descriptions and promotion and salary ladders need to be carefully planned.

Deciding on *whom* to train *when* is sometimes a trade-off. If you attempt to educate everyone, the instantaneous demand on the system may exceed its capacity, and system credibility will suffer. Training an elite will solve that problem, but this quickly creates two classes of users separated by a gap that will grow with time.

The corporate computer and large local hosts serve a sophisticated user group that knows how to command their power. Documentation is developed to a point where the administrators are satisfied about its value, and new users are expected to learn from it. Unfortunately, new users have not lived through the pioneering days and cannot wade through the documentation's jargon and massive size; it's just too formidable. The computer group wonders why people do not use their services and instead find imaginative ways to buy personal computers.

Successful companies offer continuing classes on the software tools available on the larger computer systems. Documentation is current, readable, and easily available.

Successful labs also encourage selected target groups to become educated users of personal computers. Voluntary classes and computer discount arrangements encourage scientists to buy one for work and one for home (what you learn at home you do not need to learn on the job); both are ways to encourage standardization in hardware and software tools. Since the personal computers must sooner or later become part of the network arrangement, this standards incentive plan has some important pay-offs for both parties. Data and program assurance have already been mentioned. This approach also guarantees that the rapidly developing demands made by the user can be satisfied by network connection. Tasks appropriate to the personal computer will remain at that level, but more complicated tasks will migrate upward in the net naturally.

Selling the System to Yourself and Others

The best of plans must be explained both to the people who will use it and those who must fund it.

The User. Earning the fealty of users obviously involves continuing education and listening carefully to people's needs and comments. No coordinated approach should attempt to interfere with existing custom systems having only local effect, except to attempt to provide network access if desirable. Implementors of systems should realize that building credibility in the user's eyes involves initially providing simple services that work well. Once that credibility is established, more difficult tasks can be tackled.

Yourself. In general, it is not advisable to develop custom hardware and software in-house, unless no commercial product is available. Design, fabrication, testing, and documentation costs merge with long delivery times to create credibility problems with users. Where exceptions occur, such as Searle's data system or Burroughs-Wellcome's ring network, a unique combination of personnel made in-house development feasible.

Lab managers often do not understand software development. The mysticism surrounding professional programmers and the enthusiasm of "hackers" lull managers into accepting unreal or nonexistent performance–time charts. Key developments often are allowed to proceed in the hands of one programmer, with no monitoring or intermediate documentation. What happens if the person leaves? Managers do not realize that the goal of many good software folk is to stay at the cutting edge and to remain highly visible in the marketplace. Lab problems are often prosaic and low turnover is important in the development phase. This is a real managerial dilemma requiring a job mix that has both challenging and bread-and-butter components.

Corporate Management. Corporate administrators are interested in fiscal and operational justification of new computer purchases.

Return on Investment. Computerization can improve the quality of data and the ability to extract more information from a given amount of work. Strong cases can and must be made for the increased ability to rapidly store, recover, and use the information produced by company operations. A productivity enhancement of even one percent can provide a significant return on investment, considering the large number of people affected by the electronic laboratory. It is to be hoped that your management realizes that timely, competitive innovation requires the electronic laboratory, even though such creativity enhancement is hard to quantitate.

Resource Centers. Any plan must include provisions for a well-staffed resource center to provide standard and custom software support. Users who develop a rapport with a service center will develop their own skills, and a voluntary adherence to the necessary standards develops naturally. Personnel additions will be needed to service the broader user base that is more computer-literate and demanding. Many companies are delocalizing these important service center personnel, creating interesting administrative problems. Who pays, evaluates, promotes, or terminates them?

Who Pays for What? Multidivisional, multisite, and multinational corporations can often benefit from coordinated system development and cloning, in appropriate circumstances. Such coordinated solutions spread development costs and assure intersystem compatibility. However, this requires a high-level management commitment to, and financial support of, global solutions. Individual laboratories have no financial incentive to fund longer term, higher cost, global solutions, even though the corporation would benefit. In multisite cloning the similarity of the target sites must be honestly evaluated, not assumed or dictated.

Pay and Play. In some institutions there is a trend away from the traditional funding allocation to central computer centers. Instead, they are expected to capture their operating costs. Charge-back fees for service provide a mechanism for performance and cost appraisals by group and individual users. This gives the users a powerful weapon and useful flexibility where legal requirements and corporate needs permit. For instance, they might choose to use some of their computer allocation to purchase alternative computing power.

Users could then establish their own large local computer, servicing a mixture of terminals and personal computer work stations. This facility, in turn, would be networked to the rest of the corporate computing system.

Personal computers have a well-defined role in the lab. Properly used, they provide a new form of time sharing. Poorly used, they become expensive terminals on an overused system, or isolated islands. In the best labs personal computers and the larger lab computers synergistically interact with corporate facilities. In these labs the acquisition both of large lab computers and of personal computers is rapidly increasing.

Cluster Systems. Most installations involve implementation in stages. Financial and personnel constraints, as well as prudence, necessitate such an approach. The trends of vendors to adopt communications standards, to develop families of computers, and to design computer cluster architectures make this even more attractive. Families of computers have a built-in compatibility. In cluster systems several computers can share disk and memory; they are connected by very high speed links. Job and task queue handling allocate demands to the most appropriate processor. As more power and more speed are required, hardware additions are made to the cluster in a way that is transparent to the users. Great strides in fault-tolerant methods are being made in such systems. This guarantees that hardware failures in any component will be detected and absorbed by the "organism" in such a way that no harm will occur to the user, the programs, or the data. Such cluster techniques are essential to the success of the electronic laboratory.

Our Changing Environment

The electronic laboratory is a servant that we must master. We now have the ability to make computers and robots perform tasks that humans cannot. This has extended our world. We must also be willing to change the way we do things as these servants provide us with the resources to alter the way we can conceive and solve our chemical problems.

Acknowledgments

The author would like to thank the many industrial laboratories and individual scientists who provided the exposure that led to these philosophies. In particular, many new insights were provided by Steve Benezra (Burroughs-Wellcome), Keith Caserta (Procter & Gamble), Ian Chapple (Virginia Tech), Steve Heller (EPA), Howard Moraff (Merck, Sharp and Dohme), H. Randall Munson (A. H. Robins), Michael Starling (Union Carbide), John H. Taylor (E. I. du Pont de Nemours) and Vincent Venturella (Hoffmann-La Roche).

Part 2

General Motors Research Laboratory
Executive Department
Warren, Mich. 48090

Contributor: Clay Snyder

General Motors Research Laboratories (GMR) is the central research facility of the General Motors Corp. In April 1982, an Office Systems Task Force was appointed to develop a five-year plan to address internal issues surrounding office and interrelated computing equipment. The task force was represented by members of the Budget and Cost Control, Computer Science, Engine Research, Executive, and Technical Information departments. Implementation of an integrated office and research work station network began in January 1984.

Several specific objectives were defined for this program:

- Increase the productivity of all labor groups, including managerial, professional, technical, administrative, and secretarial.
- Improve the quality of research. The new computer-based tools will result in increased and effective access to research data.
- Reduce future staff costs. GMR expects to handle an increased volume of research with little or no increase in current administrative and secretarial staff. The expanding professional and technical staff will yield greater per capita productivity and higher quality research through the automation program.
- Improve the research environment by providing a variety of advanced tools that will attract top professionals.

The first priority of the task force was to examine the status of GMR's existing information processing systems. The first question, of course, was what systems were we talking about? The answer was not so simple. It depended on your perspective and on your definitions of "information," "processing," and "system." The task force discovered there were five information processing systems in operation at GMR: computation and analysis, data acquisition, office business and administrative, information acquisition, and technical publications and presentations. With the exception of some equipment that handled work

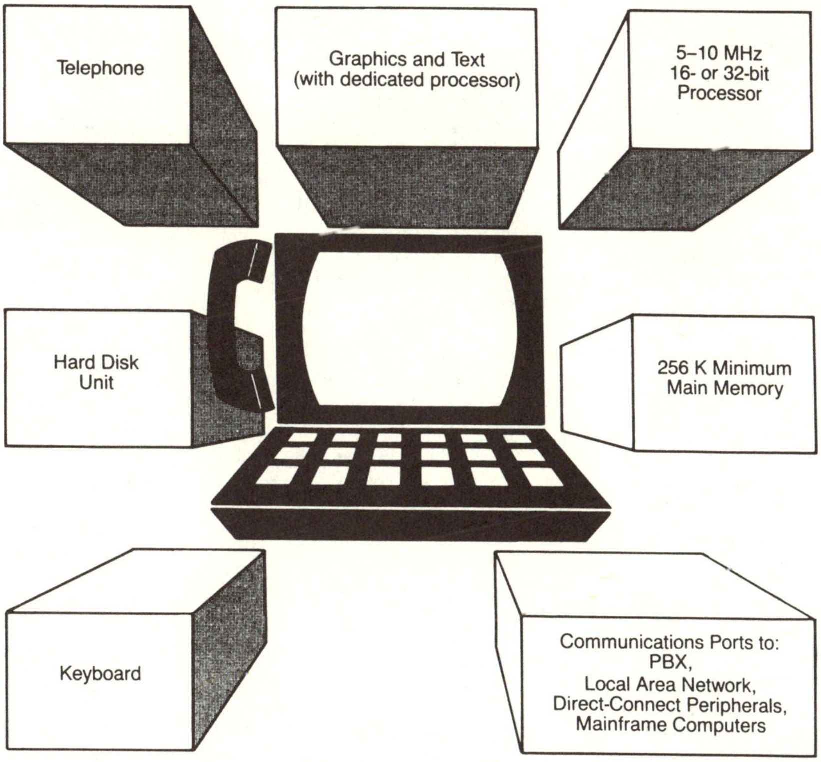

Figure 1. Specifications for professional work stations at GMR

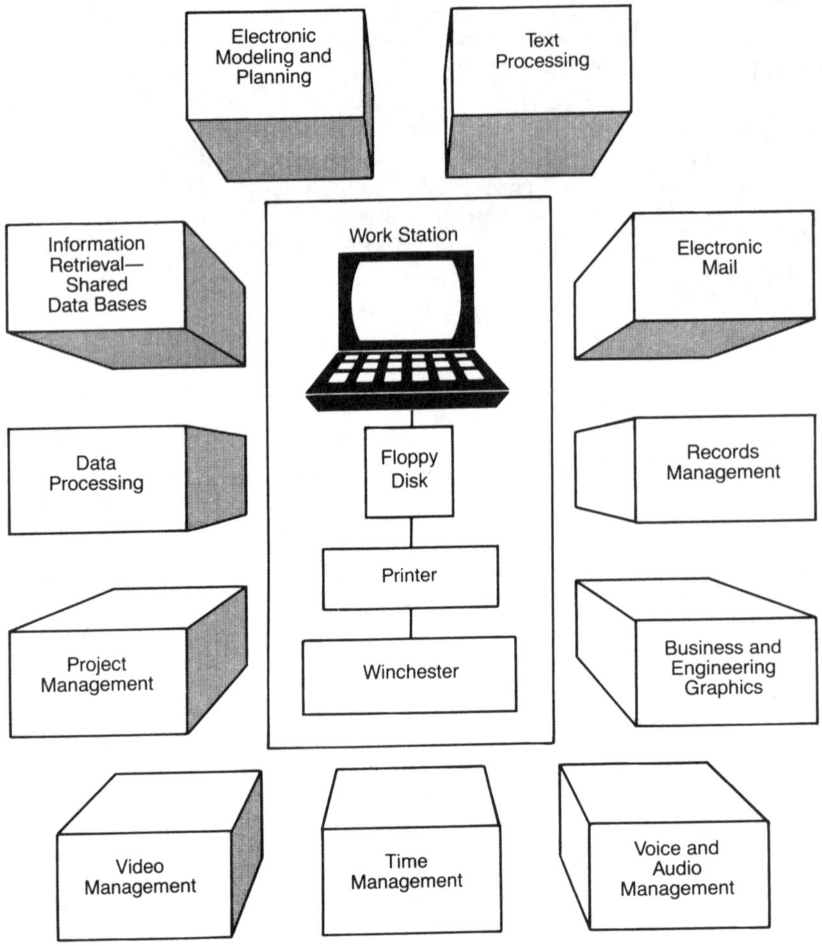

Figure 2. System applications of GMR work stations

for two or more systems, these five systems operated almost completely independently of each other.

Let's look briefly at each of the systems.

At present, a major computer facility involving a number of mainframe computers supports the engineering and scientific work of the laboratories. A data acquisition system collects data from laboratory instrumentation. Providing access to these systems by any professional work station is of prime importance.

Office systems consist primarily of the telephone system, the mail system, approximately one word processor per secretary, and an assortment of typewriters. The formal focal point for the acquisition of outside scientific and technical information is the research library. An automated library system called DATALIB is in place. All of these systems should be accessible from professional work stations.

The generation of reports and publications by GMR professionals and managers functions as a separate information processing system of its own. The Technical Information Department operates a report-processing center, as well as the publications, graphic arts, and visual communications sections necessary for the creation and distribution of such materials. These systems all need to be integrated into the professional work station environment.

In the fall of 1982, 20 "focus sessions" were conducted to determine user requirements for a wide range of GMR personnel. In total, 192 employees attended these meetings.

The sessions served the following purposes:

- identifying the requirements and developing application specifications for office systems throughout GMR,
- finding out about the use of current systems,
- allaying fears of the unknown and generating excitement about the proposed changes and their effects on productivity,
- gaining participation in, and awareness of, the plan to implement integrated office concepts at the labs, and
- creating an awareness among personnel of the feelings and opinions of people with whom they work.

The sessions were regarded as an information-gathering mechanism only. There were no preconceived notions about what people thought or what they needed. The task force's aim was to generate, rather than test, hypotheses. Our thinking was that group involvement was preferable to the isolation of answering a questionnaire; that peer and social interaction could help gain acceptance for new ideas; and that a feeling of participation would aid in the implementation of new ways of doing things.

As a result of the focus sessions, a profile of the systems needed for GMR was developed. It was decided that the GMR office environment would consist of multifunctional work stations on the desks of most employees. Generic equipment characteristics of those work stations are shown in Figure 1.

The focus sessions identified 11 applications that determined most of the salient features and capabilities needed by GMR personnel to increase their effectiveness (Figure 2). These eleven applications became the driving force behind development of the work station network.

Management approved the task force recommendations, and a new group was formed in April 1983 to implement the new system. Its first task was to prepare a request for proposal (RFP). This document was sent to more than 50 system vendors, and was structured to permit equal comparisons of vendor capability to provide the required functionality.

Twenty-six vendors responded, and in June 1983 an evaluation team analyzed proposals. Based on technical and functional requirements, the team quickly narrowed the vendor list to 11. At this point in the evaluation process, an outside consultant reviewed the remaining proposals with the team, and the list was narrowed to five.

The five finalists were invited to make presentations. In addition, the evaluation team started visiting vendor sites to review future products offered as part of the proposal. The finalists were ranked according to the weights assigned to the 11 functions, and a multivendor solution was chosen. Before making the final choice, another series of focus sessions was held to obtain user input.

To avoid the problems inherent to a multivendor solution, a second RFP was issued soliciting bids for a vendor to act as an "integrator" and be a single point of contact for the following activities and services:

- Physical installation of all hardware and software, as well as the delivery, testing, and maintenance of these items.
- Training for all applications packages provided with the work stations.

The GMR implementation for 1984 consists of 250 professional work stations capable of providing for the applications shown in Figure 2. Multiplexers will provide coaxial connection to the mainframes, while asynchronous communications will provide access to outside data bases. Integrated phone dialing and directory management also will be provided by the work station.

Our Computer Science Department is concurrently evaluating other network strategies. It is anticipated that by 1985–86 a total communication integration will be in place using broadband and private branch exchange (PBX) technologies.

E. I. du Pont de Nemours & Company, Inc.

Central Research and Development Department
Wilmington, Del. 19898

Contributor: John H. Taylor[a]

The management of Du Pont's Central Research and Development Department commissioned a study team in 1983 to examine research computing needs at the Experimental Station, Du Pont's research complex outside Wilmington, Del. The study team was composed of four laboratory heads, two computing specialists, and three outside consultants. They worked full time on the study for seven weeks during the summer of 1983, concluding with a final report and plan.

The recommendations contained in the report were to:

- Form an expanded and responsive central research computing organization to provide the majority of the computing support at the Experimental Station.
- Establish end-user management control of the central organization by forming a board of directors of laboratory heads to which this organization is accountable.
- Authorize a major expansion in the 1984 capital budget (approximately double the 1983 budget) for computing equipment.
- Approve a major expansion in the operating budget for the central organization—approximately tripling it in 1984.
- Make a major commitment to provide training and consultation in computing applications.

A fivefold increase in computing resources and a twofold increase in staffing were recommended for the coming five-year period.

The study team interviewed a cross section of research staff and management covering all departments and disciplines to assess needs. They found a high interest in additional and more available computing tools, and a real need for training programs. They concluded that there were urgent problems placing severe limitations on R&D effectiveness. Major needs identified were:

- The existing demand for computing resources far exceeded what was available.
- Automated data collection facilities were needed in laboratories and semiworks.
- Data manipulation tools were needed to screen large quantities of data, draw statistically significant conclusions, and gain insight through graphic presentations.
- The information explosion had created a need to store, search, and retrieve information of all sorts—experimental results, compounds, synthesis methods, purchased chemicals, etc., with a common format for shared accessibility.

To assess justification the team analyzed the benefits from meeting those needs. Qualitative benefits were:

- More efficient and effective research—products to market faster, processes to plant sooner.
- Increased creativity—more ideas, better conceptualization, better analysis of results.
- Better planning decisions—planning based on more thorough access to past results.
- Improved employee satisfaction—less frustration, better tools, more excitement from using modern tools.

Methodology

During the first phase of the study, the team made a detailed assessment of the R&D activities at the Experimental Station to determine the tasks performed. In addition to operations within a group, they noted significant interactions with external groups. The analysis looked at generic tasks rather than at specific job requirements related to a specific organizational structure. This was done to ensure that the recommendations from the study would not be made obsolete by any changes to the structure.

The next phase of the study involved interviews with 31 people representing a cross section of the user community, to determine their problems and concerns in conducting research. The interviewees were asked

---

[a] Adapted from a Du Pont study report coauthored by J. H. Taylor, F. J. Darnell, F. F. Huppe, H. M. Loux, R. D. Nelson, and F. G. Pater.

to describe the effects their problems have on research and what benefits would be gained from eliminating the problems. Some interviewees suggested possible solutions.

Not all problems identified were computer-related. The final phase of the study fused the computer-related results into functional requirements. The remaining problems were appended to the study report for the attention of research management.

Operating Philosophy

As a corporation, Du Pont has recognized an increasing overlap among the four sectors of computing: business data processing, technical and scientific computing, process automation and control, and electronic office technology. Our future ability to communicate among systems and organizations will require coordination of growth and avoidance of incompatible proliferation.

A central research computing organization offers the vehicle to focus the needed coordination. The central organization should capture all costs associated with its activities and should be accountable for the cost–benefit ratio of its achievements. Cost recovery from the user organizations should be on a *use basis*, rather than by allocation. Distribution of computing costs within user organizations should be at the discretion of end-user management.

Computing and information systems have become a critical resource in effective research. The central organization requires clear direction from the end-user regarding needs and priorities, and must be responsive. Research management must control the computer resources, just as they do the other resources critical to their success. This includes justifying levels of commitment (money and manpower), setting priorities, and allocating limited resources.

The central organization should therefore be answerable to, and take its direction from, a board of directors composed of laboratory heads from the research community. The board of directors establishes policy and sets priorities.

Solutions

General Approach. Meeting the needs identified in the study will require a fourfold increase in computer power and an associated doubling of support personnel. This power increase will occur in increments of 200%, 100%, and 100% over the next three years and will change the nature and distribution of computing power.

Although growth in the central computer facilities at the Experimental Station has been effectively restrained over the past two years, growth in facilities use has increased nearly 50% each year. A Sloan School of Management study shows 50–90% growth per year in end-user computing in many organizations. Thus the planned expansion is reasonable.

Software Tools. Last year's introduction of RS/1 had a tremendous effect on many researchers. This electronic notebook program provides data storage, statistical analysis, and graphics capabilities in an easy-to-use format; the most-cited need in the study was for more computer resources (DEC VAX) to support RS/1 use. In the longer term we will seek the ability to run RS/1 on other systems, both larger and smaller.

With the development of smaller and more powerful personal microprocessors or work stations, we anticipate strong growth in personal computers. These will gradually replace "dumb terminals," and many programs now on a central computer will migrate to personal computers.

Creation of an effective data transfer network and establishment of standards, formats, and protocols for files, transfer, and archives are essential to the development of common data bases and effective communication. A central responsibility appears necessary to accomplish this.

A solution with distributed resources and responsibility usually lacks adequate software documentation, standards, protocols, back-up, and security. The user or manager usually is not in a position to carry out these tasks effectively. A centralized system can provide such services efficiently without taking control of resources away from end-user management.

Hardware. A number of hardware alternatives were considered, including simply expanding the existing "VAX farm" or buying a supercomputer. The team concluded that neither approach adequately met the needs expressed.

They therefore designed an integrated system that provided DEC VAX capability for RS/1, data acquisition and analysis, electronic mail, and report drafting; but also provided an IBM/3081 with an FPS-164 array processor for number crunching, data bases, and archiving. The plan recommends purchase of supercomputer time from a service bureau where appropriate.

A network linking all terminals and computers, providing direct file transfer between DEC and IBM components, is a critical element.

Implementation

The implementation strategy has four major components: the network, hardware and software, consultation, and training. The research computing organization must grow by about 70% in 1984. Finding and acquiring appropriately skilled people to fill these positions will be the greatest challenge. Many may have to be recruited from outside Du Pont.

Consultation and Training. Consultation and training were major needs expressed repeatedly in the interviews. The need for guidance and advice ranged from needing "hotline" assistance and guidelines for selecting laboratory automation hardware to consultation on algorithms and techniques. Fifty percent of the planned personnel increases in the central computing services organization are in consulting and training positions. Effective training has a high multiplication factor, both by avoiding the many frustrations encountered in the learning process and by quickly bringing scientists and managers to effective use of computer resources.

Training materials and user documentation must be prepared. Training programs and schedules must be tailored to achieve end-user management goals. Guidelines and standards for hardware and software must be developed and promulgated. Relationships must be established with the user community so that the trainers and consultants are recognized and valued.

Custom Software. Computer applications that are specific to a certain research organization or research project and that do not have broad utility within the community are "custom applications." The study concluded that custom applications development should be the responsibility of the end-user organization, with consultation or advice from the central organization. Staffing and implementation should not be the responsibility of the central organization. The fraction of user research personnel required to provide applications support appears to be 5–10%.

Benefits

The study team concluded that the potential benefit from effective computer support for research is a 15–20% increase in productivity. This is a "soft" economic benefit and difficult to measure. As another approach to evaluating the reasonableness of the plan, numerous other large industrial research organizations were contacted. These contacts indicated that they were typically spending 3–5% of their research budget for computing support; the Du Pont plan is in this range.

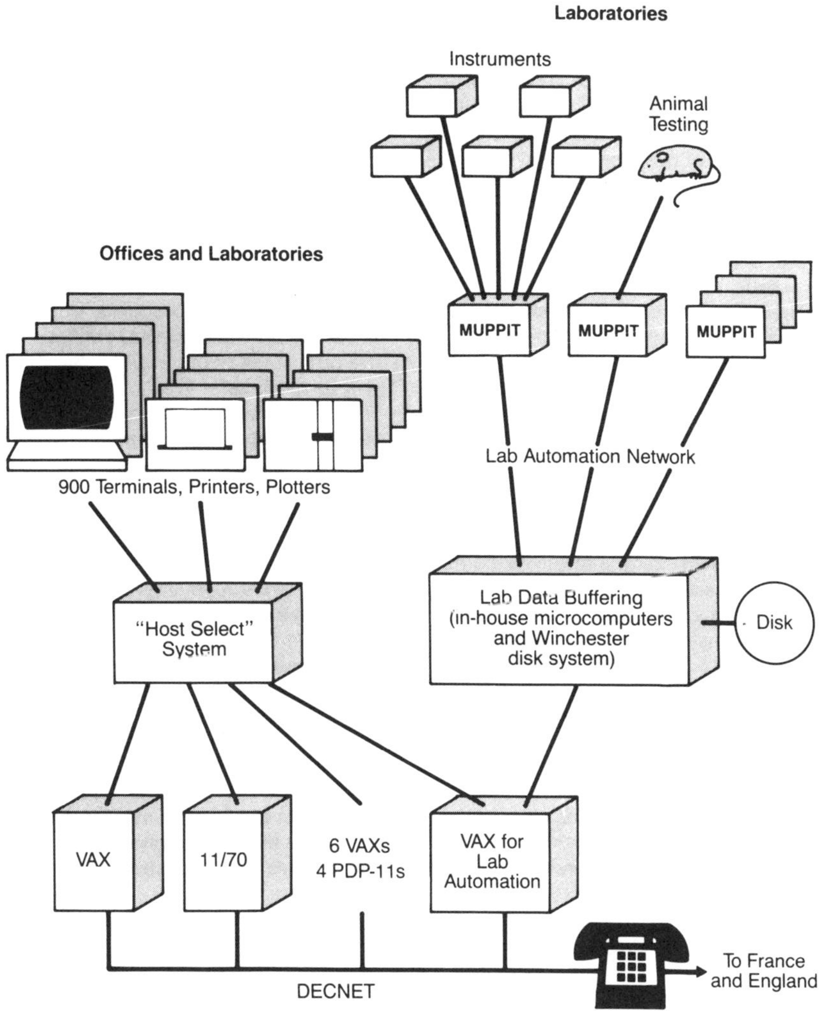

Figure 3. Searle R&D computer networks

Searle Research and Development
Research Computer Services
4901 Searle Parkway
Skokie, Ill. 60077

Contributor: Frank Schweingruber

The Searle R&D site in Skokie, Ill., is composed of eight major buildings that are closely connected on 10 acres of land. About 1100 employees are involved in pharmaceutical research and development activities. The analytical laboratories process about 250 000 chemical analyses each year. Office automation supporting the large amounts of documentation that must be submitted to regulatory agencies is well developed. The paperless office has not arrived at Searle, but electronic mail is heavily used and more than 900 terminals are scattered throughout the buildings to support information management, word processing, and scientific needs.

Computer services to support these activities have evolved along two tiers. The first tier is closely associated with the laboratory instruments, and provides equipment control, data acquisition, and initial analysis. It employs a network of in-house-developed microcomputer-based components. The second tier, providing long-term data bases, extensive graphics, statistical services, LIMS, word processing, and electronic mail, is supported by a family of six DEC VAX computers.

DEC Computer Network

For user access to the DEC computers, a "host select" system from Digital Communications Associates (DCA) allows any of the 900-plus terminals to access any of the computers. The DEC computers at Skokie, and those in France and England, are linked by DECNET. The Skokie computers will soon be connected via a "VAX cluster." This hardware family will then be easily expandable to support growing needs (Figure 3).

Personal computers (PCs) are not forbidden, but only a few are in use since most data must be accessed by a large number of people, and the state of the art of reliably interfacing PCs to central computers is in the expensive embryonic stage. The software tools currently available on most PCs also are not very different from those on a central computer. However, the Apple Lisa computer is perceived by us to be a major step forward in terms of ease of use, and this computer will become of even greater interest when a highly reliable mainframe connection becomes available. Integrity and security of data are very important in the pharmaceutical industry, and an error-free, reasonable cost link between PC and host is mandatory.

Lab Automation Network

Chromatography support for 144 gas (GC) and liquid chromatography (LC) instruments has been the major responsibility of a Xerox Sigma 3 system. This computer also supports 15 other types of analytical instruments and many unique in vivo animal monitoring systems.

When Xerox abandoned this part of the computer business in 1975, we concluded that no single computer was a suitable replacement. Individual microcomputers could more appropriately do the job. This required the design of an inexpensive multipurpose microcomputer system that could be variably configured to carry out required tasks. Specifications for these microcomputer boxes included the following:

- Interface to a host computer.
- Integrity of data shipped to the host must be guaranteed and its storage must be confirmed; i.e., through error-checking protocols and message acknowledgment.
- The experiment must be able to be run whether the host computer is available or not (built-in buffering).
- Sufficient analysis should be done at the microcomputer so that only small amounts of data need be shipped to the host.

The systems are based on the Intel multibus, over which several CPU boards can access the same memory. This permits segmenting of the software, with the separate CPUs communicating with each other.

Our microcomputer systems (called MUPPITs for *m*ulti*u*nit *p*re*p*rocessing *i*n*t*elligence) can support several terminals simultaneously, interface to lab equipment in a variety of ways, provide high-resolution graphic displays, and carry out very CPU-intensive activities. Standard units accommodate RS-232 input from instruments, binary-coded decimal input from electronic balances, as well as numeric keypad entry.

This has led to the development of a chromatography network with a small microcomputer box at each GC or LC to perform the analog-to-digital conversion and send data to a host computer for analysis.

The host computer is a VAX 11/780. A redundant microcomputer-based Winchester disk buffer is inserted into the transmission lines between the various lab units and the host. This assures that the lab units can run for 24 hours without risk of data loss in case of a host failure. The analyzed data end up on a disk within the DEC network and can be integrated into the appropriate LIMS or other data bases. It was an essential design feature that the data eventually reside on communally accessible data bases.

Who Does What

Management and operation of these computer resources is the responsibility of the Research Computer Services (RCS) group, with a staff of 50 people.

Within this group is Real Time Systems (RTS), a group of 25 people with a charter to support automation in the laboratories and, in some cases, serve as an in-house "vendor" for customized applications. The role of RTS ranges from giving advice on the purchase of commercial equipment to building unique systems such as the microcomputer network. All microcomputer network programming is done by the RTS staff.

High-level languages are available to the user community. However, most programming and data base design is carried out by the RCS staff. Heavy involvement by one or more individuals from the requesting laboratory is essential in defining and developing a successful software system.

User participation is probably the single most important factor in the success of a project. To maintain communications, RTS is accountable at a weekly meeting of the users' group, which is attended by interested, irate, or ecstatic users.

Education of users at the beginner and novice levels is provided through a large selection of regularly scheduled courses often taught by the appropriate system support person.

Another important element is the degree of responsiveness by system development people to the individual questions and needs of users. RCS management has asserted that support of installed systems is of higher priority than new system development, and an open-door policy is highly encouraged. Since programs are maintained, in general, by the same people who create them, this policy can frequently be detrimental to meeting scheduled dates for new systems. The advantage, however, is that it gives a continuous reading of the suitability of the software and educational tools.

Software Validation

The GLPs (good laboratory practices) are FDA-defined regulations for carrying out much of what goes on in pharmaceutical R&D. The FDA and the industry are actively working on the question of how these must be changed to cover the substantial effects that new computer technology is having on industry practices. Computer validation and software validation are important items that eventually will be defined for GLPs.

Our emphasis on the guaranteed integrity of data transmission and storage was influenced by these GLP regulations and expectations of what computer validation would mean. Software guarantees are much harder to give. The best way to create reliable software is through careful control during the design and coding phases. Regularly scheduled design reviews and examination of the code by someone other than the author are important. To make software testing as automatic as possible, command files are created, along with large collections of test data, so that repeated stressing of the system can be performed easily. Testing is invoked whenever any changes are made in the system.

Originally published in
ANALYTICAL CHEMISTRY **1984** *56*(11) 1200A
and **1984** *56*(12) 1312A.

Expert Systems

Part I

The concepts of science, and our ability to use them, are hard won and easily lost. Years of training, followed by exposure to many and varied problems, are required to produce an expert. These specialists have an uncanny knack for dissecting a problem and proposing appropriate solutions. They are eagerly sought after as consultants. It is now possible, under appropriate circumstances, to capture the skill of the expert in the form of computer software. Such *expert systems* can be used by scientists as another tool to help them solve the challenges they face and also to provide a training tool for the young expert.

Expert systems are currently operational in mass spectral (MS) and nuclear magnetic resonance (NMR) interpretations, neutron activation analysis, X-ray powder diffraction, real-time optimization of quadrupole mass

A. Industrial Applications

Diesel Locomotive Repair	*General Electric*
Configure Computers	*Digital Equipment Co.*
VLSI Design	*Daisy*
PCB Fabrication	*Hazeltine*
Steam Plant Control Training	*Bolt, Beranek and Newman*
Telephone Cable Maintenance	*Bell*

B. Exploration and Technical Applications

Oil Well Drilling	*Elf-Acquitane*
Oil Well Logging	*Schlumberger*
Mineral Exploration	*SRI International*
Nuclear Reactor Control	*EG&G Intelligentics*
X-ray Powder Diffraction	*Amoco*
Soy Bean Diseases	*University of Illinois*

Figure 1. Artificial intelligence systems and their developers

spectrometers, nuclear power plant control, core drilling analysis, repair of computers, cars, and diesel locomotives, configuring computers, telephone cable maintenance, integrated circuit and PC board manufacture, steam plant operation, evaluation of earthquake-induced building damage, and a variety of legal applications. Expert systems diagnose pulmonary conditions, prescribe antibiotic treatments, track digitalis therapy, identify a disease in internal medicine, analyze ophthalmic disorders, and diagnose infectious diseases. Figures 1a–d show some representative systems and their developers.

Although unrestrained problems may easily represent hundreds of thousands of dollars in software engineering costs, there is evidence that isolated problems can be successfully attacked by a skilled programmer in a few months. An expert system for the repair of complex electromechanical systems, written in FORTH, can be run on a DEC PDP-11/23 microcomputer. Small expert systems will become more common in the laboratory, servicing separation sciences, spectral interpretation, and methods development.

Imagine being able to ask a computer-based expert system the following questions:

What chromatography methods would best separate the following homologous series . . . ?

What structures are compatible with this pair of MS and NMR spectra . . . ?

How could we analyze for the following constituents in a matrix composed of . . . ?

What is the best way to configure our GC/MS/MS to assay for . . . ?

The realm of artificial intelligence (AI), to which expert systems belong, is filled with numerous intimidating words and constructs. Exaggerated or premature claims by developers, distortions by the popular press, and rejection by potential users because of fear have further clouded the development and application of AI. The function of this tutorial is to explore what expert systems are, how they function, and to examine their construction and maintenance. The goal is to leave the reader with a working vocabulary and some basic concepts of the area.

Professor Carl Djerassi, the eminent synthetic and structural organic chemist, has commented in his foreword to "Applications of Artificial Intelligence for Organic Chemistry—The Dendral Project" as follows: "The use of computers in the manipulation of symbolic rather than numeric in-

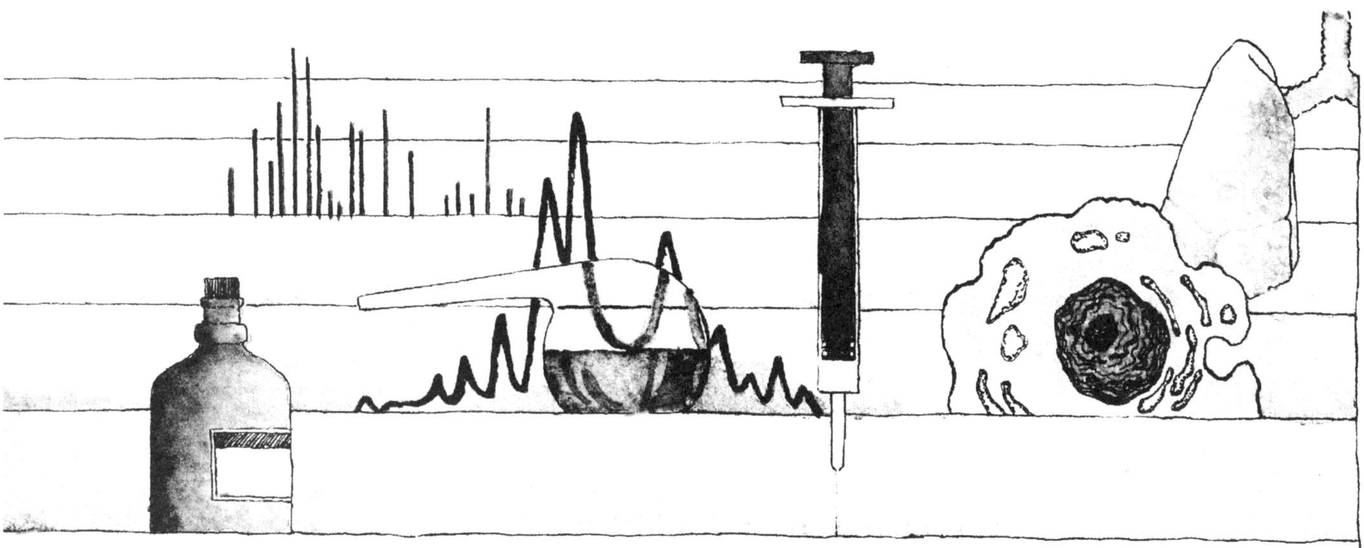

C. Chemical Applications

Mass Spectrometry	*Stanford University*
Molecular Biology Experimental Design	*Stanford University*
Organic Syntheses	*University of California*
Separation Science Planning	*Varian*
Neutron Activation Analysis	*Schlumberger*
Quadrupole MS–MS–MS Optimization	*Lawrence Livermore Laboratories*

D. Medical Applications

Blood Diseases	*Stanford University*
Cancer Treatment	*Stanford University*
Lung Ailments	*Stanford University*
Glaucoma	*Rutgers University*
Internal Medicine	*University of Pittsburgh*
Electrolyte Imbalance	*Massachusetts Institute of Technology*

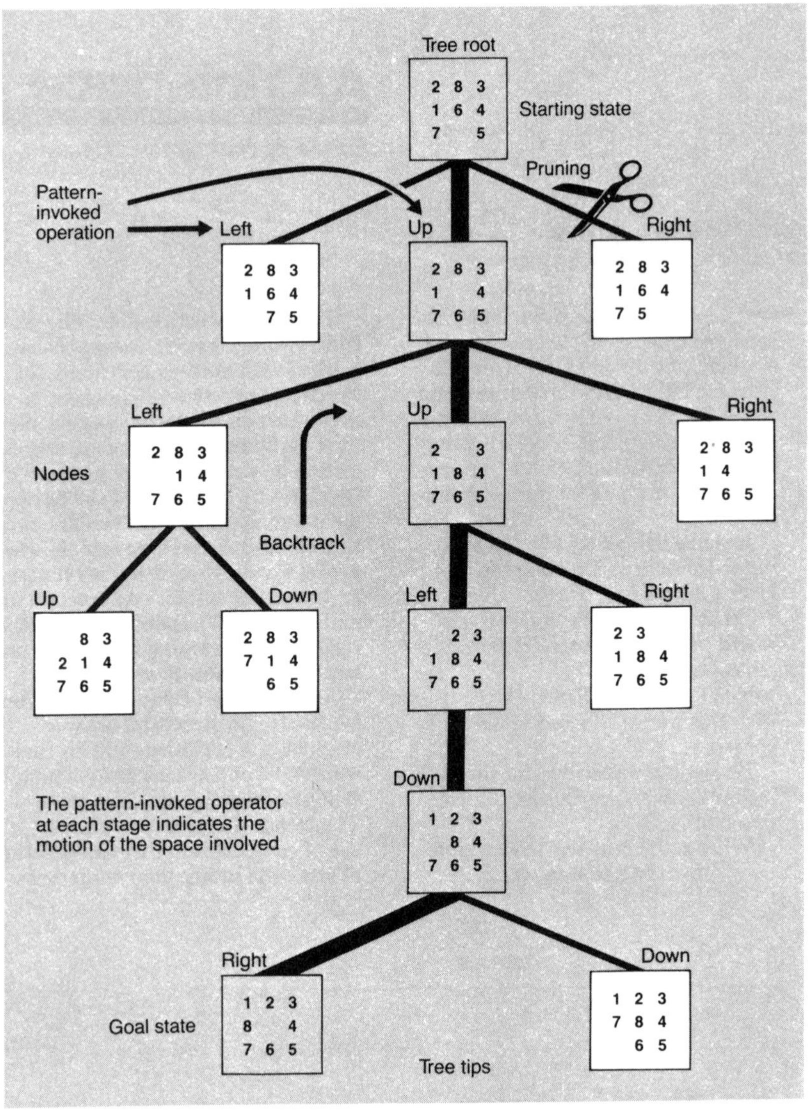

Figure 2. A portion of the state space for an "8 puzzle"

puts is of . . . recent origin and until lately has been ignored, and for psychologically understandable reasons even opposed . . . because of the fear that the thinking man will simply be reduced to an appendage of a machine . . . " (*1*). These attitudes must be changed. Expert systems, properly used, complement human abilities, rather than substitute for them.

Defining an expert system

The traditional software system consists of a program and the data it manipulates. Any attribute that a human might call knowledge is implicit in the program. However, expert computer systems normally provide separate formats for explicitly representing declarative knowledge and procedural knowledge. Discrete, separate control structures exist for manipulating that knowledge and providing a user interface.

Knowledge base. An expert system begins with a *knowledge base.* It contains information extracted from many human experts, usually by exhaustive interviews conducted by "knowledge engineers." These engineers require an unusual blend of capabilities in human communication skills, fundamental science, learning psychology, and computer programming. Attacking a problem from the knowledge base usually involves a *starting state* and *operators* that can connect that state to a *goal.* Problems are like games—chess or checkers for example. These have a starting state (the initial board layout) and a goal (to capture all the opponent's pieces or lead to a checkmate). A different and less complicated game is the "8 game" in which the numbers one through eight are inscribed on movable blocks held in a frame containing nine spaces. The goal is to arrange the numbers in a predefined order by moving the pieces through the single empty space. The knowledge base for each game describes the allowable rules of movement.

State space. The number of possible states or conditions that might exist from starting state to goal defines the *state space* for the game or problem. For checkers the state space is about 10^{40}; for chess it is about 10^{120}! Each move in these games produces a branch path that grows toward a potential goal; as the number of possible pathways increases a *combinatorial explosion* takes place. Novice players can explore the chess state space just a few moves ahead; even masters can only see the consequences for 15–20 moves ahead. Medical diagnosis involves a state space size near that of checkers. Making a complete search of a state space this size (10^{40}) is an impossible task—even for a computer.

Inference engine. Therefore, it is obvious that a useful computer-based expert system will have to incorporate a control structure to perform the same function as the reasoning processes of the human expert, although it usually approaches the problem quite differently. This is the *inference engine.* Its function is to first reduce, or constrain, the extent of state space that must be explored in connecting the starting state to a goal. It must then make inferences from declared and procedural knowledge that produce solution(s) to the problem.

User interface. The final element is a *user interface*, which allows the scientist to easily enter data pertinent to the problem, validate the conclusions, and possibly alter the system's function.

The knowledge base, its state space, the inference engine, and the user interface comprise an expert system. Modularity is desirable, since it makes possible the substitution of one knowledge base by another, permitting creation of new expert systems.

Functioning of an expert system

The skeleton and control structures. A portion of state space describing a solution to an "8 puzzle" (Figure 2) suggests a tree structure. As one moves about this upside-down tree from the root to the tips a strategy for movement must be chosen. Each branching point is called a *node.* One could randomly pick a path and follow it in depth to its very end; if this proved undesirable, another path could be chosen, again at random. This implies the existence of a mechanism for *backtracking*, or moving back to a node where an alternate search can begin. Such *depth-first, blind searches* are too time-consuming.

A more time-efficient approach involves *problem reduction.* Strategies might be devised that ordered the first row (or column) of an "8 puzzle." A separate strategy, building on this, could order the second row (or column). Finally, the last row (or column) would be attacked. The problem has been simplified by reducing it to a set of simpler problems.

Alternatively, the control structure could prune away some branch paths which are obviously unproductive, based on available data.

Then, ad hoc rules, rules of thumb, or heuristics would suggest some ordering of the remaining main branches, indicating which ones should be pursued first. This type of depth-first approach would be useful in cases where multiple acceptable goals exist or multiple paths to a single goal state occur and it is not important which acceptable goal or path is used (Figure 3).

With other problems it might be appropriate to move to the same level along all branch paths first and then examine the costs of getting to each equivalent node. The distance from each node to the potential goal(s) would also be evaluated. This *breadth-first* strategy allows the control structure to move toward a goal on a *best-first* or *least-cost* basis. This requires a complete ordering of pathways, which may be costly in machine execution time. An alternative is to select a path based on local gradients of improvement, called *hill-climbing.* The term is derived from the observation that one way to attempt to reach a summit quickly is to choose the path with the steepest gradient.

Another strategy invokes froglike leaps when an acceptable solution is far away. Like a rash, intuitive guess this uses any route that appears capable of rapidly getting near the goal, a *means–end* approach.

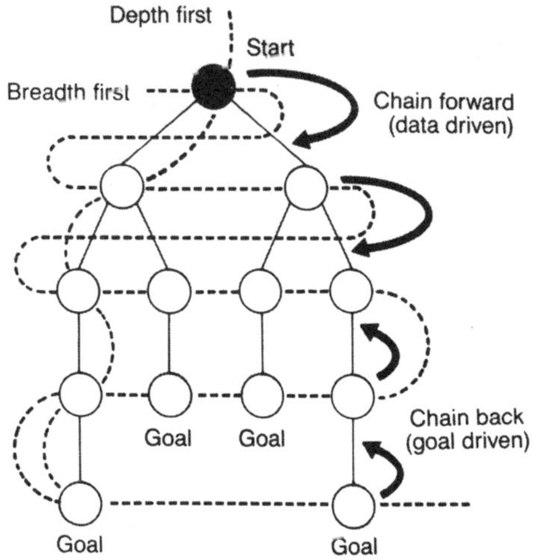

Figure 3. Various search methods can be used to reach goal states

All of these processes constrain the problem and/or make it amenable to a reasonable solution, as evaluated by time and computer resources, such as memory and disk space.

Reasoning about problems. There are two general ways to reason about problems. Human experts sometimes begin with available data, which help define a starting state, and *chain forward* to a solution by induction. Many other problems, like medical diagnosis, start with the goal (a tentative diagnosis) and *chain back* to see if all symptoms and test results are consistent with the diagnosis. These two approaches are also respectively termed *data driven* and *goal driven* (Figure 3). Many human experts rapidly develop a subset of solution hypotheses consistent with available information via a forward chaining process and then test all of them by a backward chaining process. Other problems are best solved by a *bidirectional search*, which simultaneously begins at the goal state and starting state and chains toward the middle. All of these techniques are used in various computer-based expert systems.

Each problem area has an approach that suits it best. Games like the "8 puzzle" are commutative, i.e., it is always possible to reverse an operation. Chess is noncommutative since pieces are lost in the process. Different strategies are needed.

Trees vs. nets. Simple tree structures often fail to correctly or efficiently represent problems. Interconnections between nodes are required, and this leads to a network, where each node may have more than one parent (Figure 4). An example is the causal network used in CASNET, an expert system used to diagnose diseases of the eye. Clinical signs, pathophysiologies, and diseases form a complex network. Recent extensions to an internal medicine expert system, CADUCEUS, link its causal data base to another one that classifies diseases by organs and tissues. Links and bridges between these two structures are required. In complex systems 85% of the problem may be data base management, and only 15% artificial intelligence.

The memory. How do we really store declarative knowledge?

Semantic networks organize declarative knowledge around the objects being described. The objects are represented by nodes. Relationships between objects are represented by labeled connections called arcs (Figure 5).

Frames. The need to meticulously specify each node in a semantic net, its inability to handle generic objects efficiently, difficulties in interconnecting object attributes, and problems handling unspecified attributes or default conditions have led expert system developers to the concept of storing declarative knowledge in *frames* (Figure 6). Frames are data structures that hold all knowledge about an object, event, or condition and are composed of *slots.* For example, a frame for a chemical would contain slots for formula, structure, physical constants, manufacturer, etc. Information in slots can be shared between frames by a process called *inheritance.* Slots may be further subdivided into *facets.* To describe a particular chemical a copy of the "chemical frame" would be *instantiated* (a specific instance of it is created) and the slots would be filled with data on the compound. Default conditions can be specified, e.g., it is carcinogenic until proven otherwise. Connections to other frames are easily made. Slots may also contain hypotheses that relate to the expert system's functioning, rules about program situations, or pointers to other frames. All of the information about the item is in one place.

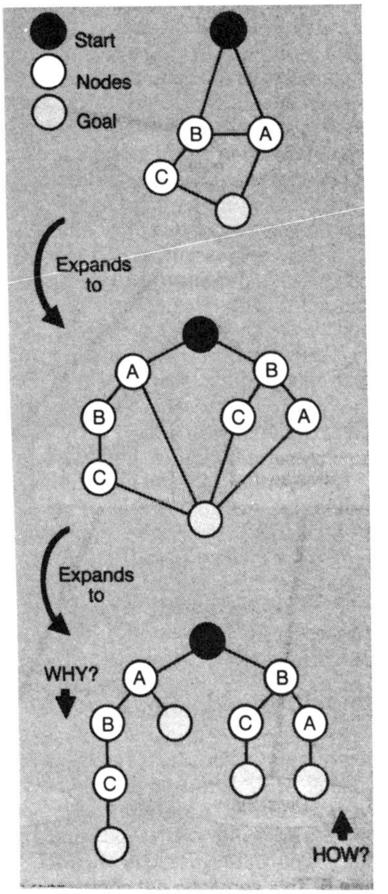

Figure 4. Basic data structures: trees and networks
All of the above structures are equivalent

First-order predicate calculus rep-

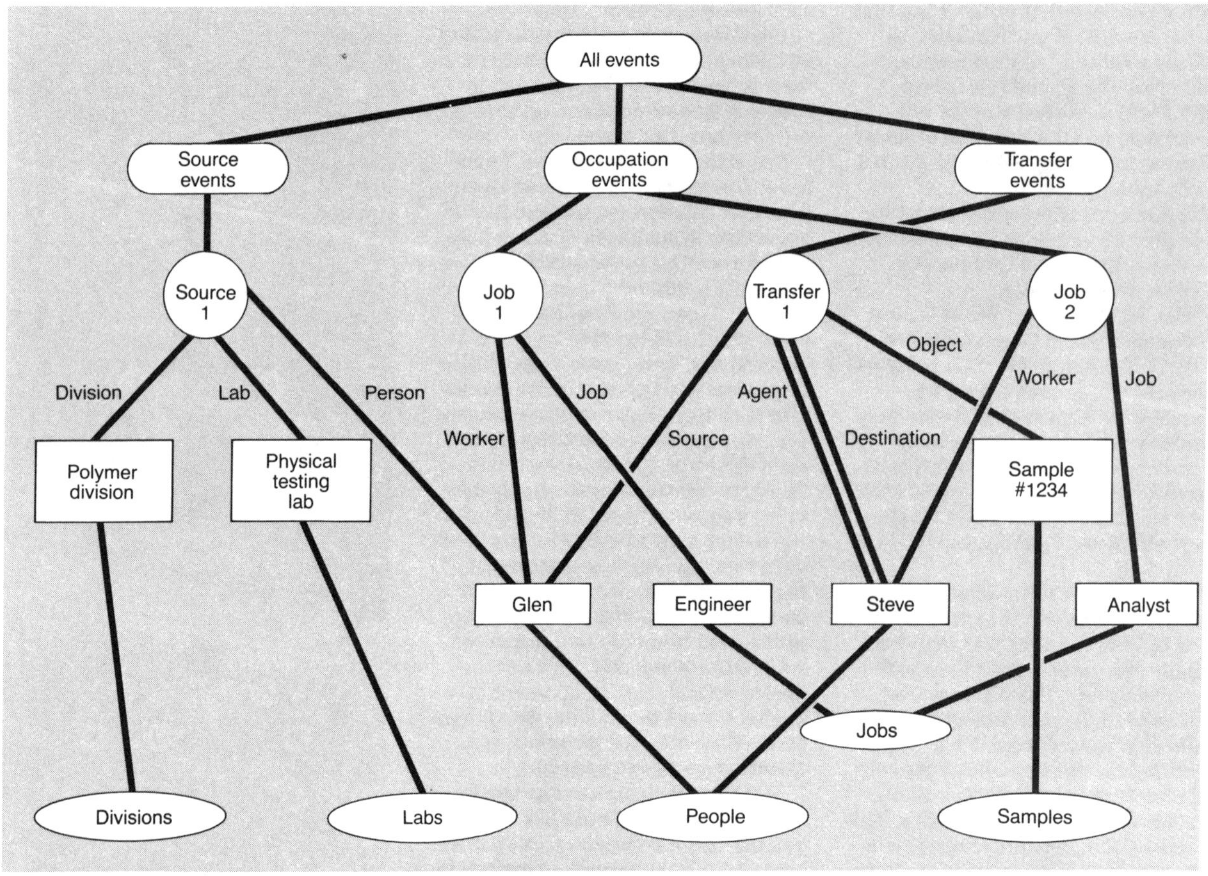

Figure 5. This semantic net connects objects in a laboratory information network
A sample is being submitted from the polymer lab by engineer Glen to analyst Steve

resents information by means of formulas. Function, variable, and constant symbols are set off with parentheses, brackets, and commas to form statements. These propositions, which are either true or false, express relationships between specific and generic objects. The sentence "Michener wrote 'Poland' " would become WRITE (MICHENER, POLAND). Connectives, such as ∧ (logical AND), ∨ (logical OR), → (implies), and ∃ (there exists) allow more complex expressions;

WRITE(MICHENER, POLAND)
∧ WRITE(MICHENER, CENTENNIAL).

We can formulate declarations such as,

MANAGER(ANALYTICAL_LAB, COFFEY)

TECHNICIAN(ANALYTICAL_LAB, FRESNEL);

then define a new function, called SUPERVISOR, that is based on the MANAGER/TECHNICIAN relationship;

[TECHNICIAN(x,y)
∧ MANAGER(x,z)]
→ SUPERVISOR(y,z)

and finally ask a question like "Who is Fresnel's supervisor?"

(∃X)SUPERVISOR(FRESNEL, X)

This, incidentally, is the nucleus for a relational data base.

The muscles. The knowledge base also tells how the data can be manipulated to solve the problem. This is procedural knowledge. If the problem is well understood, conventional programming techniques can be used. But, in many cases, complex alternative paths must be explored.

Pattern-invoked programs. In many situations pattern-invoked programs, or operators, may be defined. In the "8 game" these would be:

UP: if the empty space is not at the top of the board then move it up;

DOWN . . . RIGHT . . . LEFT

These operators (Figure 2) are activated by the control structure whenever certain conditions hold.

Production rules/systems. One important form of a pattern-invoked program is the *production rule*, which takes the form

IF (condition) THEN (action).

Often, the production rule involves a series of sequential predicate logic statements (a series of logically ANDed statements, each requiring the return of a TRUE or FALSE answer). For example:

IF (ANIMAL IS A MAMMAL)
(ANIMAL HAS HOOFS)
THEN (ANIMAL IS UNGULATE)

IF (UNGULATE IS RUMINANT)
(UNGULATE HAS FOUR HOOFS/FOOT)
(UNGULATE MALE HAS ANTLERS)
THEN (UNGULATE IS DEER)

This kind of knowledge base is called a *production system.* This is the most common approach used to develop expert systems today. The prototype expert system developed at General Electric for repairing diesel locomotives contains 530 rules. The final system may contain over 2000.

Production systems do not require that programmers carefully order the sequence of rules. Thus expert system developers can add, modify, and delete rules easily.

Predicate logic can also be used to represent procedural knowledge. Other techniques involve the use of conditional probabilities, i.e., various events will occur IF other events have occurred. This is called a Bayesian approach, after its developer, Thomas Bayes.

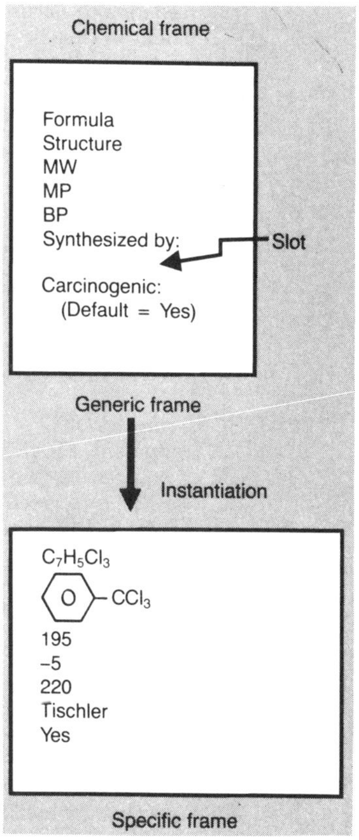

Figure 6. Data structures that hold all knowledge about an object in one place are called frames

Let's explore some difficulties expert systems face.

Decisions, decisions. Expert systems work poorly in cases where no understanding of the field exists. They are redundant in areas where theory is fully mature. Some problem areas that have been addressed well by expert systems are based on *empirical* observations, others on *conceptual* understanding, while others have some basic theory in place. Many of the problems are well structured, and singular paths lead to singular solutions. Other problems are ill structured, and many paths lead to a variety of acceptable solutions.

The information input to the expert system may or may not be quantifiable. It may be incomplete or inexact. Some systems can handle *fuzzy sets* where the quality of assignments to a variable might be "reliable, partly reliable, unreliable" or "poorest, poor, good, better, and best." The user employs subjective techniques to assign values to his or her input.

Relationships within the knowledge base are also subject to fuzziness. For example, THIS condition always/often/sometimes/never is associated with THAT condition.

One medical expert system (MYCIN), associated with antibiotic treatment, assigns values to inexact inferences ranging from −1 to +1. A −1 represents complete confidence that a proposition is false, while a +1 represents complete confidence that the proposition is true. When a production rule succeeds, the *confidence factors* of its component clauses are combined, yielding a value that may be used in weighing the validity of a solution.

In some systems, such as medical diagnosis, solutions are often reported only if their confidence factor "score" exceeds a certain threshold. In other systems more weight is placed on the difference between the highest ranking conclusion and its second-ranking neighbor. Advanced systems short-list potential candidates and then begin to rule out certain items based on conflicts with existing or newly requested input data. The remaining candidates are subject to discrimination, which attempts to separate the contenders by a more complex search strategy. Finally, the best solutions are pursued to see how well they survive a detailed examination. The analogy with the human process is obvious.

In diagnosing human ills, or malfunctioning electrical and mechanical equipment, multiple problems are often concurrent. Once the initial phase of diagnosis or troubleshooting has occurred, some expert systems explore the remaining symptoms to see if another abnormal condition can be characterized. This pursuit begins by subtracting all symptoms consistent with the primary diagnosis and using the remainder as the starting state for a second search.

The flesh. The expert system is used by many types of people. Each should see only what he needs or desires.

The simple user only wants a solution to the problem. From time to time, the scientist will want to know WHY the expert system needs to know certain information. This not only guides the user in providing the correct amount and type of information, it builds a sense of confidence. The scientist will also ask HOW the expert system arrived at a certain solution. The skeleton, muscles, and control systems described make it possible for the expert system to handle HOW and WHY requests easily (Figure 4). This will become even clearer when we explore the language structures that are used.

Students using expert systems need prompting and often guidance if they are forcing the system to pursue an unrewarding avenue. GUIDON is an example of a medical diagnostic system that provides this service. It can detect when a dead-end branch is being explored and construct terse but human-acceptable comments and questions to direct the student. The best approach does not use "canned" messages. It builds comments from the symbolic information associated with the nodes in the neighborhood of the problem area using list-processing techniques to be described.

Validation of an expert system tests whether correct answers are being produced. Improvement of the system requires a human expert to interact with it and pursue why incorrect answers are being generated. HOW and WHY questions are an essential part of this process. Additions and changes to existing production rules are then implemented. Modification of the production rules must be followed by testing to ensure that new anomalies have not been created. TEIRESIASIS is a system that provides a user-cordial improvement system. Once again, the success of the system lies in its ability to process symbolic information that is very close to the user's natural language.

Many scientists believe that the solution to programming problems in general lies in natural language processors which will allow us to command a computer in the same way we communicate with colleagues.

How are expert systems programmed?

Implementing expert systems

The tongue and voice. It is now obvious that our skeleton and muscles are probably made up of symbolic knowledge stored in frames, with procedural information expressed in predicate logic statements. The symbolic items can be most easily created and manipulated by a computer language designed to handle such information. LISP (LISt Processing) is one such language.

LISP. LISP is used to define and manipulate irreducible objects associated with an alphanumeric label (an *atom*). An atom is much like a FORTRAN variable. Atoms may be assigned *values*. These values are combinations of atoms and operators arranged in some list structure.

EXAMPLE_OF_AN_ATOM
(THIS IS ITS ASSOCIATED LIST
OR VALUE)

atom	*value*
MARBLE	(SPHERICAL GLASS BLUE)

LISP is a pointer-oriented language (*2, 3*). Atoms are stored in memory wherever there is space, and they are located by means of *pointers*. These are numbers representing an address. Lists are represented as a collection of memory cells whose contents are pointers to other memory cells (Figure

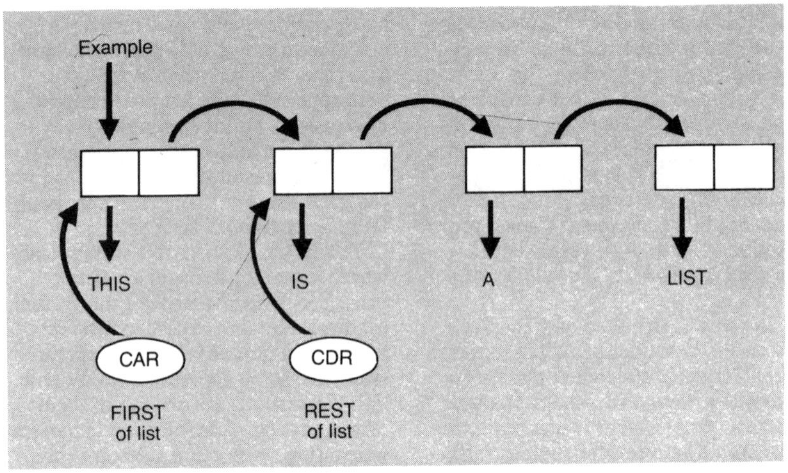

Figure 7. A list is a collection of memory cells which act as pointers to other memory cells

7). *Box and arrow* notation represents a pointer by an arrow, and memory cell(s) as a box. We begin with an arrow drawn from an atom that leads to the atom's value, a list. A right arrow connects successive elements in the list. When a list element is an atom, a down arrow is drawn to that atom's name. The end of the list is marked by a null.

Manipulation of lists merely involves manipulation of their pointers and application of a few simple logic processes. The objects themselves need not be "handled." Such lists need to be manipulated in a variety of ways:

- Concatenate List 1 with List 2.
- Create new list by connecting List 1 and List 2.
- Are two lists the same?
- Is an atom in question a member of a list?
- Perform a logical AND between two lists.
- Perform a logical OR between two lists.
- Evaluate predicate statements.

LISP contains primitive commands that allow it to interrogate, identify, and separate the FIRST member in a list from the REST (*4*). Tradition has assigned these FIRST and REST commands the names CAR and CDR, respectively (Figure 7).

These also provide the basic operators necessary to explore trees. One operator, FIRST, gives the left subtree (Figure 8), while REST gives the right subtree. Incidentally, the expansion of the tree in Figure 8 shows the *prefix* nature of LISP. Operators come before operands; +AB, rather than A+B.

Exploration of trees and nets can easily be done by creating new lists representing pathways to be explored. These are pointers to pointers. This provides not only the mechanism for tree exploration, but the method for reporting that a desired goal has been reached along a path; the successful path list is returned to the user. This is a "map" of the nodes that were passed on the way to the goal. The path list provides the route back when failure of a search occurs. The path list approach also permits users to ask questions about HOW and WHY certain steps were/are taken. We can move forward toward the goal and ask WHY the system wants certain facts; or we can move back and find HOW the system made a decision (Figures 4 and 8).

LISP-type languages and path list approaches allow one to explore a net (Figure 4) and detect that an infinite loop is being approached (a node about to be traversed already occurs in the path list). The net can even be redrawn to a tree structure by list replication, as shown in the figure. This makes manipulation and exploration simpler.

There are many dialects of LISP. Examples are basic (or common) LISP, PLANNER, CONNIVER, QLISP, SAIL, POP-2, MACLISP, ZETALISP, INTERLISP, INTERLISP-10, and INTERLISP-D. Compatibility packages available from several vendors allow programmers to run their own software in different LISP environments.

INTERLISP-10 runs on DEC PDP-10 and -20 equipment. FRANZ LISP runs on VAX equipment, under VMS or UNIX. MuLISP is targeted to 8080 and Z-80 CPUs. TLC-LISP and P-LISP run on IBM PCs and Apple IIs, respectively.

At a slightly higher level, rule-based languages which operate on top of LISP and compile into LISP have been developed. OPS5 (Carnegie Mellon University) and KS-300 (Teknowledge) are examples which can be used to create expert systems.

SMALLTALK and FLAVORS are higher level languages that are ideally suited to describing and manipulating objects. In conventional programming, procedural operations (programs) are considered primary, and the data or objects they manipulate are secondary. In object-oriented programming the roles are reversed. Here, the program sends a message *to the object* indicating the operation, or method, that is to be performed.

LISP and LISP-like languages are

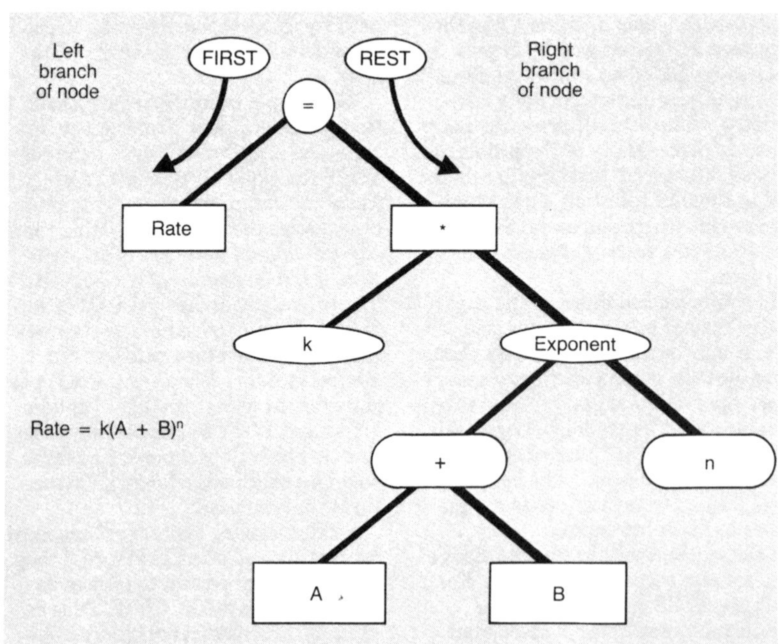

Figure 8. LISP pathlist: RATE = (∗k(exponent(plus A,B)n))

typified by their interactive nature, the emphasis on symbolic expressions, and a tree-oriented approach to data structures.

PROLOG. There is an alternate language approach which relies heavily on predicate calculus structures. PROLOG (PROgramming in LOGic) is the best known example. Advocates of PROLOG argue that it is more readable, provides a more suitable natural language interface, and has the ability to execute in a parallel fashion that improves speed of execution. PROLOG supports the easy construction and use of relational data bases. It has many long-time proponents in Europe and new ones in Japan.

PROLOG is available for Z-80-, 68000-, and PDP-11-based machines, as well as DEC-20's and VAX equipment. As with LISP, both compiler and interpreter forms exist.

Development systems. At least four companies provide full application development tools. All operate on top of a LISP environment. System-1 (Teknowledge) and LOOPS (Xerox) are rule-based. KEE (Intelligenetics) and ART (Inference) are frame- and rule-based systems. These systems consist of the control structures to which the user must add his or her own knowledge base. Other examples are EMYCIN (*MYCIN E*mptied of its specific knowledge base, Stanford) and ROSIE (Rand Corporation).

Other firms, such as Arthur D. Little and Applied Expert Systems, use KEE or LOOPS to help build systems for clients.

Many of these high-tech firms feel that it is essential to have system expertise and a good platform on which to build. They therefore will not sell the system tools separately, but contract to work directly with a client's application experts.

RuleMaster (Radian, Intelligent Terminals Ltd.) offers a system that runs on widely available personal computers.

Knowledge engineering programs that run on IBM PCs are M.1 (Teknowledge) and EXPERT-EASE (Expert Software International). A small expert system is being developed by Microsoft in collaboration with Expert Software International. A PROLOG-based expert system shell, requiring a UNIX host, is available from Nixdorf.

Hardware. Expert systems will run on traditional computer hardware. However, special LISP-oriented CPUs give better results. These are called LISP engines. Vendors such as Xerox, Lisp Machine, and Symbolics can provide such LISP engines. Costs range from $20,000 to $100,000 including software. These machines are usually characterized by long word length (to hold pointer addresses), a large virtual memory address space (the software is often quite big), and primitive LISP commands implemented in microcode (for speed). High-speed cache memory, pipelining, and separate processors for input/output and graphics complete these packages. These concepts have been introduced in previous tutorials dealing with languages and operating systems for the laboratory (*2, 3*).

PROLOG adapters for such systems are becoming available.

A Japanese government–industry–university consortium has been organized to develop a fifth-generation AI system. Some U.S. scientists feel this might well place U.S. technology and industry in a disadvantageous position, since machine intelligence can cut down on product development time and cost and improve product quality dramatically (*5*).

The present and future

Over 50 expert systems are in active use (*6*). More are under development. Many pharmaceutical firms, consumer product companies, and petrochemical labs have contracts with the high-tech firms that can provide the required tools. Others are developing their own from LISP engine platforms. Estimates by marketing analysis firms indicate that expert systems will grow to more than a $25-million market by 1985 for hardware and software. Services such as custom programming and education will be an equivalent industry. Small systems will become increasingly important.

Next month's capsule reports will describe some of the smaller chemical systems that have been developed.

Acknowledgment

The author wishes to thank Marc Tischler, a member of the Laboratory Automation and Instrument Design Group at Virginia Tech, for the discussions that helped to develop some of the concepts presented.

References

(1) Lindsay, R. K.; Buchanan, B. G.; Feigenbaum, E. A.; Lederberg, J. "Applications of Artificial Intelligence for Organic Chemistry—The Dendral Project"; McGraw-Hill: New York, N.Y., 1980.
(2) Dessy, R. E. *Anal. Chem.* **1983,** *55,* 650–62 A.
(3) Dessy, R. E. *Anal. Chem.* **1983,** *55,* 883–92 A.
(4) Winston, P. H.; Horn, B.K.P. "Lisp"; Addison-Wesley: Reading, Mass., 1981.
(5) Feigenbaum, E. A.; McCorduck, P. "The Fifth Generation"; Addison-Wesley: Reading, Mass., 1983.
(6) "The Handbook of Artificial Intelligence"; Burr, A.; Feigenbaum, E. A., Eds.; William Kaufman: Los Altos, Calif., 1982; Vols. I and II.

Part 2

Four examples of expert systems designed for chemical laboratory use follow. The terms employed have been defined in the previous tutorial on this subject.

Varian
611 Hansen Way
Palo Alto, Calif. 94303

Contributor: Joe Karnicky

Problem area. The Systems and Techniques Laboratories of Varian Corporate Research have undertaken a program to study the modern programming methods collectively referred to as artificial intelligence (AI), with a view to applying these techniques to the computing capabilities of Varian products.

The writing of an expert advisor program for liquid chromatography (LC) was chosen as a prototype project. The system is being developed in cooperation with the LC Research and LC Applications groups at the Varian Walnut Creek Instrument Division. The system is envisioned as assisting the chromatographer in developing a separation method to fit a sample via iteration of a cycle that includes experimental design, execution, and analysis. Use will be made of stored information relating to previously published separation methods and physical properties of analyte chemical groups.

This expert system is being written with the intent of making the domain knowledge readily separable from the rest of the structure. Thus, consultation systems for applications in other domains can be constructed easily by replacement of the knowledge base.

Development approach. The expert system is being written directly in MacLISP, as opposed to using a more fully developed set of tools. A number of advantages are associated with this approach. Writing in a programming language provides maximum flexibility in the choice of system architecture and implementation. The MacLISP system provides a well-maintained, well-documented, comprehensive environment for programming and debugging. It has been possible to create pattern matchers, rule interpreters, and frame manipulators as extensions of simplified functions written in MacLISP, as described by Winston and Horn (*1*). The system is completely transportable. Of greater importance, this development environment is closer to the expected delivery environment for an instrument expert sys-

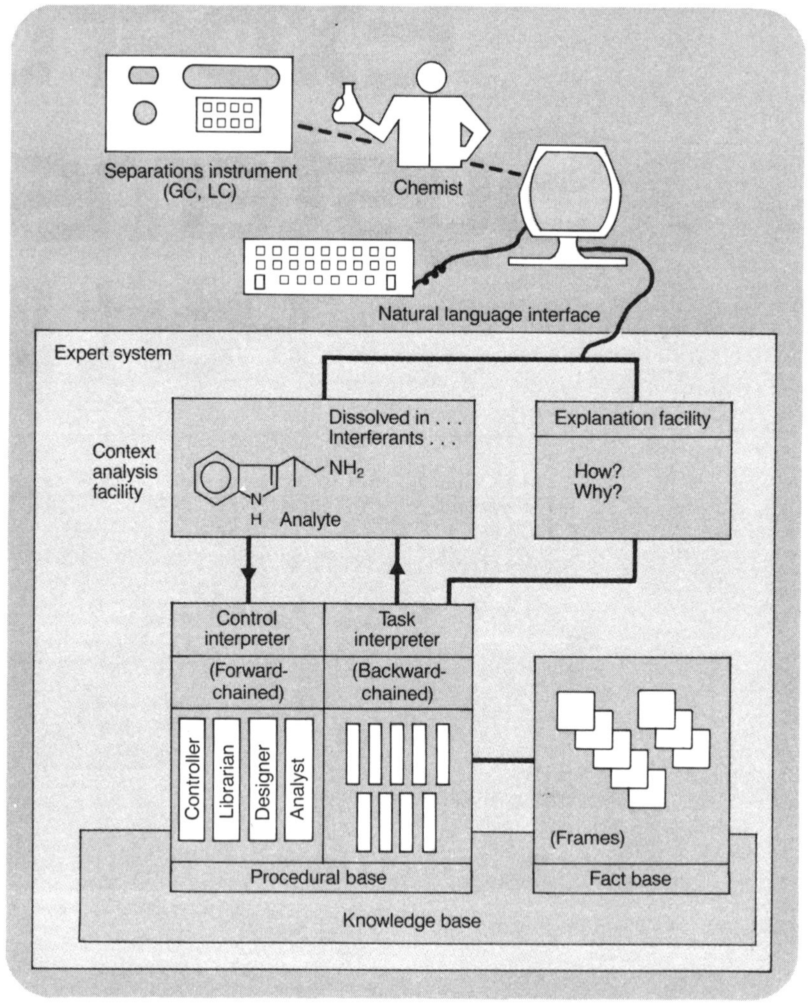

Figure 1. Expert system architecture for separations science

The user begins by entering the analyte structure, sample solvent system, and possible interferants via a terminal. The context analysis facility evaluates the problem. The procedural base consists of two parts. The control interpreter contains procedural operations determining the actions of a librarian, an experiment designer, an analyst, and the master controller. The task interpreter contains modules reflecting available procedural information on agents and situations. The fact base consists of literature methods and chemical information on analyte compounds. The knowledge base is made up of the procedural and fact bases. The explanation facility interacts with the user, providing information on how decisions have been reached and why questions have been raised. The system generates the preferred analytical method for an analyte in the specified matrix

```
FIRST. ENTER THE GENERIC NAMES OF THE COMPOUNDS YOU WISH TO ANALYZE.
WHEN YOU HAVE NO MORE COMPOUNDS TO ENTER. TYPE NIL
FIRST COMPOUND:  THEOPHYLLINE
NEXT COMPOUND:   DESIPRAMINE
NEXT COMPOUND:   NIL
NEXT. ENTER THE SAMPLE MATRIX:  SERUM
NEXT. ENTER THE INTERFERENCES TO BE RECKONED WITH
COMPOUND:        PLASMA
COMPOUND:        NIL
THANK YOU
NU. OF SOLUTIONS FOUND = 2
I RECOMMEND THAT WE USE THE METHOD IDENTIFIED AS LCAW-87 AS A
STARTING POINT FOR YOUR SEPARATION
DO YOU WANT TO:
    (A)  PROCEED UPON MY RECOMMENDATION
    (B)  CHOOSE A (DIFFERENT) LIBRARY METHOD AS A STARTING POINT
    (C)  STUDY THE LIBRARY OF METHODS ON YOUR OWN BEFORE DECIDING
    STUDY
(LCAW-87  (AKO (VALUE LIBSEP)))
          (TEXT (VALUE (LCAW) (107)))     /LITERATURE REFERENCE TO METHOD
          (AUTHORS (VALUE (LARSON-L)))
          (TITLE (VALUE (DETERMINATION)
                        (OF)
                        (SOME)
                        (TRICYCLIC)
                        (ANTIDEPRESSANTS)
                        (IN)
                        (SERUM)))
          (ANALYTES (VALUE (DOXEPIN (TR 8.6 MIN))      /INVOLVED ANALYTES
                           (DESIPRAMINE (TR 10.3 MIN))
                           (NORTRYPTYLINE (TR 12.4 MIN))
                           (IMIPRAMINE (TR 13.6 MIN))
                           (AMITRYPTYLINE (TR 16.6 MIN))))
          (INTSTAND (VALUE (P_CHLORODISOPYRAMIDE (TR 6.0 MIN))))
          (NOINT (VALUE (SERUM)))                      /CONDITIONS
          (PROGRAM (VALUE (ISOCRATIC (2.0. MLMIN))))
          (MOBILEA (VALUE (ACN (35%))
                          (KH2PO4 (0.01 MOLAR))
                          (TMACL (0.02 MOLAR))
                          (H3PO4 (1%))))
          (COLTYPE (VALUE (MCH-10 (MICROPAK))))
          (COLSIZE (VALUE (STANDARD (ID 4.0 MM) (LENGTH 30.0 CM))))
          (DETECTOR (VALUE (UV-ABS (LAMBDA 240 NM) (VARIAN-5020))))
          (SAMPREP (VALUE (SERUM (2.0 ML))          /SAMPLE PREPARATION
                          (MADE-ALKALINE (NAOH))
                          (EXTRACTED (HEXANE ISOAMYL-ALCOHOL))
                          (EXTRACTED2 (H2SO4)))))
```

Figure 2. Printout of interaction between user and an expert system

tem—i.e., running in compiled LISP on a time-shared super-mini or translated into one of the procedural languages, such as Pascal.

The dominant disadvantage in writing directly in MacLISP is the very slow rate of development of the system. Further, LISP has no direct graphics facility. To do graphics one must call up programs written in other languages from within the LISP program.

Program architecture. Figure 1 shows the architecture of the expert system. The design is an example of a system based on a model of independent intelligent agents manipulating a data structure at various levels of abstraction. The four agents are:

- a librarian who assists with searches through the fact base,
- an experimental designer who assists in the specification of the next set of separation conditions to be employed,
- an analyst who converts data (chromatograms) into information, and
- a controller who determines what task needs to be performed next and reconciles inconsistencies.

Knowledge representation. Two of the major design objectives chosen were to represent as much domain-specific knowledge as possible in declarative form and to have the results of any decisions invariant to changes of the order in which rules and assertions are listed within the knowledge base.

Knowledge for control of each intelligent agent is represented in a forward-chained rule-based production system whose interpreter is data driven. Task-performing knowledge is kept in a collection of separate knowledge bases segregated according to agent, task, and situation. The task interpreter is a modified version of a PROLOG interpreter (using depth-first, backward chaining) embedded in LISP. The modifications allow for the use of predicates that can be evaluated outside of the knowledge base and representation of uncertain knowledge.

The fact base of stored knowledge (the library) consists of frames containing structured representations of separation methods taken from the chromatography literature and chemical information related to the analyte compounds. The frames are linked by inheritance properties.

The complete expert system described is a fairly ambitious project. Fortunately, because of the incremental nature of expert system development it is possible to make use of the system even when incomplete.

At present, the various LISP functions have been written to implement frame building, searching, and manipulation; forward and backward chaining; and rule base construction. An explanation facility has been written, and the system has been exercised with minimal rule bases. A printout of a fragment of a session between the system and a user is shown in Figure 2.

Future. It is interesting to speculate on the future applications of expert systems to chemical analysis. For the near term we can expect systems to:

- routinely perform consultative and educational functions (e.g., automated methods development and management of structured data bases); and
- implement intelligent response optimization strategies that use rule-based methods to control starting conditions, process parameters, and termination conditions.

In the middle term we might see the routine use of expert systems that coordinate results from different analytical techniques to effect identification and quantification, as well as intelligent detectors (flames, plasmas), which use rule-based computation to manage their operating conditions.

In the long term we might see the creation of cognitive analytical instruments, possibly mobile, which could function autonomously.

Acknowledgment

The contributor wishes to thank Phillip Hornung of Varian for several of the concepts presented here.

Merck, Sharp & Dohme Research Laboratories
P.O. Box 2000
Rahway, N.J. 07065

Contributor: Hugh B. Woodruff

Theoreticians can calculate exactly the frequencies at which infrared radiation will be absorbed for a small molecule, assuming the spatial arrangements of the atoms and the strengths of the bonds are known. However, the vibrational characteristics and interactions of the vast majority of molecules are too complex for **adequate theoretical treatment.**

Problem area. This complexity also makes the infrared spectrum one of the most specific molecular properties that can be determined. The infrared spectrum essentially provides a fingerprint of a material, so a peak-by-peak correlation of an unknown substance with an authentic sample provides excellent evidence for identity. Unfortunately, manual comparison of spectra from an unknown substance with a library is very tedious. However, computer techniques have greatly reduced this problem.

Search–match. Collections of spectra can be recorded on magnetic bulk storage media. Comparisons between an unknown spectrum and this library can be made by search–match programs. As computer technology progresses, the time and cost required to search these libraries also decrease.

At present, computerized search-and-compare schemes are still the most commonly employed techniques and are available on all commercial vibrational spectrometers equipped with microprocessors. However, in the event that an exact match is not found, identification of the substance is not achieved, and the scientist must either interpret the spectrum or an alternate computer approach must be used.

Pattern recognition. Pattern recognition has been used by a large number of research groups to attempt such spectral interpretations. Pattern recognition attempts to find a relationship between a characteristic set of features (e.g., a spectrum) of an object (e.g., a compound) and some property of the object (e.g., presence or absence of a particular functional group) without using chemical knowledge or chemical prejudices.

Pattern recognition techniques also require the presence of a collection of digitized spectra. If the problem is to determine whether or not a particular compound contains a carboxylic acid functionality, then a collection of representative acid and nonacid spectra is needed.

One study employed a data set consisting of 2600 spectra (200 per class for each of 13 classes, such as acid, alcohol, or aldehyde). The resultant predictive abilities often ranged between 70% and 90% accuracy.

Such results are satisfying and encouraging. But to be truly useful, many more classes of compounds would have to be added. Unfortunately, the addition of more classes means increasing the size of the data set, and much of the speed and manageability of pattern recognition is lost. In my opinion, the use of pattern recognition as a tool to solve actual problems by practicing spectroscopists will be limited. However, during the next decade, expert systems may provide the assistance needed.

Expert systems. Unlike pattern recognition, where a relationship is sought from a representative collection of spectra without using chemical knowledge, expert systems make use of a knowledge base that does contain the chemical knowledge a human expert would use to solve a problem. A good knowledge base also contains other information that adds up to the intuition a human expert employs.

To be successful, an expert system must give useful information and advice to a human trying to solve a problem, and it must also be able to explain how and why it reached its conclusions. To achieve this, the knowledge base used by the expert system must be modular, allowing easy modification of the "rules."

Based on these requirements, no true expert systems currently exist to interpret vibrational spectra. Perkin-Elmer does have a combination of an interpreter and a library search on its infrared data station. However, whatever rules are used to effect the interpretation are proprietary and hence unavailable to the practicing spectroscopist.

Our research effort, based on work initiated at Arizona State University, has resulted in a package called PAIRS (program for the analysis of IR spectra). The inference engine is a Fortran program that analyzes a spectrum and any available supplemental information (such as sample state) by using the rules available in the knowledge base.

The rules were initially extracted from the experiences of scientists through textbooks, journal articles, and verbal information. PAIRS tests for a total of 168 different functionalities. The rules for each functionality (class) are stated on paper in the form of hierarchical trees. With certain exceptions, the order in which functional groups are considered during an interpretation is of little consequence.

Figure 3a shows a small section of the rules used to determine the likelihood of an aldehyde being present; Figure 3b depicts these same rules in the form of a hierarchical decision tree. The first determination is wheth-

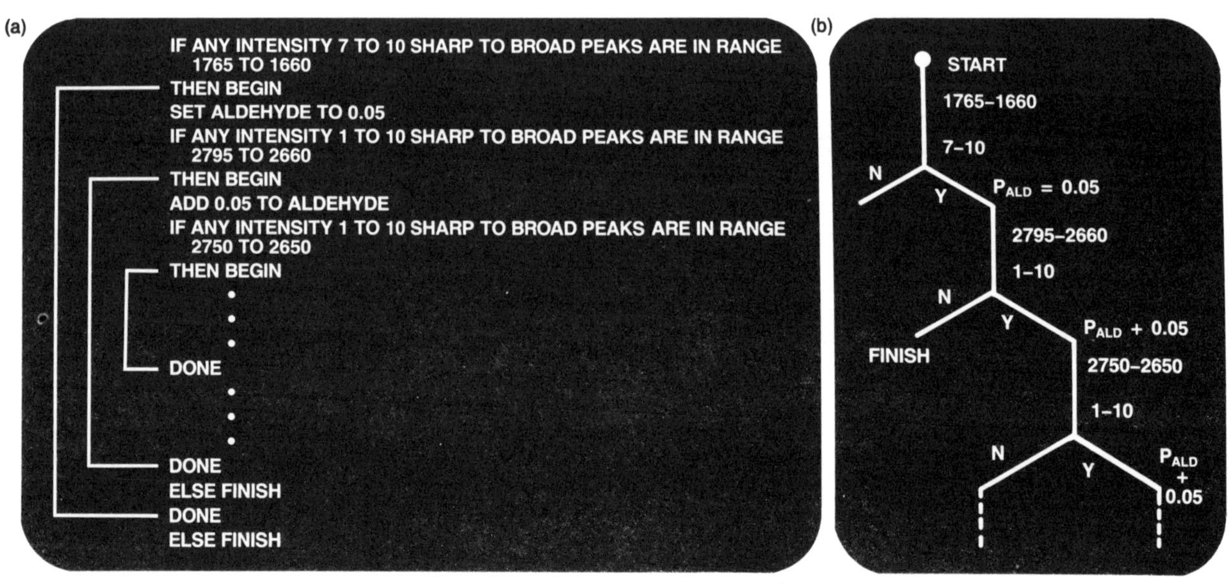

Figure 3. Small section of aldehyde rules
(a) in program form, **(b)** in tree diagram

er a carbonyl peak is present between 1765–1660 cm^{-1} and is of strong intensity (intensities range from 7 to 10). If a peak is present in the correct range of positions and intensities, the probability of the compound being an aldehyde is set to 0.05. If no appropriate peak is present, the aldehyde tree is finished. If a carbonyl peak is present, queries about the C–H stretching region are then posed. "True" responses to these questions would lead to a further increase in the probability index.

Rather than continuing to lump all aldehydes together as a single class, three subclasses have been created: saturated aldehydes (no alpha–beta unsaturation), unsaturated aldehydes (in at least the alpha–beta position), and aldehydes containing an alpha-electronegative group. A scientist would attempt to make this distinction; PAIRS should do so as well. A real aldehyde would yield sufficient true decision tree responses to result in a probability index greater than 1.0.

PAIRS has been tested on many spectra of varying complexity. In one study a total of 180 entries were drawn from the bulletins of the International Center of Information on Antibiotics. Many of the 56 classes involved were interpreted very successfully by PAIRS. Four classes—acetal, ketal, methylene, and nonaromatic olefin—were handled poorly by PAIRS. Each of these functionalities is difficult to interpret accurately using infrared spectroscopy. PAIRS made a total of 10,080 predictions (180 spectra × 56 functionalities) and was correct 88% of the time. Eliminating the four most difficult classes led to predictions that were more than 90% correct.

As software tools that expedite the development of expert systems become more prevalent, the availability and usefulness of these systems in analytical chemistry and spectroscopy will increase markedly.

Note: PAIRS, including the knowledge base, is available for unrestricted distribution as Program 426, Quantum Chemistry Program Exchange, Indiana University, Bloomington, Ind. 47401.

Lawrence Livermore National Laboratory
Livermore, Calif. 94550

Contributor: Carla M. Wong

Problem area. Lawrence Livermore National Laboratory (LLNL) is implementing an expert system for instrument control of a totally computerized triple-quadrupole mass spectrometer (TQMS). This system is an ideal candidate for self-adaptive, or knowledge-based, control for several reasons. The spectrometer is a very complex instrument with more than 30 computer-controlled parameters, and it produces vast amounts of data. Most important, the skill required for tuning or optimizing operating conditions is the kind of knowledge that can be represented as procedures and rules (*2*).

TQMS is a multistage instrument consisting of an ion source, a quadrupole mass analyzer, an rf-only quadrupole collision gas chamber, a second quadrupole mass analyzer, and an ion detector (*3*). Figure 4 shows the basic operational modes. It also shows the opportunities for the varied experimental data that can be collected and the increased operational complexity inherent to such systems. The problem is how to optimize for the most significant data, as opposed to merely collecting the most data. To do this one must encode into a knowledge base a tuning procedure for the TQMS that includes heuristics to describe what will eventually become a self-adaptive, feedback control process for real-time optimization of the data acquisition procedure.

Data structure and language. The TQMS tuning program is implemented in the knowledge representation language called KEE (knowledge engineering environment) developed by Intelligenetics. This sits on top of the InterLISP-D language running on a Xerox LISP processor. KEE includes an integrated, object-oriented knowledge representation system and a rule-based reasoning system. The knowledge base includes a set of related *objects* or *units*. The units describe a concept in terms of its attributes, the actions it can perform, and its relations to other units. The units have a hierarchical relationship.

Units can pass their attributes to other units by a process called inheritance. Inheritance allows economical information storage because an attribute that is shared by many units is stored in only one. These attributes either can describe a unit, such as the optimum voltage, or delineate the procedures that the unit can perform. One of the procedures that a unit can perform is to display a graphical representation, or icon, of itself, such as the schematic diagram shown in Figure 5.

Search strategy and examples. Production rules of the form:

IF ⟨premise⟩ THEN ⟨conclusion⟩

are used. The control structure in-

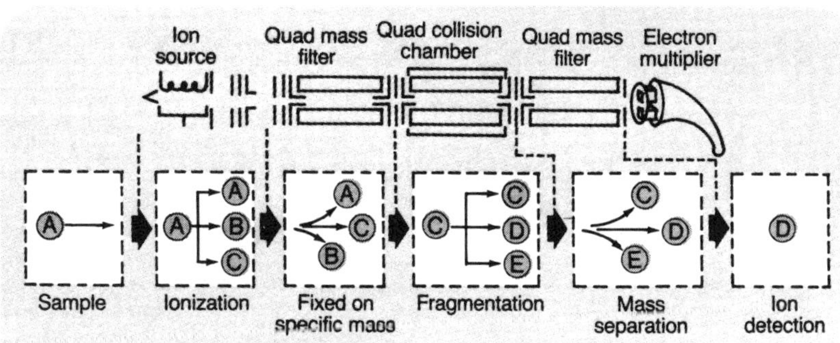

Operation mode	Quad 1	Quad 2	Quad 3	Results
MS	Separated by mass	All masses passed No gas	All masses passed	Normal mass spectrum
Daughter	Fixed on specific mass	All masses passed Collision gas	Separated by mass	Spectrum of all daughter ions from the selected parent ion
Parent	Separated by mass	All masses passed Collision gas	Fixed on specific mass	Spectrum of parent ions that fragment to give specific daughter ion
Linked	Separated by mass	All masses passed Collision gas	Separated by mass	Fixed mass difference between two scanning quads gives specific neutral mass loss
Monitor	Fixed on specific mass	All masses passed Collision gas	Fixed on specific mass	Single- or multiple-reaction monitoring

Figure 4. Basic operational modes of triple-quadrupole mass spectrometer to be controlled by expert system

volves a backward-chained search strategy. A conclusion is viewed as an initial hypothesis, which is assumed to be true. The initial hypothesis that is stated explicitly in a slot of the unit called CALIBRATION is: "The device is tuned or calibrated." The rule interpreter then tests whether there is evidence for the truth of the hypothesis.

In addition to back-chained rules, the tuning program uses instructions that are interpreted sequentially. An example of the rule structure and sequential instructions is shown below; these are described in more detail elsewhere (*4, 5*):

```
(RULE D)
(IF (DETECTOR OUTPUT IS MAXI-
    MIZED AFTER VARYING LENS1-Q1
    VOLTAGE)
    (DETECTOR OUTPUT IS MAXI-
      MIZED AFTER VARYING LENS2-
      Q1 VOLTAGE)
    •
    • Other ANDed statements
    •
    (DETECTOR OUTPUT IS MAXI-
      MIZED AFTER VARYING Q3 BIAS-
      VOLTAGE)
    (THEN (DETECTOR IS COARSE-
      TUNED)))
```

Sequential instructions have the same syntax as rule preconditions:

```
(RESET CALIBRATION WORKING-
  MEMORY)
(INITIALIZE LENS1-Q1 BIAS-
  VOLTAGE)
(INITIALIZE LENS2-Q1 BIAS-
  VOLTAGE)
  •
  •
  •
(ALLOCATE DETECTOR NET-PEAK-
  WIDTH)
(CLEAR STATE CALIBRATION-
  MODE)
(TEST-HYPOTHESIS CALIBRATION
  IS SUCCESSFUL)
```

This overall procedure is described in the top-level TQMS rule shown below:

```
(RULE D2)
(IF (DETECTOR IS COARSE-TUNED)
    (DETECTOR IS FINE-TUNED)
(THEN (CALIBRATION IS SUC-
    CESSFUL)))
```

Strengths and weaknesses. The primary strength of using a system such as KEE is that it provides a development tool that enables one to encode knowledge very rapidly and to modify and change it very easily. KEE's primary weakness is that in the object hierarchy one can define only specific objects, not classes, with instantiation. The rule structure also cannot invoke itself recursively.

Future. The TQMS tuning procedure can be done much more rapidly via an expert system. The system can be operated by a novice rather than an expert mass spectroscopist. However, the most important aspect is not typical instrument tuning over the entire mass range, but the ability to do on-line, real-time optimization of the data selection, acquisition, and interpretation.

An issue that has surfaced repeatedly in mass spectrometry focuses on the number of GC/MS instruments available to be used vs. the number of skilled mass spectrometrists available to operate them. This problem becomes even more critical when dealing with the operation and data interpretation of MS/MS and hyphenated instrument systems such as GC/MS/FT–IR. This problem cannot be solved by normal university training programs. Therefore, it is essential to capture the expertise of instrument operators and data interpreters in an expert system. With such systems the operation of the complex instrument, the acquisition of the most relevant data, and the ultimate interpretation of that data will be readily available to all analytical laboratories, rather than just the few lucky ones having live experts.

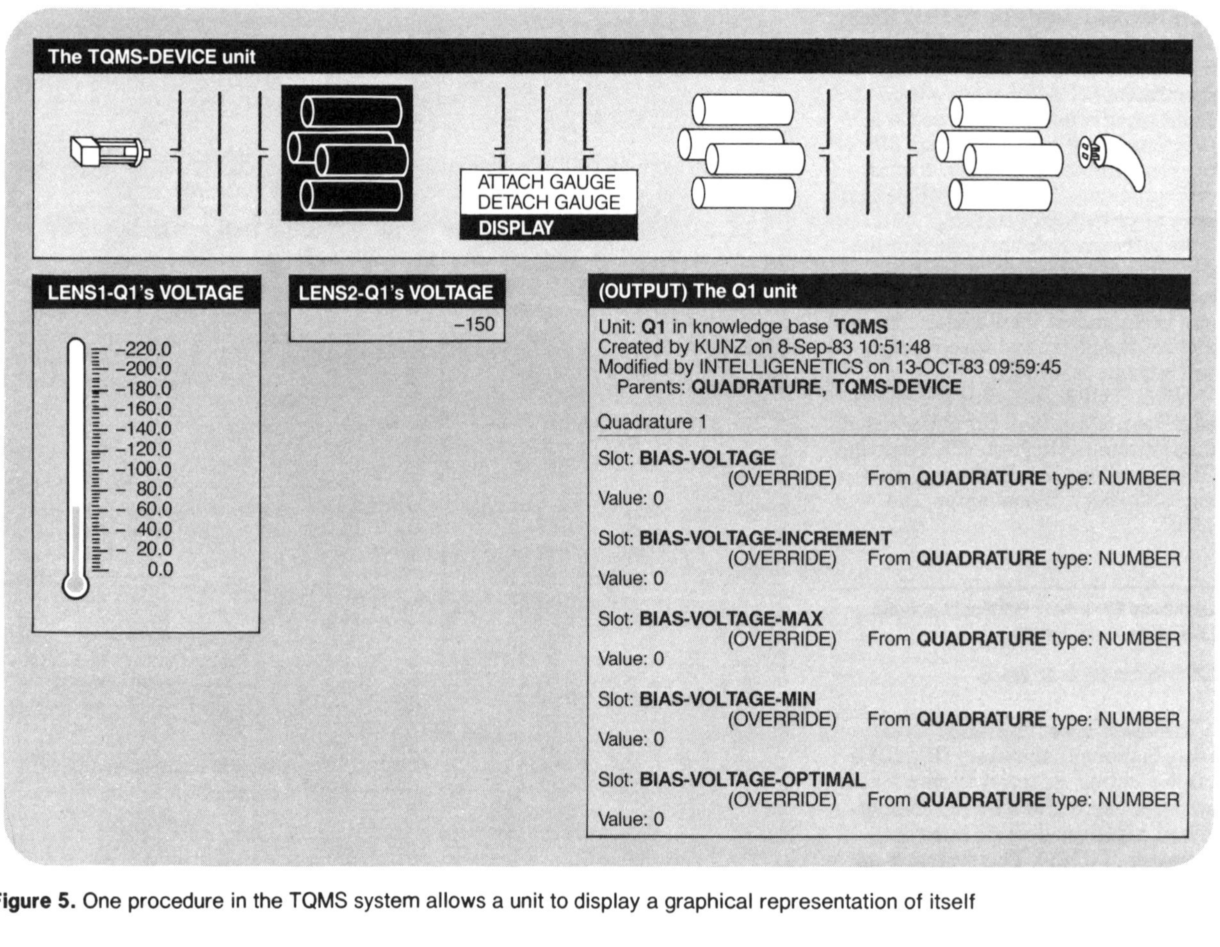

Figure 5. One procedure in the TQMS system allows a unit to display a graphical representation of itself

With the rising public concern for toxic and hazardous waste monitoring and the generation of new organic materials daily, the problems of rapid, reliable, complex organic analysis must be approached by the expert system method because it is not possible for enough people to become experts soon enough.

This work was performed under the auspices of the U.S. Department of Energy by the Lawrence Livermore National Laboratory under Contract W-7405-ENG-48. This document was prepared as an account of work sponsored by an agency of the U.S. government. Neither the U.S. government, nor the University of California, nor any of their employees makes any warranty, express or implied, or assumes any legal liability or responsibility for the accuracy, completeness, or usefulness of any information, apparatus, product, or process disclosed, or represents that its use would not infringe privately owned rights. Reference herein to any specific commercial product, process, or service by trade name, trademark, manufacturer, or otherwise does not necessarily constitute or imply its endorsement, recommendation, or favoring by the U.S. government or the University of California. The author's views and opinions expressed herein do not necessarily state or reflect those of the U.S. government thereof, and shall not be used for advertising or product endorsement purposes.

Amoco Production Company
Research Center
Tulsa, Okla. 74102

Contributor: Susan Ennis

An expert system in its most desirable form can be described as a program that allows the knowledge about a particular application area to be coded in a form very close to the English language, makes it possible for the expert to examine and modify that knowledge, and provides explanations on how the knowledge is applied for any particular result.

Each of these characteristics has implications of vast importance to the user of an expert system:

- It should not be necessary for any user to understand programming or computers to build or modify the system.
- The expert ought to be able to examine the knowledge and test the system to determine if the knowledge set is complete and correct.
- The expert has to be able to see how the knowledge was applied to a particular problem and also to see why any particular part of the knowledge was *not* applied to the problem.
- The key to expert systems is that all of this ought to be able to be done without rewriting software programs for each application. There should be a single software base that can be used for problems in different domains. In this way, programming costs can be reduced and the explanation and query facilities can be enhanced beyond the cost effectiveness of a single application.

To my knowledge there is not a software system in existence today that performs all of these tasks in a sufficiently robust manner. Many research institutions and some industries are working on the creation of this type of software. At Amoco we have attempted to implement several systems with varying degrees of success (*6, 7*).

Problem area. Amoco's X-ray system was written to experiment with expert system techniques. The general problem is to determine the mineral constituents of a rock sample given the powder diffraction spectrum of that sample. The system recognizes a limited suite of about 20 minerals found in sedimentary rocks.

The problem involves three phases. First, the possible existence of a mineral has to be recognized, usually by the presence of a characteristic peak. Second, the presence of the mineral has to be confirmed by the presence of subsidiary peaks. Third, the spectrum has to be modified or reduced to eliminate that portion attributable to a mineral that was found. The system cycles until the entire spectrum is reduced to a *noise* level.

A system has been written in the LISP language to perform these tasks. The form of the production rules was determined by talking with an X-ray diffraction specialist. A prototype system to examine and apply these rules was developed in about one month. In another two months the external interface (getting data from real X-ray runs) and the query interface (being able to ask the system how it found or did not find a certain mineral) were in place. The entire system was written in less than six man-months.

An X-ray expert was able to run the process, examine the results, change the rules, rerun the data, etc. The system required a 4-M address space and took about 15 min to run for each sample. Because of the rapid turnaround the expert could see the results of any change made within 20 min. The rule base was completed in another two months.

The system has been rewritten in Fortran, without access to the English form of the rules and without the query facility. It is now a small subroutine, and the run time is so fast that it can no longer be measured with a wall clock.

The X-ray system was later used as a test problem to investigate the usefulness of several generalized expert system development tools (*8*). It took about a month each to show that the system could be recoded to run on EMYCIN (*9*), UNITS (*10*), and OPS-5 (*11*). These systems are available from Stanford University, Intelligenetics, and Carnegie-Mellon University, respectively.

The rule forms in each of these three systems were not as "nice" for the expert, who probably could not have completed the checkout of the knowledge base if the process had originally been written with these tools. In particular the UNITS and OPS-5 systems had rule forms that were more computer science oriented.

Warnings to potential users. Amoco has been involved in the creation of several other expert systems. This experience prompts the following comments.

In general, problems can be classified by two orthogonal criteria. First, some problems can be isolated (like the X-ray system), but other problems have to be integrated into a much larger data processing system. Second, some problems have been studied to the extent that the data required for their solution are known (like the X-ray system), but other problems are currently being solved by experts who can't say what data they use to reach their conclusions.

If the expert system must exist as a part of a larger data processing system, development is *much* harder. Such an effort has to be considered a research task and probably cannot be cost justified at present unless some of the costs are written off as research. And if a totally integrated system is envisioned it is best to get AI included at the beginning of the design. Retrofitting a good existing system can be done, but it involves much recoding.

There has to be a critical mass of people involved in a project. In addition, some of the failures at Amoco have involved projects where it was difficult or impossible to secure the cooperation of the experts.

If the expert system requires integration, or if the domain is not well understood, the time to build it will be long. In such circumstances there must be a long-range research commitment to the system's development. There will be false paths followed in the development of any system, and the transition from nothing to a full system will not be a smooth, advancing line of increasing expertise. Schlumberger has documented this for its Dipmeter Advisor. There were periods when the software was redesigned, and no progress was made on the system. There were several short periods when the addition of rules to the system actually degraded the value of its answers.

Schlumberger has estimated that only 15% of its effort in creating a Dipmeter system was AI (*12*). The remainder was in the design and development of the user interface. Expert systems are not a substitute for data processing. They represent an evolution from current programming practices, and they embody a different way of looking at information.

Learning. There has been much written in the popular press about systems that can learn from experience and can upgrade their own performance. The advent of practical systems of this type will not occur in the immediate future. There are a few university learning systems, but they operate on very small problems that are well understood, where there are a very small number of parameters, and where all answers are immediately known to be right or wrong.

Many industrial engineering problems have no absolute answers; there are only good, better, and best answers. Much of the needed data is simply unavailable at the time that the problem has to be solved, and the number of needed variables is both enormous and currently undefined.

The potential benefits of expert system technology are enormous, but there is a great amount of work that has to be done to realize these benefits in the integrated industrial environment.

References

(1) Winston, P. H.; Horn, B.K.P. "LISP"; Addison-Wesley: Reading, Mass., 1981.
(2) Wong, C. M. et al. *Rev. Sci. Inst.* **1983,** *54,* 996–1004.
(3) Yost, R. A.; Enke, C. G. *J. Am. Chem. Soc.* **1978,** *100,* 2274–75.
(4) Wong, C. M. et al. *IEEE Trans. Nucl. Sci.* **1984,** *NS-31* (1), 805–10.
(5) Wong, C. M.; Lanning, S. M. "Energy and Technology Review," (UCRL-52000-84-2), 8–15, Lawrence Livermore National Laboratory: Livermore, Calif., 1984.
(6) Tompkins, J. W. In "IBM Engineering and Scientific Study Conference Proceedings"; IBM: Poughkeepsie, N.Y., 1982.
(7) Weiss, S. et al. In "Proceedings of the National Conference on Artificial Intelligence"; William Kaufmann, Inc.: Los Altos, Calif., 1982.
(8) Ennis, S. P. In "Proceedings of the National Conference on Artificial Intelligence"; William Kaufmann, Inc.: Los Altos, Calif., 1982.
(9) Van Melle, W. et al. Heuristic Programming Project Report HPP-81-16, Stanford University: Stanford, Calif., 1981.
(10) Smith, R. C.; Friedland, P. Heuristic Programming Project Memo HPP-80-28, Stanford University: Stanford, Calif., 1980.
(11) Forgy, C. E. "OPS-5 User's Manual," CMU-CS-81-135, Department of Computer Science, Carnegie-Mellon University: Pittsburgh, Pa., 1981.
(12) Smith, R. C.; Baker, J. D. In "Proceedings of the Eighth International Joint Conference on Artificial Intelligence"; William Kaufmann, Inc.: Los Altos, Calif., 1983.

Originally published in
ANALYTICAL CHEMISTRY **1985** 57(1) 77A
and **1985** 57(2) 310A.

Workstations in the Laboratory

Part I

Workstations are responsible for dramatic changes taking place in the electronic offices of business and engineering personnel. As laboratory assistants, workstations can increase productivity by providing easy access to tools for acquiring, filing, retrieving, correlating, reporting, and plotting data. As para-scientific colleagues they provide modeling and simulation support (Figure 1). This tutorial will acquaint analytical chemists with the vocabulary of workstations, why they are needed, how they provide productivity leverage, and potential pitfalls of using them.

Workstations are interactive computer-based devices that allow the user to control the world of laboratory operations—from data capture through information release. At the personal computer level *integrated* software allows exchange of data and information across the boundaries that exist between the various types of programs. At the network level integrated software and hardware allow interchange across the laboratory environment.

The bottom line

Today's analytical chemist must play many new roles. Pressures are mounting to provide more, better, and quicker service. The extraction and communication of information from analytical data has traditionally been a labor-intensive process. Workstations in the electronic laboratory can change that as management and the scientist recognize the time and money these new servants can save.

Studies indicate that it is fiscally justifiable to add a $5000 workstation to an engineer's desk. These surveys show that workstations can dramatically increase productivity, with a typical payback period of less than 18 months. The analytical chemist's tasks involve a similar bottom-line analysis.

Stand-alone vs. mainframe

There are two extreme views regarding the proper hardware approach to workstations. One side advocates the use of intelligent terminals connected to a mainframe computer. The other advocates stand-alone workstations. Mainframe critics point out the cost in time and frustration to access services on large computers with overloaded facilities. Software costs also drive users to stand-alone workstation environments. The per-user licensing fee for a multiuser software package on a large computer is usually high compared to single-user workstation packages. But critics of stand-alone workstations point out the inability to intercommunicate results.

The obvious solution is an approach in which workstations are embedded in a good local area network (LAN) facility. The goal is the ability to import and export data and information between programs running on the station and between other workstations and larger computers on the net. The network connection allows the individual scientist the freedom necessary to accommodate individual needs and work habits. It ensures the laboratory that his/her findings can be exchanged with others and that they are available for future use.

The beginning

The number of requirements for laboratory workstations is immense. A basic package must provide for word processing, spreadsheets, data base management, graphics, and telecommunications. This marketplace is currently the focus of fierce survival struggles among software/hardware vendors. Such workstations require a computer with memory mapping to handle the large programs that provide intertask communication in integrated software. Acceptable graphics require at least moderate-resolution screens. Many personal and professional computers have one attribute but not the other.

Initially, hardware limited the capabilities offered; recently software has become the limiting reagent as computer designers leapfrogged over existing programs. The growing trend toward 8-Mbyte addressing and 640 × 400 pixel bit-mapped display screens will help stabilize the market, but the current technical lifetime of products is only about 18 months. The survival rate for vendors will be low. Success is not always a question of the best product; it is often a question of astute marketing.

Since most laboratory packages are

Figure 1. Integrated software in laboratory workstations provides the chemist with a number of servants that can increase productivity

being built around software and hardware originally designed for the business or engineering office, our tour of workstations begins with the integrated software currently available for personal computers.

Integrated software

The goals of integrated software are straightforward: to allow the insertion of data into a variety of programs with the same data format and command keystrokes, facile arithmetic and statistical manipulation of the data, and automatic transfer of information between the various software modules without specific operator intervention. Various vendors interpret the basic vocabulary and implementation tools quite differently. This makes it difficult for the prospective buyer. A description of the essential elements of integrated software and a common vocabulary are as follows.

Multiple tasks—at a minimum word processing, data base management, spreadsheet, graphics, and statistics. Switching tasks in unintegrated systems requires returning to the operating system every time. An integrated system that runs all tasks concurrently is much more efficient.

Unified command structure—similar functions asserted by the same keystrokes in each task. In unintegrated software every task has its own command structure. For example, ⟨Ctrl⟩-K,D saves a document in WordStar (MicroPro); a spreadsheet uses ALT-S. This requires the user to remember each conflicting set of commands and to shift gears at the beginning of each new task. An integrated system that has one set of command structures is much more logical.

Import–export—intratask and network transfer, as well as telecommunications support. But how are data transferred between tasks? Often, the data from one program are incompatible with another program's requirements. One package might require that TABs be replaced by SPACEs and might ignore special printer control functions needed by another program. Even worse, this package might place a different interpretation on a control character. Some programs can take alphanumeric input derived from foreign programs if it is in a neutral format. Although each system may have its own private format for storage, it does recognize a neutral *lingua franca* for exchange purposes. The .DIF format made popular by VisiCalc (VisiCorp) and the .SDF protocol of dBASE (Ashton-Tate) are examples. Both 1–2–3 (Lotus) and dBASE can import–export information in .DIF format, but this is not an easy task. Some systems allow import of neutral formats, but not export—apparently to encourage sole use of that product. Much more convenient is an integrated system that uses a common data format.

Live vs. dead transfers. Data and information in integrated packages must flow easily from a source task (transmitter) to destination task(s) (receiver). If the transmitter does not remember what tasks it has conveyed items to, and receivers do not store what tasks gave them data, then subsequent changes in the data require explicit retransmission of the items by the human operator. Have you ever set up a meeting, decided to cancel it, and forgotten to inform *all* of the prospective attendees? Wouldn't it be prudent to build an integrated system in which any change in a datum was reflected in updates in all affected areas?

Truly integrated software links all transactions. Transfers that occur without automatic linkage are termed *dead* transfers. Live transfers are very desirable, but the most difficult to implement correctly and consistently. Most of today's systems link only the simplest connections, e.g., spreadsheet or data table-to-graphic functions. Even these usually link in only one direction, so that tracking a point on a plot back to its data table is difficult.

Implementing integrated systems

Integrated software can be implemented in several ways:

Tight integration—a single memory-resident program that can effect all the required tasks. Examples include Symphony (Lotus) and Framework (Ashton-Tate).

Managed integration—specially written programs running under a manager, such as VisiON (VisiCorp) and Smart (Innovative Software), or a coordinator, including the Mac series (Apple) and Windows (MicroSoft).

Imposed integration—rewritten independent programs running under a coordinator, like Intesoft (Schuchardt) and IT (Martin-Marietta).

Loose integration—completely independent programs running under a coordinator. Examples include Concurrent PC-DOS (Digital Research), DesQ (Quarterdeck), and Topview (IBM).

In Symphony the data management and graphics package sit on an underlying master spreadsheet. Symphony implements live, linked transfers by consolidating the data in one large file. Passing of information is intra-file. Alternatively, a multiple-file system can be used, with points establishing the linkages (Framework). The tight integration is easy to learn and use, but may pose difficulties if your needs do not match the skeleton or if you want to add on other programs.

A less tightly coupled integration may make it possible to add other programs. VisiON provides a rigid set of integration rules; added programs may adapt to them. The operating system (OS) is so submerged that it is almost invisible. On the other hand, DesQ uses the OS to provide input/output (I/O) services for the programs you select to run under it. It is often a simple matter to add your favorite program. But, if the programs you add perform the I/O themselves, peculiar results occur. Windows is somewhere between these two extremes.

The Macintosh takes a hardware approach, placing essential software tools in ROM. It is a delightful machine to learn on, but adapting other software to this environment is not a trivial task.

In most of these systems, dead transfers are implemented using a *clipboard* buffer as an exchange medium. A program writes data into a clipboard area and *hangs* it up for access by others. Since source and destination are unknown to each other and may require different formats, some systems allow only alphanumeric ASCII exchanges. Others place multiple versions of the data on the clipboard, allowing the destination to choose from among the offerings. Where concurrent task execution is not employed, there is little alternative (Macintosh).

At another level of integration, software is available that will run on microprocessor workstations *and* the hosts they are connected to via a network. Examples are Goldengate (Cullinet) and Millenium (McCormack-Dodge).

But how can we take advantage of these integrated tools? We need a mechanism for looking at spreadsheet and text at the same time, just as we view a pile of papers on our desk.

Windows

A CRT screen can provide several windows on the data being manipulated by integrated software. Windows may be overlapped, so that proximal screens obscure a portion of remote ones (Figure 2). This allows more images on the screen at one time, but can easily confuse the view. Tiling, like a mosaic, allows each window to be viewed in its entirety. Poorly laid out, it can also be bewildering to the eye, and the small window sizes may reduce the amount of information to uselessness. Most CRT screens are not well designed for viewing typescript information. Reflect on how you scan a printed document or the newspaper, and it is obvious that today's medium-resolution, small screens are inappropriate. The cathode ray tube face needs to be larger and have a vertical

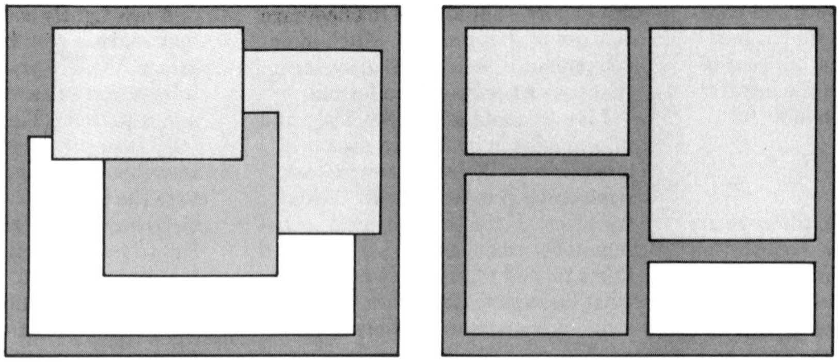

Overlapped Tiled

Figure 2. Multiple windows provide simultaneous views into the various facilities of integrated software

aspect ratio. Above all, screen update needs to approach the subsecond level.

Reviews of current integrated packages uniformly focus on one frustrating aspect—the slowness of the system to respond. In some systems the keyboard locks as a request is processed; in others the type-ahead feature captures what you type, but there is no visible evidence of your efforts until the executing task terminates. Does this sound reminiscent of your contact with large mainframes? Even if you are not interested in real-time data collection, subsecond screen update is not just desirable, it is a necessity. Users' time is valuable, and hardware design should permit work habits that are built around the way our senses work most effectively.

Concurrency

Many integrated packages claim to be multitasking. At first appearance multitasking is a simple word meaning the ability to handle multiple tasks. The question is *how?* Some integrated systems handle multiple tasks by a master–slave relationship; one program is preeminent. Often termed a foreground task, it has high-priority access to machine facilities. Other tasks run in background whenever the foreground task is idle. Spooling of printer output, spelling error detection, or even data base sorting might be handled in this fashion.

This approach fails when multiple users, multiple tasks, network communication requirements, and interactive graphics become involved. CPU time must be allocated in some way. This involves concurrent execution of tasks. Many systems advertised as "multitasking" are not concurrent.

Often, scientists want their workstation to perform real-time data acquisition. Most personal computer OSs handle only one task at a time. It would be impossible to have them collect data from an analog-to-digital (A/D) converter at fixed intervals and "time out" a second task concurrently. Let's explore time in more detail.

Time allocation. In time slicing the operating system allocates discrete units of time to each task on a rotating basis. More complex systems allocate multiple quanta of time to bigger, more difficult programs.

Round robins pass control of the machine from task to task via a scheduling algorithm that views all installed tasks as a circular linked list. This control can be invoked from an external control program, or the control mechanism can be built into the executing routines themselves. In this approach all tasks are considered as equals and are called coroutines. Coroutines are like subroutines that can call each other, a member dropping out only when its task is finished.

Schedule-driven systems create a work list at every tick of a master clock. This list contains all tasks that need service as determined by interrupt requests, time of day, or system demand.

Concurrent PC-DOS uses fixed-time slicing, although suitably modified programs can set their own priorities and invoke variable time slicing. Windows employs a round-robin coordinator. However, most PC systems have a response time that is highly load dependent, making them unpredictable in high-speed real-time situations.

Examples of OSs suitable for real-time work are Forth (Forth Inc.), a coroutine-oriented systems language; RTOS (Micro-Way), a reconfigured version of Intel's RMX-86; and QNX (Quantum), a Unix-oriented approach.

Data collection also requires careful choice of an A/D subsystem that is matched to the OS.

Interrupts. Most OSs today use interrupt architecture to service the needs of A/D subsystems. Interrupts involve allowing a device, by special hardware, to cause the equivalent of an immediate *jump to subroutine* when service is needed (Figure 3a). If serviced quickly enough, and if interrupt load is not high, the other users and tasks are not affected by this diversion of machine resources. With interrupt-driven devices, each time a data point is available a processor interrupt occurs. This requires the CPU to execute a series of instructions that stores the data and keeps track of how many points have been taken. This software overhead consumes a few hundred microseconds for each point. Interrupts can handle about 20,000 transactions/s at a maximum. The rate may be much lower if OS or language overheads are excessive. Actual values may be as low as 30 to 120 Hz.

The contretemps. With coordinating mechanisms and interrupts available, why is concurrency such an elusive goal? One answer is that the software tools used, and their manipulation by programmers, often lead to inefficient packages. Another answer lays the blame at the feet of hardware designers who have crippled machine performance by clumsy input–output handling. Even well-handled interrupt service can limit real-time response to request rates of 5 kHz. Poorly designed systems fail at slower rates. The IBM PC running a Concurrent PC-DOS fails at network speeds of 1200 bits/s (8 ms/character, event rate of 120 Hz). A much slower CPU, the DEC LSI-11, running a Forth operating language, can easily handle five serial lines at 9600 bits/s (event rate of 5 kHz). The reason for the difference? Software quality!

Thus, although the speed and memory size of current microcomputer hardware appear awesome, it is more telling to observe what the simultaneous increase in OS memory requirements and inadequate optimization can do to these impressive figures. The combination of a Unix-like OS and graphics–floating point support can easily bring a 68000-based processor to its knees, permitting two simul-

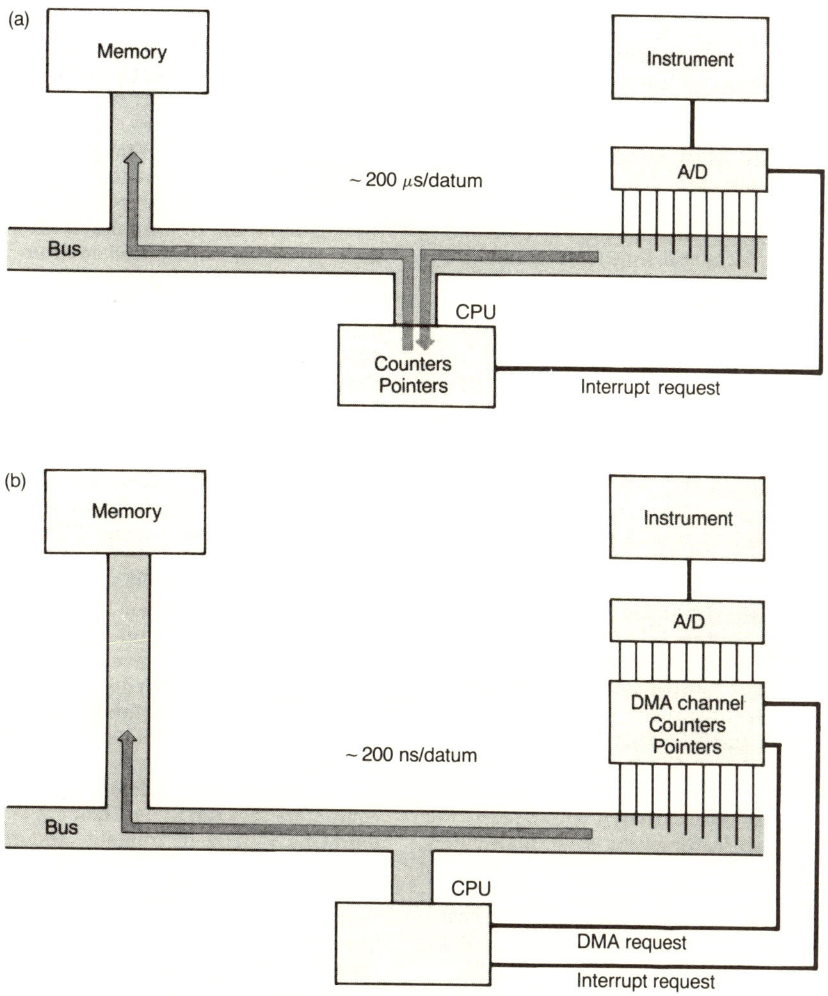

Figure 3. Interrupts and direct-memory access devices
(a) Interrupts momentarily halt execution of the main program and cause execution of a software program that moves the data from the A/D to memory. This program also updates a pointer to memory that directs data to the proper storage area and decrements a counter that keeps track of how many points remain to be acquired. These values are held in main memory or in CPU registers
(b) Direct-memory access devices use their own special high-speed hardware registers to keep track of pointer and counter values. Only when data acquisition is complete is a single interrupt request made

taneous users at best and very, very slow system response. Such systems often have 30-Hz A/D converters simply because that is all the system will support! In such cases, what alternatives exist?

Direct memory access. In direct memory access (DMA) transfers, hardware keeps track of where to put incoming data and how many points are required. The CPU operation is affected for only the few hundred nanoseconds required to transfer the data directly to memory. DMA channels can achieve 100 kHz to 2 MHz throughput to memory (Figure 3b). Hardware DMA subsystems are available from a number of vendors for the most popular PCs (e.g., Data Translation, Tecmar).

Transient recorders—A third approach to real-time data capture is a separate CPU–A/D–memory subsystem, connected to its host via a serial or parallel link. This decouples the host OS from the actual process of collecting data; it merely sends commands to the subsystem, which does the actual work. The required information is returned to the host's memory when the experiment is over. Such subsystems are available from Transiac, Teknivent, Analog Devices, and Cyborg. Potential users should be aware that some implementations solve the speed problem, but are still not concurrent.

Graphics and data bases

Graphics introduces severe speed and format problems. Integrated software should allow mixed text and graphics. CRT interface hardware limitations may preclude this (e.g., the IBM PC without a Hercules-type graphics card). Most systems cannot embed chemical structures in text. Packaged hardware systems such as the Star (Xerox) and the Lisa/Macintosh (Apple) have abilities in this area. Graphics–text software packages are available, such as MACCS (Molecular Design), Tekalike (Creative Solutions), and Spellbinder (Lexisoft). The chemist should realize that many of these systems do not adhere to a graphics standard that allows easy use of other display devices, ready modification of the software, or porting of other software packages to the system.

Many vendors are offering products that adhere to the Core graphics standard or the Graphics Kernel Standard (GKS). These companies include Precision Visual, Nova Graphics, Apollo, DEC, FutureNet, Ramtek, and Tektronix. The engineering community is fostering an Initial Graphics Exchange

Specification (IGES). GSX, a graphics tool kit from Digital Research, makes different screens and hard-copy devices look the same to the programmer.

To plot data you need to find it! Well-developed relational data bases are available in the engineering–business workstation field. In a distributed environment, such tools allow sharing of data between software packages and with other workstations. Popular relational data bases include micro-Ingres (Relational Technology), Informix (Relational Database Systems), Progress (Data Language), and Unify (Unify Corporation). Nicolet uses Informix on some of its lab spectrometer systems. Oracle (Oracle Corporation), Focus (Info Builders), and IDMS/R (Cullinet) are packages that run on both mainframe and microcomputers. General Motors Research Lab uses Oracle throughout its PC-based workstation network.

Standard buses

High-speed data acquisition, multiple users, and graphics all increase the demands made on a workstation CPU. The easiest solution is to increase the number of internal pathways for communication by creating multiple buses and to add more CPUs, allowing parallel processing of the various demands. These are clearly the directions high-performance workstation designers are taking.

In addition to software standards, bus hardware standards create an open system where many vendors find it simple to manufacture board level products. In turn, system integration houses find it economical to build more and different systems. STD and S-100 buses or DEC's Q bus are typical examples. Intel's Multibus II, built on the success of the original Multibus, is a five-bus standard; the Motorola/Signetics VMEbus, a dramatic upgrade of the VersaBus, is a three-path system. Both systems have a bandwidth of greater than 40 Mbytes/s, a Eurocard form factor, and full 32-bit primary data and addressing lines.

Standard buses mean that it is possible to purchase interface elements from third parties to extend an instrument-based workstation. Yet queries suggest that few scientists know, for example, that the Perkin-Elmer (P-E) 7000 series computers use a VersaBus and that it is possible to add "foreign" cards to them.

Standard operating systems

Many workstation vendors are focusing on Unix as the OS under which the facilities will execute. The ability to transport software developed under Unix to other machines has been a highly touted feature of this OS and its companion language, C. The language and the OS have similar architecture, which could aid both developer and user.

However, most Unix systems are large and cumbersome. The basic Unix lacks real-time features, and the command line structure is obscure. Its major attribute is the large number of companies that appear to support it.

Many licensed and unlicensed versions of Unix exist. Unix was originally developed at Bell Labs. V5 was licensed to educational institutions beginning in 1973. V6 was licensed to commercial firms in 1975, and V7 was introduced in 1979. System III Unix signaled AT&T's commitment to the OS as a commercial product. Unix/Berkeley 4.2 added many nice features. System V Unix incorporates many of the UC/Berkeley enhancements. Licensed lookalikes such as Xenix (MicroSoft), Uniplus+ (Unisoft), and V/68 (Motorola) exist. Unlicensed lookalikes such as Coherent (Mark-Williams) and Idris (Whitesmiths) also exist.

Unix is, therefore, not a single entity, but some 30 distinct personalities. The degree of compatibility between these versions is complex. Xenix has source compatibility with V7 and Systems III and V, but file system compatibility only with V6. To resolve these incompatibilities the prospective user must alter ported software. This may require consultants or trained in-house support. Indeed, the Unix advantage may lie only in the development laboratories of the instrument vendor, where a standard can allow experienced software engineers to develop software more rapidly.

Chemists purchasing systems that are to be used as general-purpose workstations should insist on a standard OS with a demonstrated ability to run any needed third-party software.

Neutral data format

A standard interchange format for data would provide the mechanism to move data between different combinations of software–hardware tools. The electronics community is addressing this need through the Electronics Design Interchange Format (EDIF), developed through a consortium made up of Tektronix, Texas Instruments, Motorola, Mentor Graphics, Daisy, and the University of California. Other neutral-format protocols in electronic manufacturing are already ANSI standards. Varian and DEC have suggested a similar neutral data format for analytical instruments and laboratory workstations. Unfortunately, it has not been warmly received by other instrument vendors.

With the observation that integrated software, neutral-data-format protocols, and concern about standards for languages and OSs are prevalent in the engineering workstation environment, one questions the retarded development in analytical instrumentation. Perhaps the analytical laboratory does not need such sophistication. This is doubtful. More likely, it does not recognize the need for such capabilities or that it is possible to achieve such standardization in a competitive industry. Standards will become a high priority among instrument vendors only when buyers demand them.

A wave of workstations

Some workstations have been designed from the ground up specifically for the target group involved. Others have been developed as add-ons to popular personal computers. Some larger microcomputer manufacturers are beginning to provide workstation extensions to their "professional" computers that run under their standard OSs. Finally, instrument vendors are extending the function of embedded computers to include workstation functions. Regardless of the approach, the buyer should carefully examine the ease of installing third-party software.

For many scientists, the first attachment to the workstation environment begins with their discovery that small data management programs can improve their access to a personal literature file and address/phone/contact information. Personally oriented literature reference files have been created around dBASE, Rbase (Microrim), Knowledgeman (Micro Database Systems), and Condor (Condor) running on conventional personal computers. A "full-text" search approach is available from Sci-Mate. Inexpensive address/phone programs and other support utilities are available from the new IBM Personally Developed Software Directory.

Bolt, Beranek and Newman's recent announcement of a version of RS/1 that runs on the IBM PC opens up access to that very fine and well-accepted electronic notebook for scientists. It allows entry of tabular data, plotting, fitting, statistics, and modeling. RS/1 file structures can be easily linked to larger data base programs such as DEC's Datatrieve.

Less expensive plotting–statistical packages are available for the IBM PC. An example is the Grafmatic–Plotmatic package (Microcompatibles). 2D and 3D plots, bar and pie charts, contour plots, and cubic and bicubic spline or least-squares fitting are provided. It is compatible with a variety of standard Fortran compilers. Statpro (Wadsworth) provides regression analysis, analysis of variance, and factor and cluster analysis. Scatter, triangle, and regression plots, dendograms, and histograms are supported. DAES (R&L) is a similar package.

One of the first to appreciate the need for merging real-time data acquisition and the power of integrated software was the Electronic Notebook (Laboratory Technologies). This provided software to digitize data from experiments, and then linked the data files to Symphony or Lotus 1–2–3. Running on IBM PCs, the system uses A/D and D/A subsystem cards from Data Translation. Running on the DEC Rainbow, an Acro Systems data acquisition expansion box is used. This is an 8088-based unit that is controlled by ASCII strings sent from the host micro. The collected data are returned to the host over the interconnecting RS-232 lines.

Isaac (Cyborg) is an IBM PC add-on that provides the hardware for data acquisition and control (A/D, D/A, and parallel I/O) as well as the software that links the captured data to Lotus 1–2–3 and Symphony. The related IBM PC Engineering/Scientific Services package weds a data acquisition and control adapter to an IBM PC/XT or AT coupled to an advanced graphics display supporting the GKS graphics standard. Asyst (McMillan) also provides for data acquisition, graphics, statistics, and mathematical transformation of data on the IBM PC. It is stand-alone. Typical functions provided by such packages include eigenvalues, eigenvectors, polynomials, curve fitting, convolution, smoothing, and fast Fourier transformation.

Some instrument companies are offering GC/LC-type packages running on IBM PCs that offer workstation services (e.g., Nelson and Spectra-Physics).

DEC's Pro-300 series of PDP-11/23-based computers have a real-time hardware option that services parallel I/O, serial I/O, and IEEE-488 devices. An analog I/O subsystem has been shown, but not delivered. Software support packages include CT*OS scientific word processing, RS/1 electronic notebook, Pro/Datatrieve data base management, and Supercomp-Twenty spreadsheet/modeling (Access Technologies), all running under a subset of the RSX-11/M OS.

The Perkin-Elmer 7500 Professional Computer is a 68000-based system running under Idris. Built to support a number of P-E's instruments, its spectroscopy data acquisition and manipulation packages have been augmented by word processing software. This includes P-E's Pecowriter and Lex (Ace Microsystems), a full-function system. Supercomp-Twenty is also available. It is encouraging to see other vendors' software supported on an instrument company's "somewhat dedicated" computer. Such a workstation approach can bring software tools to the bench more rapidly and at an acceptable cost.

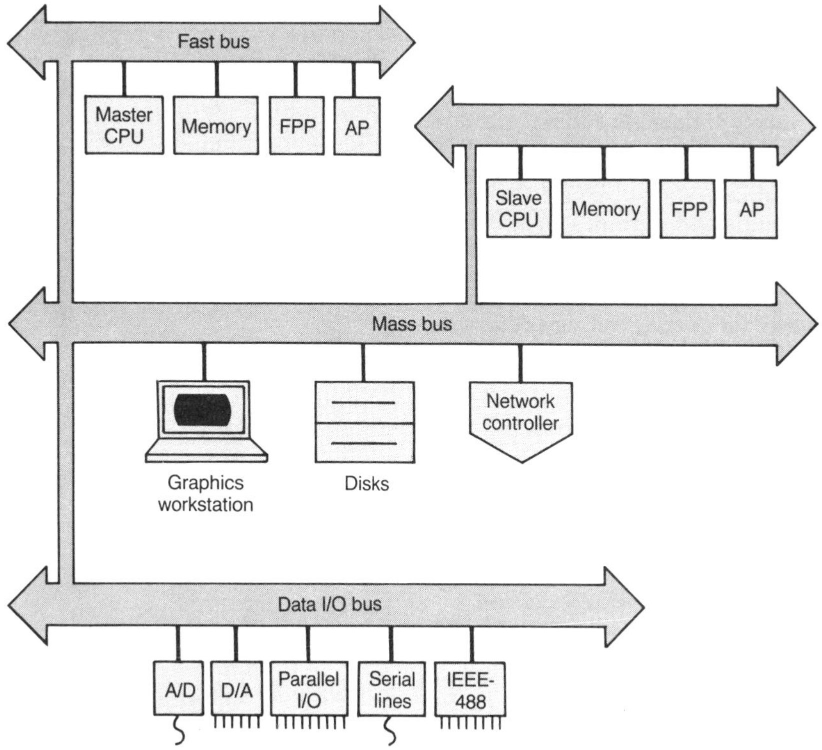

Figure 4. Multi-CPU, multi-bus configurations are essential for high-performance workstations

Where real-time data acquisition is supported, it is important to request specific details on the A/D subsystems supported and the maximum throughput of the system, including software and OS overhead. Some systems support only slow data acquisition rates. Most of the add-ons are not truly concurrent.

Higher performance workstations

Moderately priced higher performance workstations specifically oriented to rapid data acquisition and manipulation are available. Typical is the SPT-871 (Burcon), which uses a Forth-like language. Laboratory Technologies has a similar higher performance system built around the Intel real-time operating system. Both systems are crafted with hardware and software tools to meet high-speed needs.

StarLab (Mattson) provides a network-based integrated package oriented toward spectroscopic analysis laboratories. It provides sample-tracking functions and spectral searching in addition to workstation facilities. The system is C and Unix based. Real-time data acquisition and control are supported by modifications to the standard Unix, providing an *automation shell.* StarLab supports the relational data base management system called Informix.

High-performance workstations

How can one computer system handle the diverse requirements of instrument control and integrated workstation software support? Hardware architecture provides a solution to a number of the challenging areas mentioned above.

Data acquisition. High-performance workstation environments that can support multiple users, graphics, and very high speed data acquisition transcend single-CPU/single-bus/interrupt-driven architecture (Figure 4). In such workstations, such as the MC-500 series (Masscomp), the main CPU needs the support of both a floating-point processor (FPP) and an array processor (AP). High-speed FPPs can handle 75-bit precision math, or more, at speeds of over 500,000 Whetstones (a measure of math capability). APs can address over 16 Kbytes of 32-bit high-speed memory, allowing storage and manipulation of large arrays. APs often have their own CPU implement virtual memory addressing, increasing dramatically the array sizes that can be handled in relevant time.

All of this activity increases traffic on the bus, so that it is common to use multiple buses. The MC-500s have three! A Multibus links the main CPU with the disk, tape, and graphics processor controllers. An STD+Bus links the main CPU to the analog and digital I/O peripherals, as well as to higher speed communication interfaces. Finally, a special in-house bus links the main CPU to its memory. It is even possible to add a second slave CPU to the Multibus. A/D task operation involves DMA. These extremes are necessary to support 2-MHz data acquisition rates in an environment that supports multiple users and graphics under a Unix operating system.

Graphics. Graphics workstations for chemistry impose awesome demands on hardware and software alike. The amount of information for static displays is high, and the bandwidth for moving images is extremely large. The PS-300s (Evans and Sutherland) are 68000-based processors with FPP and AP. The FPP manipulates all coordinate data and transformation functions rapidly. The AP transforms the data structures in main memory and applies the function networks needed to dynamically modify these objects. With this support, refresh and update can occur at the same rate, and smooth temporal motion can be effected. The vector display gives an image free from the step-functions associated with raster displays. Systems for molecular modeling require the ability to handle 15,000 to 20,000 vectors and can use some 1800 programmable "colors" to properly display data. Silicon Graphics and Sun also provide excellent raster graphics workstations.

Network integration. An interesting approach to meeting the diverse needs of the scientific laboratory is the Domain family from Apollo. These workstations have proven very cost-effective at Monsanto. The Apollo graphics workstation provides 2D or 3D viewing and image transformation capabilities using the Core graphics software protocol. Computational workstations possess 32-bit processors with pipelined architecture and integrated FPP hardware. Each of 24 tasks can support 256 Mbyte of virtual address space. More traditional workstations in the Domain network invoke windowed, icon image screens for access to a variety of word processing, data processing, and programming tools. All of the workstations can intercommunicate over a 12-Mbit/s baseband LAN. A relational distributed data management facility allows sharing of information between stations.

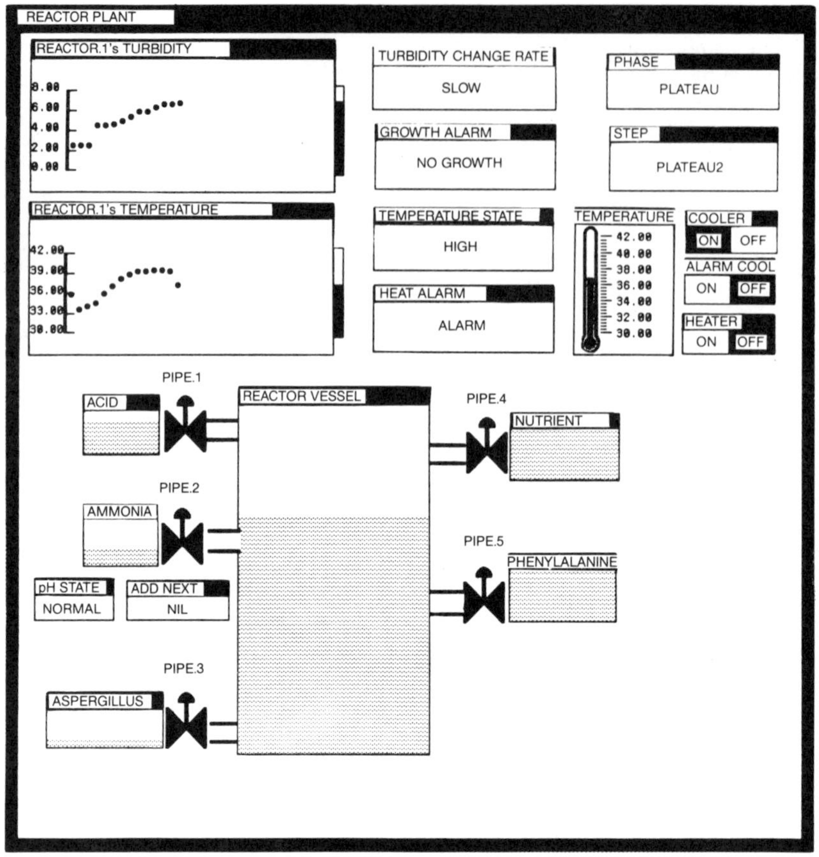

Figure 5. Bioreactor simulation display from an artificial intelligence workstation (IntelliCorp's Knowledge Engineering Environment)

Aspergillis niger growth (citric acid synthesis) is being monitored by turbidity. Temperature and pH are being controlled. The real-time simulation is seen via a "mimic" control panel. Photo courtesy of IntelliCorp Knowledge Systems Division, Menlo Park, Calif.

This type of integration, layered over the integration of software tools at the individual workstation level, will be the hallmark of the modern electronic laboratory.

AI workstations. Finally, a new generation of workstations is in the offing. Based on artificial intelligence (AI) constructs, these workstations are intended to provide the user with an intelligent but well-mannered assistant that can provide annotated notebook support, an intelligent encyclopedia, discrete event simulation with direct manipulation of the "mimic" objects, as well as planning and communication support. The systems can even "kibitz," so that the interaction with the user is pleasant, not presumptive or preemptive. The systems are ideal for methods development in the analytical laboratory. They have already been used successfully in bioengineering fermenter design, chemical engineering process control, and laboratory time–motion analysis (Figure 5). Such workstations are characterized by large (2–10 Mbyte), high-speed (<200 ns) memories and an associated FPP. Typical examples are Xerox's 1108 Interlisp-D machine, TI's Explorer, and the computers made by Symbolics and Lisp Engine.

Some excellent menu-driven natural-language approaches to the human–machine dialogue are already available. Natural-Link, which runs on the TI PC, proves that such tools can run on small machines. This AI contribution allows users to select phrase components from a menu system to create sentences that, in turn, are used as commands to the system. This ensures that all queries are acceptable by the machine (good linguistic coverage) and that an answer exists for the question (good conceptual coverage). The advantage? How often have you asked a computer system a question it did not understand and then found, after several attempts at rephrasing the question, that it had no data on the subject?

The local area network

Experience in the engineering arena suggests that workstations work best when they are connected to mainframes and clustered in multiple-user networks. Perhaps the best implementation exists at General Motors (GM), where some 1000 workstations will be

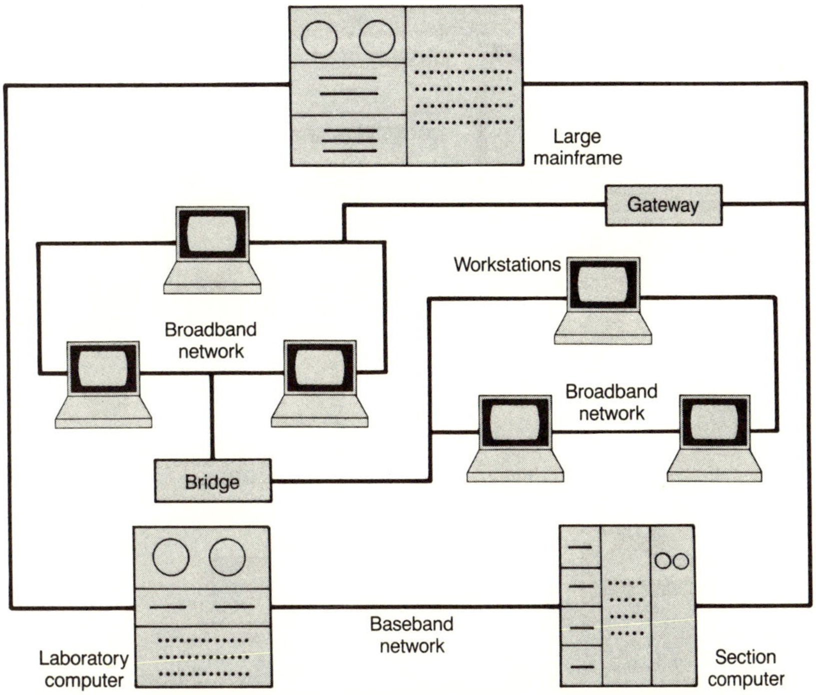

Figure 6. Networking provides another level of integration for the electronic laboratory.

Here, two broadband nets, with a bridge joining them, interconnect a number of workstations. A baseband link ties larger computers together. A gateway couples the two dissimilar networks

networked by the end of 1986. A dual baseband–broadband net interconnects the larger host computers and microbased workstations, respectively. Word processing (Mass-11, Microsystems Engineering) and data base management (micro- and mainframe Oracle) are available to all scientists.

The mailman. As workstations become more prevalent, intercommunication will become the most important factor—and quite possibly the bottleneck that prevents the system from serving its users properly. Some type of high-speed LAN is required. A number of vendors—for example, Sytek, Ungerman-Bass, Corvus, Network Systems, and Nestar—have responded.

A LAN provides a conduit over which information can flow. It does not establish, per se, a protocol for message format. LANs have been likened to a postal service that delivers letters (Figure 6) but does not constrain what is in the envelope, either in form or language. All that is required is that a reasonable address be on the face of the envelope. Convention has established a form for the letter inside, and a language common to sender and receiver must be used. Electronic communication also requires a rigid format that must be adhered to at both ends for effective communication (Figure 7).

The letter. Data transmission is usually effected by one of two coding schemes (the language): The American Standard Code for Information Interchange (ASCII), a seven-bit code; or an IBM eight-bit code called EBCDIC.

IBM mainframe connections may be binary synchronous (bisynch, or BSC). Advanced network requirements led to IBM System Network Architecture (SNA). Most other companies use ASCII conventions and binary asynchronous (asynch) communication protocols. Fortunately, a number of vendors make ASCII/BSC, ASCII/SNA converters that can be installed in workstations. They work like voltage converters, in that their operation should be transparent to the user. Examples are Irma (Digital Communications) and Cleo (Phone 1). GM employs Forte 3270 converters for its IBM PCs and Avatar units for its DEC Rainbows. Many of these converters emulate IBM 327X terminals.

Unfortunately, today there are five major protocol choices for message architecture (the letter's form): Xerox Network System (XNS), the U.S. Department of Defense's Transmission Control Protocol/Internet Protocol (TCP/IP), IBM's SNA, DEC's Digital Network Architecture (DNA), and the open system implementation of the International Standards Organization (OSI/ISO). In addition, there are smaller, special-purpose protocols such as Microcom Network Protocol (MNP). For example, VisiCorp integrates three protocols into its VisiON products: For point-to-point connection over modems MNP is used; for peer-to-peer communication over a LAN users choose XNS; and for mainframe connections the SNA protocol must be used. Most engineering workstations and graphics terminals al-

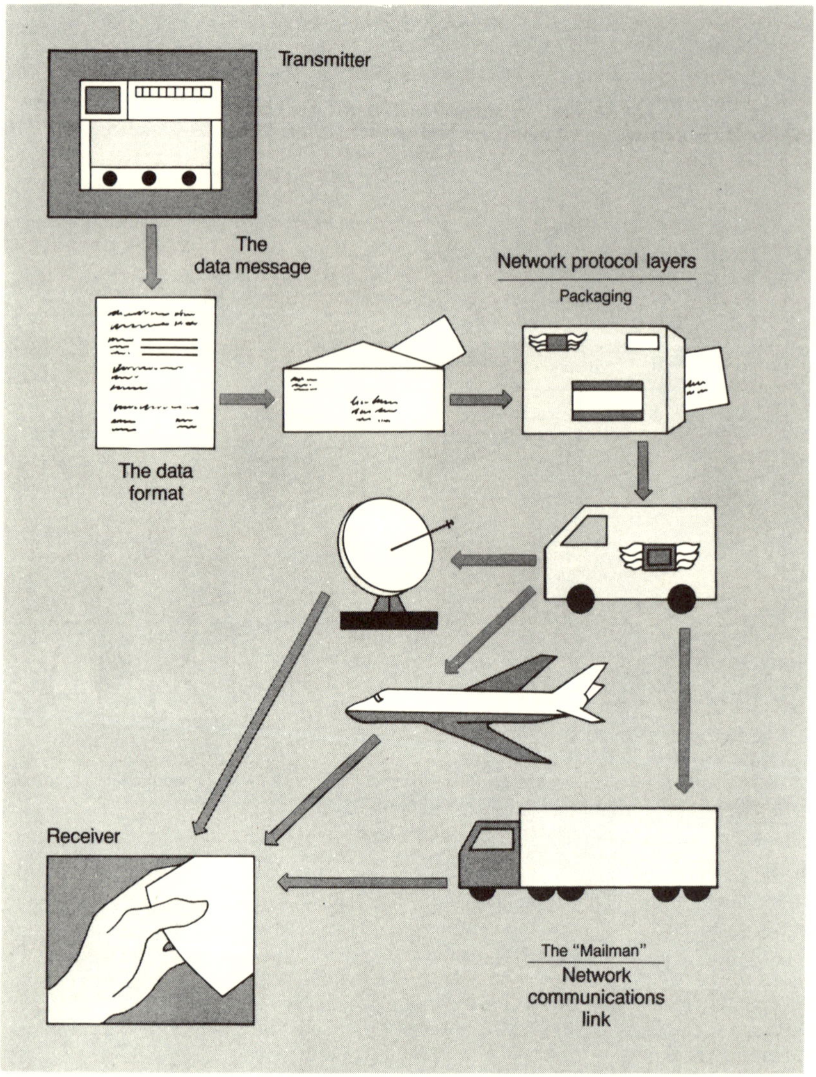

Figure 7. Network conventions involve a common language between transmitter and receiver, an agreed-upon neutral data format, and a standard packaging protocol for transmission

The proposed ISO standard involves seven layers of protocol. There are four other popular conventions. Electrical transmission standards, such as the IEEE 802.X baseband and broadband carrier sense multiple access/collision detection (CSMA/CD) or token-passing protocols, provide the delivery system for the packet

ready incorporate the TCP/IP standard, since many Unix versions have this network control already built in. As the OSI/ISO protocol becomes an international standard more companies are adopting it. DEC, Data General, Hewlett-Packard, Burroughs, Honeywell, Xerox, and Tektronix have commitments to the OSI/ISO standard. An incredibly potent force behind adoption of this standard is GM's MAP (Manufacturing Automation Protocol) Network; it endorses the ISO standard and has widespread industrial support.

Purchasers of workstations for the lab should examine the availability of, and support for, micro-to-mainframe connections. They are essential to provide access to the data services available on mainframes via products from Cullinet, MSA/Peachtree, Applied Data Research, Cincom, Computer Associates, McCormack-Dodge, Information Builders, and On-Line Software, as well as DEC, IBM, or Hewlett-Packard.

Workstations will require a major change in work habits and attitudes. The electronic office has welcomed this new servant. The laboratory scientist will find workstations useful as well. Using workstations does require learning a new vocabulary and accepting an assistant on your desk and workbench. Most people find it fun, profitable, and expedient. What more can we ask?

Part 2

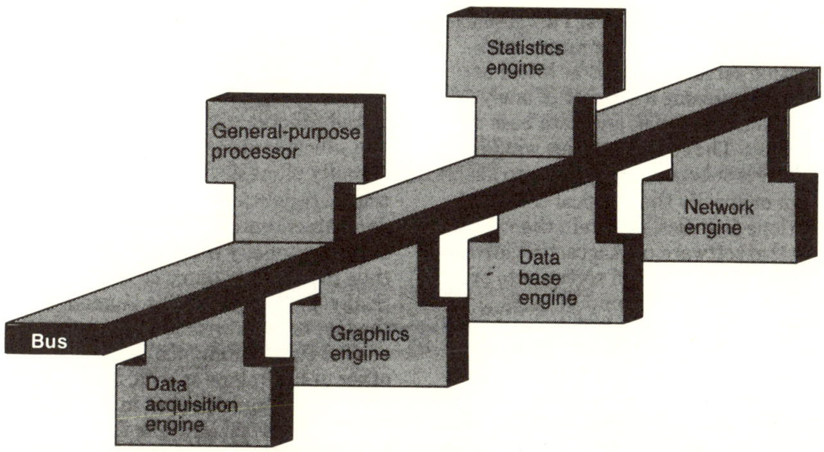

Figure 1. A multiengine workstation

Special-purpose computers, or engines, provide data acquisition, graphics, statistics, data base, and network services. The engines are connected to a common bus. Such separation of tasks eliminates the competition for time so common in single-processor, standard operating system configuration

The authors of the following capsules were asked to discuss the design philosophies of the scientific workstations with which they are involved. Each capsule is an example of a different approach, addressing different needs. The intent is to provide material that, in conjunction with last month's tutorial, can provide the scientist with an ability to critically evaluate the numerous commercially available scientific workstations.

Union Carbide
Chemicals and Plastics
South Charleston, W.Va. 25303

Contributor: Michael Starling

An Introduction to Many Points of View

Laboratory scientists have had many new tools presented to them over the past few years. Many of these devices are the results of including computers in existing instruments.

The first applications of embedded processors to instrumentation really served the instrument vendor rather than the scientist. Ease of design, potential flexibility, parts count reduction, and field serviceability were among the early motivations; but the cost to repair a modern instrument has not dropped, and custom networks and software have tended to restrict purchasing freedom, rather than enlarge it.

Scientists have therefore begun to look for new capability as well as the "ease of use, or repair, and increased consistency" originally touted for the new generation of instruments.

Important questions that have arisen include the following:

- How does one ensure the timing accuracy of data acquisition?
- How can the validity of the basic data be ensured or even assessed after undergoing intensive numeric manipulation in what is often a "black box" provided by the vendor?
- How are data from a number of instruments to be gathered together while maintaining their integrity?
- How can these data be arranged in a manageable and understandable form and then displayed in such a way that the underlying principles can be perceived?
- Finally, how can what is learned be fed back into the experimental design?

These concerns, which roughly follow the steps in which data are gathered and analyzed, are fundamental to the scientific method in a cost-conscious world. Yet they are difficult to implement. The solution is a proper workstation environment.

Current solutions. The concept of the laboratory workstation goes beyond the single instrument and, in fact, the single computer. The best current solutions involve networks of computers. Each computer performs that portion of the overall task to which it is best suited.

Data acquisition applications are placed within the instrument or in small to medium-size machines dedicated to specific instruments. The question of timing accuracy has gone largely unnoticed, but the question of validity of data reduction algorithms applied by the embedded or dedicated processor has attracted public concern and comment. Scientists and regulatory agencies are struggling with the definition and implementation of "software certification."

Transfer of data from instruments to larger computers, which are capable of graphic and data base manipulation of these data, is essential. Due to lack of any consistent data structure or transmission specifications, this step can be quite challenging.

Standards that address instrument control as well as instrument–computer data interchange format are essential to the effective establishment of a laboratory network.

Such standards would help designers and would prevent scientists from having to learn obscure and inconsistent interfaces to the various software packages used in accessing, reducing, and displaying data. In addition, standards would help simplify the establishment and maintenance of the audit trail required by regulatory or patent imperatives.

Required resources. This scenario is one in which the laboratory workstation serves as a window into a network possessing a number of analytical, computational, and data base resources. The support of this window involves substantial expertise to design or specify the physical communications facilities, maintain the various system software packages, provide interface hardware and software to specialty devices, and support users through problem analysis, correction, and training. This is an expensive endeavor.

The level of difficulty in establishing the laboratory workstation is certainly a problem for most organizations whose primary goal is science. A solution to this problem is certainly an opportunity for the interested, capable vendor.

Some suggestions. Recent technological advances allow an attempt at class solutions to some problems. An example of a simple class solution that has already been achieved is the integrated spreadsheet, data base management, and graphics package designed to aid in the manipulation and presentation of business data.

The generic problem of the laboratory scientist is somewhat more complex. The workstation must support at least the following functions:

- initialization and initiation sequencing for all instrument and sensor subsystems;
- real-time data acquisition in those situations where intelligent instruments are not present;
- real-time control where required;
- data file management;
- data base management;
- user interface management;
- data reduction management;
- graphics display management;
- network communications management; and
- virtual terminal management.

It becomes immediately obvious that time management in a single-processor based system will be a nightmare. Timing is just not predictable. Until recently, this would have been sufficient to cause one simply to dismiss the concept.

Fortunately, there is now no reason that multiple processors cannot be used to solve this group of problems. Several 16/32-bit microprocessors that have facilities for the construction and coordination of multiple-engined general-purpose computers are available. The National 16032 and 32032 as well as the Motorola 68000 and 68020 processors are good examples. Language and operating system environment constructs are available that do not limit the execution of programs to a single machine. Most notable of these are Forth and Occam.

The goal is the construction of a "network in a box," in which each required functionality is allocated to its own processor (Figure 1). Each functionality alone is demanding, but relatively straightforward. Time-dependent functions can be separated from computationally intensive functions, thus reducing resource contention. Such a product would be layered, but the layers would reside in several processors that communicate to one another at backplane speeds. Such a design can be implemented in stepwise or in parallel fashion. The final product can be expanded by the addition of processors addressing newly needed functions.

Given that such a thing is possible, what would be the most significant concern during the design of such a device? Simply, will the user get a friendly, usable, and consistent interface to the capabilities of the system?

Perkin-Elmer
Main Avenue, ML 75
Norwalk, Conn. 06856

Contributors: Mark Thompson, David Binkley, and Peter Passalacqua

A General-Purpose Workstation Design

Designing a workstation suitable for the analytical instrument laboratory requires the fusion of available technology with the expressed needs of the scientist and anticipation of the future demands that the analyst will make of the system as his or her interests, needs, and abilities change.

Surveys conducted by the Computer-Aided Chemistry group of instrument workstation users indicate that a large percentage (~50%) are interested in word processing, data base management, graphics, data acquisition, terminal communication, statistical analysis, and BASIC and Fortran-77 programming. Fewer than 25% of the scientists interviewed expressed strong interest in spreadsheet capabilities, a spelling dictionary, or Pascal and C programming. Little interest was shown in an integrated package of these software tools.

With these guidelines in mind the Perkin-Elmer (PE) Series 7000 professional computer was designed as a workstation to service the needs of both scientist and instrument. The philosophies behind its configuration may be useful to scientists developing their own attitudes toward the workstation environment. Design considerations were driven by the need to provide a workstation that is simple for scientists to use, yet convenient for engineers to build and modify (Figure 2).

The nucleus. From previous experience with an 8-bit microprocessor-based workstation, it was evident that laboratory requirements could be met only with a 16/32-bit microprocessor with extended addressing capability. Industry momentum surrounding the Motorola 68000 processor made it a natural choice. The 68000 has the interrupt structure, instruction set, and speed required for scientific applications. It is part of a growing family of more powerful processors that are upwardly compatible. The growing necessity to incorporate standards in intelligent-instrument products suggested the Motorola Versabus.

Increasing evidence exists that the Unix operating system (Bell Labs) and Unix-like operating systems (OS) reduce software development costs and assist in porting of software from system to system by the designer. Idris (Whitesmiths), a Unix lookalike, was chosen for the OS. The program development tools that this supports, such as the C language, have permitted software engineers to develop a broad array of application programs in a short period of time.

Language support facilities for the user include an extended BASIC interpreter and a Fortran-77 compiler. In addition to data base management, statistics, graphics, and data acquisition software packages developed by PE, the workstation supports third-party word processing (Lex-68, Ace Microsystems) and spreadsheet (Supercomp-Twenty, Access Technology).

Unix systems are not inherently real-time. It was necessary to enhance

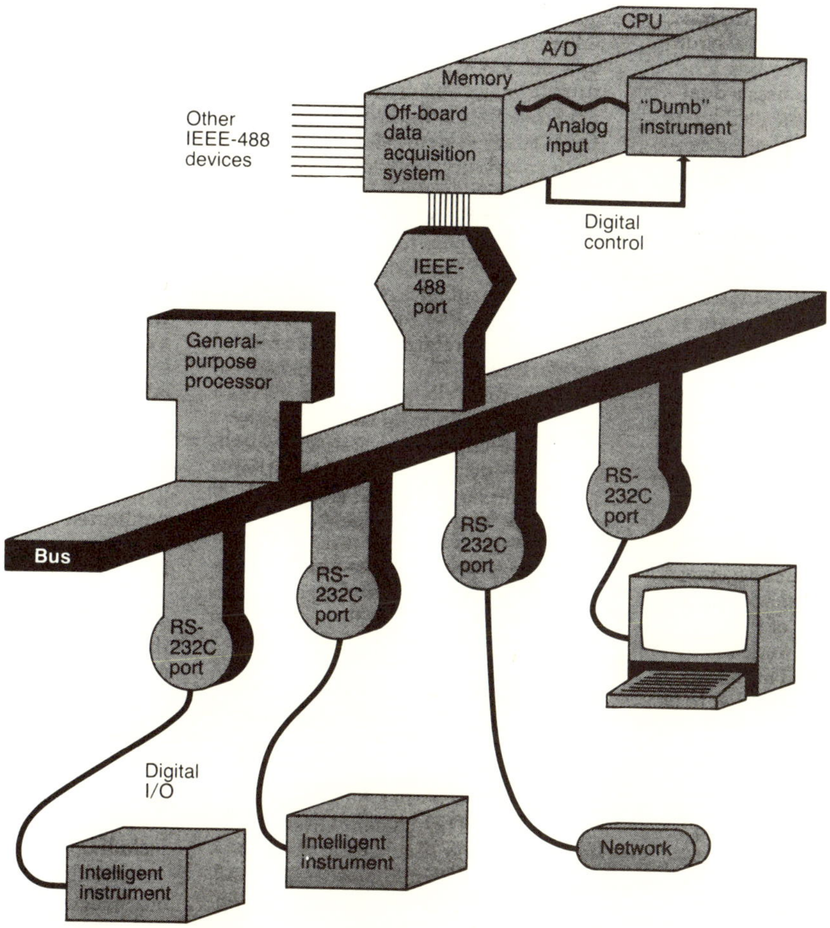

Figure 2. A general-purpose workstation

The combination of intelligent instruments with standard digital I/O capabilities and a communications-oriented workstation off-loads data acquisition and control from the host. Each instrument is connected to the workstation over an RS-232C serial asynchronous communications link. Instruments with only analog output can be serviced by an off-board A/D converter and control subsystem. A parallel IEEE-488 bus interconnection links the workstation and the subsystem

Idris to ensure optimal performance in instrument applications where the response time must be guaranteed. Sections of code can be locked into memory as required.

Idris supports partitioning of the rigid disk into six logical disks. In normal operation, applications software, data files, temporary files, and the operating system are located on individual logical drives. Since each logical drive can be assigned protection individually (read/write, read-only, etc.), applications software and the operating system can be protected from accidental changes or deletions.

Finally, standard RS-232C and IEEE-488 I/O ports were installed. The serial ports operate at up to 9600 bits/s, whereas the parallel IEEE port can communicate at up to 100 Kbyte/s. Most instruments designed today have one or the other of these digital communications standards to provide computer compatibility.

The environment. Such a generic workstation is easily adapted to the specific needs of such diverse intelligent instruments as infrared (IR) spectrometers, energy spectrometers, ultraviolet–visible spectrometers, chromatography equipment, atomic spectroscopy units, fluorimeters, and thermal analysis instruments. Hardware personality modules that use the communications ports of the workstation adapt it to a particular analysis method. Appropriate software is added to the Winchester disk.

It is possible to timeshare the workstation between such units merely by activating a switch or swapping cable connections. Laboratories routinely use one workstation to service various combinations of separation instruments and spectrometers.

Customization. For instruments that do not have a standard digital I/O port, a data acquisition module is available that communicates with the workstation over the IEEE-488 port. Many laboratory instruments require sampling rates of 30 Hz or less. The data acquisition system uses a dual-slope A/D converter controlled by its own computer; such integrating converters are immune to the 60- and 120-Hz noise common in lab environments. The host workstation uses calls from BASIC or Fortran to activate the converter. The converter stores up to 1000 values in its own RAM memory, so that the host can read the data at its leisure. This decoupling of data acquisition from the main processor and its OS prevents the unpredictability usually associated with nonconcurrent system managers.

Communications. The workstation was designed to be compatible with the 3200 series of PE minicomputers. The latter can support a full-function laboratory information management system (LIMS). By means of an ASCII link, import and export of files can take place between the workstation and the minicomputer. This allows information from the workstation to be passed to the LIMS, where it can be integrated into the data base. LIMS information can also be passed to the workstation, where data management, spreadsheet, and word-processing software can manipulate it.

Data transfer between the workstation and a variety of other computers is supported by the terminal emulation software provided. These computers include IBM PC and mainframe, Harris, and DEC environments.

The user interface. A full set of function keys has been added to the keyboard. Plastic overlays are available for standard applications, and special ones can be prepared by the user. A single keystroke of the proper function key can be substituted for the complex series of keystrokes required by the OS. For example, a function key labeled *graph* can retrieve data, scale it, and display it on the bit-mapped color graphics terminal.

In addition, eight *soft keys* are mounted on the bezel of the display terminal. These keys are labeled by changing displays at the bottom of the CRT. They are used in the same way as the function keys. However, the use of software-displayed labels makes it easy to change both the operation performed by the key and its accompanying label during the course of an experiment. Soft keys dramatically reduce both the amount of training required to use new programs and the recall effort involved when occasionally used programs are executed.

The goal. A laboratory workstation

associated with an instrument may, at one extreme, be seen by the scientist as a turnkey system that requires no programming experience or computer expertise to use. In the middle ground, the turnkey system can be enhanced as the needs and sophistication of the user increase. In the long term a properly designed workstation can become a general-purpose tool, supporting the personal and professional software becoming so popular in the electronic laboratory. For example, data from a spectrometer requiring kinetic analysis can be exported to a spreadsheet program for study and reporting. Analytical results from an IR spectrometer can be shipped to a statistics package for reduction. Atomic absorption results can be stored and correlated via a data base management package.

Varian Associates
Palo Alto Instrument Division
611 Hansen Way
Palo Alto, Calif. 94303

Contributor: Steve Smallcombe

A Data Analysis Workstation Design

The Varian Instrument Group makes a variety of scientific workstations for use with chromatography and spectroscopy. The development philosophy behind the ADS-4000 nuclear magnetic resonance (NMR) workstation may help scientists develop their own concepts about what a workstation should be and do.

A modern NMR spectrometer relies heavily on a powerful embedded data system. Such NMR instruments often contain multiple computers that handle data acquisition and spectrometer control as well as a variety of data-processing activities. The software available for these spectrometers is generally capable of performing a wide variety of NMR experiments and of processing the resulting data. This data processing uses a variety of techniques including the fast Fourier transform (FFT). The multiple-processor hardware architecture, coupled with a multitasking operating system, allows simultaneous data acquisition, processing, plotting, printing, and display.

With the powerful computer system and software available on the spectrometer, there may at first seem to be little need for an NMR workstation. In some cases this is true, because the multitasking capabilities of the spectrometer data system often allow the spectroscopist ample time to analyze and plot the data as new NMR spectra are being acquired. However, in an open-access lab, where many proton NMR spectra are acquired, the time needed for data processing and plotting often exceeds the time necessary to acquire the NMR data. Under these circumstances, the system throughput can be limited by the spectrometer's data-processing capacity. In these cases, an NMR workstation offers the possibility for improved throughput. Since the data system is normally a fraction of the cost of the NMR spectrometer, an NMR workstation that is a clone of the data system in the spectrometer can provide a low-cost method of doubling an open-access NMR laboratory's productivity. The parts of the instrument's computer system dedicated to spectrometer control and data acquisition do not need to be duplicated in the workstation, as these activities are strictly a spectrometer function, not a workstation function.

Unlike workstations for other forms of spectroscopy or chromatography, this NMR workstation does not add new capabilities to the spectrometer, but increases its throughput by providing an alternate site for data examination and analysis.

The workstation hardware (Figure 3) has many of the advantages of the parent spectrometer system hardware including:

- multiple processors on a multiple bus structure for efficient multitasking;
- optional array processor for faster data processing;
- 16-color graphics display;
- a variety of printers and plotters for hard copy;
- operator-assignable knobs that allow manipulation of spectral windows, gain settings, and cursors via a natural interaction mechanism; and
- Winchester disks for fast data storage and floppy disks and streaming magnetic tape for data archiving.

A major advantage of reproducing the spectrometer data system in an NMR workstation is that the operator interface and software are identical for both environments. This provides the operator with a common interface and reduces development costs. It is very important that an operator need not learn a new set of commands and parameters to perform the same function in two places.

Automated NMR analysis. It would seem that an alternate solution to the throughput problem is to simply increase the speed of the instrument's computer to decrease data-processing time. Although this is certainly valuable, it is not the entire answer. A focal point for a flexible software system is also required—one that can be added to by vendor and user alike.

The ultimate goal in an automated system is not simply the production of proper spectra, but the production of completely analyzed chemical results. NMR spectra have a very high information content, and it is often desirable to examine the data in several different presentations before committing the data to the hard-copy printer or plotter. This is especially true for proton spectra and two-dimensional spectra, where there are usually a large number of spectral regions where expanded plots might aid interpretation. The NMR spectroscopist must currently add his or her decision making to the data analysis and plotting process. As spectrometers become more and more automated, data must be acquired, processed, and plotted while the system is unattended. Workstation software will need to provide decision-making intelligence and thus approach the status of an *expert system*.

The wide variety of potential analysis methods necessitates that the user be able to easily add programs involving his or her own procedures. Thus, an important use of an NMR workstation is for methods development. In most cases, this is accomplished by writing new software or by modifying existing programs. This is an ideal activity for a workstation since the software can be developed and tested without impeding the throughput of the spectrometer. Varian spectrometers and workstations use Pascal as their primary programming language, and Pascal is available on these systems for user programming. It is important that the programming language available to the user be the same as the primary programming language used for the system. This ensures that the user-written programs will fit in with the system software and have the same access to data files, parameter files, and system peripherals as normal system commands.

Direct programming in a high-level language is an important feature to some spectroscopists, but many scientists would rather not learn a programming language at this level. Rather, they prefer a language they already know—the commands and parameters used to operate the spectrometer. To this end, Varian has installed a greatly expanded NMR-oriented macro language capability in the workstation and spectrometer software systems (*1*). Many languages have a macro capability that allows a series of system commands to be linked together to form a new command or macro. In this way the macro becomes a small program. The language of this program is made up of words related to the functions of the instrument itself, plus needed param-

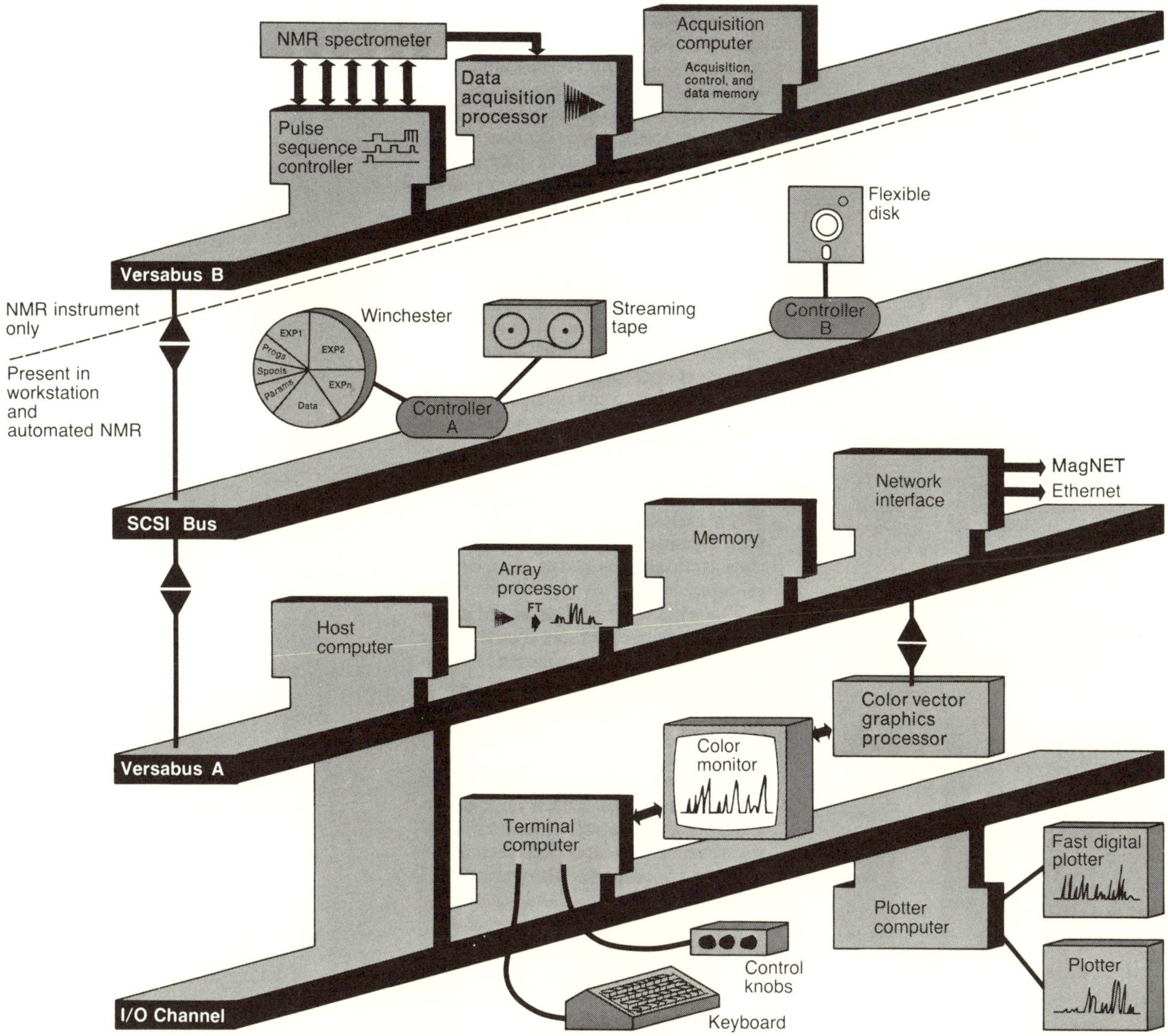

Figure 3. A multiengine, multibus NMR workstation

A workstation that replicates all of the functions of an automated instrument's computer system except data acquisition and control can service many instruments. Such workstations can increase analysis throughput, as well as serve as the site for software generation and testing. It is the logical location for future installation of an expert system. (The SCSI bus is the industry standard Small Computer Standard Interface bus)

eters. The NMR command language includes arguments; IF statements; looping; mathematical expressions; the ability to read from the keyboard; the ability to write to screen, printer, or plotter; and the ability to access an individual peak or integral within the spectra.

The user can now easily write a macro for report generation, such as a quantitative analysis of mixtures that reports the mole percent for each component. Macros can also automatically expand a spectral region. Intelligent macros can provide the decision-making capability that a chemist once had to supply.

Networking. Obviously a workstation and spectrometer can only work together if there is a convenient and fast method of exchanging data and programs between them, i.e., a network. Varian has implemented a fast parallel network (MagNET) to connect the workstation and the spectrometer. This network is invisible to the user. The scientist sees MagNET as an extension of the file system and can use all the normal system commands to copy programs, store and recall data, list libraries, etc., across the network. MagNET can be used for much more than simple data transfer since any command that accesses or writes to a disk file can do so across the network without the operator having to learn any new commands. As Ethernet (baseband) software standards stabilize, this protocol will be offered as an additional component of the MagNET system.

Future trends. As computers continue to improve their data-processing and graphics capabilities, NMR workstations will become available on a greater variety of hardware. The software available on these systems will merge the existing NMR software with a wide range of word processing, report generation, sample management, and scientific analysis software packages. This will allow scientists to perform all the related jobs necessary to solve a particular chemical problem from a single workstation located on their desks.

Masscomp
One Technology Park
Westford, Mass. 01886

Contributor: Steve Mullen

A High-Performance Workstation Design

Even with the aid of intelligent instrumentation the integration of multiple devices for high-speed data acquisition typically results in a system severely limited in speed or in its capacity to perform real-time processing while acquiring data. The following capsule describes the design philosophies required for a higher performance workstation such as the MC-500. This workstation was designed to allow predictable aggregate analog data acquisition rates of one million samples/s, yet provide for maximal use of high-level language programming and ease of use.

Triple-bus architecture

In traditional minicomputer architecture, a single high-speed internal bus transfers digital data throughout the system. The data flow bottleneck inherent in this design is avoided in a triple-bus design (Figure 4) that uses multiple industry-standard buses to distribute data acquisition, computation, and graphics among several very high performance processors.

STD+ bus. This 16-bit wide bus, capable of supporting data transfer at rates up to 2 Mbytes/s, is intended for use by data acquisition interfaces under control of a separate data acquisition/control processor (DA/CP). Standard STD interfaces may be connected to a single half of the bus. Such a bus offers the user a convenient means for adding a wide variety of simple, inexpensive, commercially available modules for application to specific problems.

Multibus. This bus is intended to be used by system-level peripherals that normally operate in direct memory access (DMA) mode. It is a superset of the industry Multibus (IEEE-7996) standard. Many I/O options are commercially available, aiding the designer. However, because timing problems may arise on a bus running near capacity, detracting from the predictability and the net transfer rate of the system, the normal Multibus bandwidth must be extended to 6 Mbytes/s to meet the 2-Mbytes/s design goal for throughput to disk.

500-Series bus. The 500-Series bus is a proprietary bus used primarily to connect the main CPU to memory. It allows the CPU to operate at its maximum speed while handling the expected traffic among the CPU, memory, floating-point, and array processors. It is required to avoid the lost time that would be spent on CPU/Multibus arbitration if memory were placed directly on the Multibus.

Multiple CPUs

The CPU module includes a 68000 processor for instruction execution and a 68010 processor for memory management. The module also contains an instruction and data cache, an address translation buffer, and three serial ports.

Cache memory attempts to store the next series of instructions required by the CPU, buffering the high-speed CPU from the lower speed memory. When instructions or data are found in the cache by the CPU no "wait states" for memory fetch are incurred. The CPU typically receives 85–95% of its words directly from cache rather than having to access slower main memory.

Virtual memory (*1*) allows the user to treat disk storage space as an extension of main memory, but disk access is another potential bottleneck. The address translation buffer is a high-speed RAM that holds the most recently referenced accesses into virtual memory. If the desired information is resident in main memory no CPU wait states are incurred in accessing it. Automatic disk–memory swapping is invoked only if the information is not memory resident.

The addition of a second CPU, additional memory, and a second floating-point processor (Figure 5) can almost double the computing power of such a system. The second CPU operates as an asymmetric slave processor.

Data acquisition and control

Hardware. The DA/CP is an intelligent, front-end interface that manages transfers between devices on the STD+ bus and memory. The DA/CP is connected to the host processor via the Multibus and can perform DMA data transfers directly to or from system memory. The DA/CP converts inexpensive, interrupt-driven devices into high-speed DMA devices. To ensure that Multibus DMA and interrupt latencies do not affect performance, all data transfers are buffered by hardware first-in first-out (FIFO) registers. A multibuffering method is used for continuous transfers. Since the DA/CP executes 24-bit arithmetic, it can perform simple data reduction

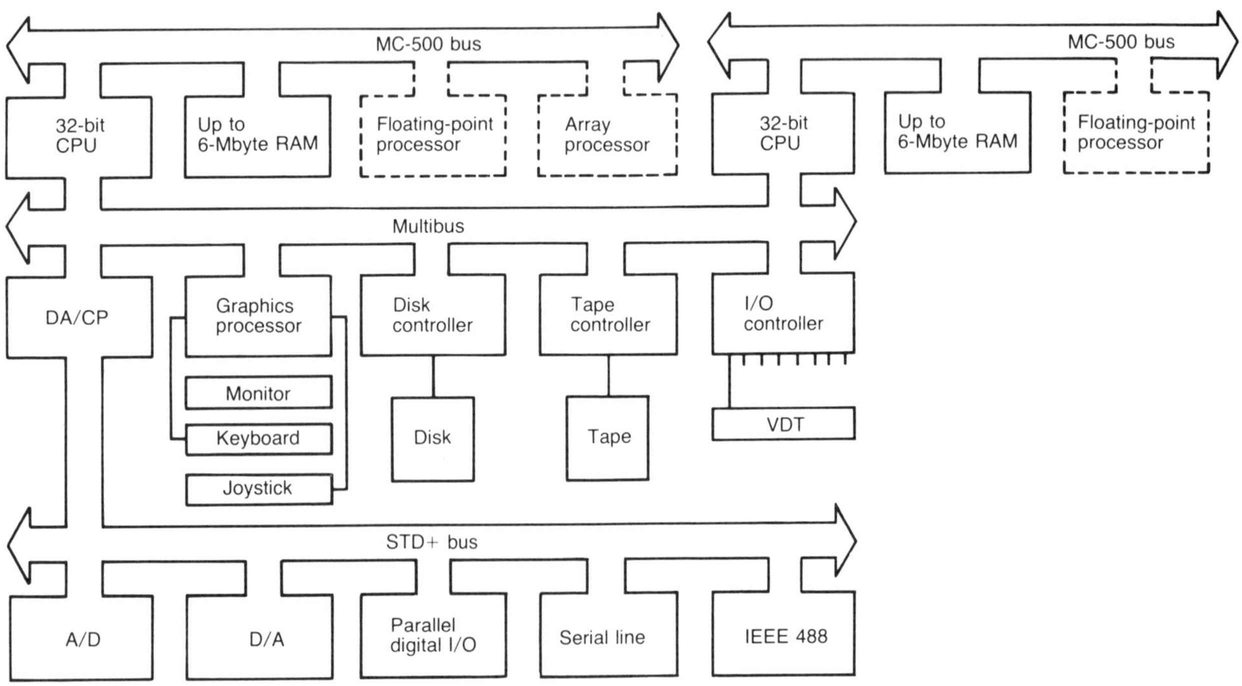

Figure 4. A multiengine, multibus, general-purpose, high-performance workstation

This triple-bus configuration permits master–slave processors to be installed on the proprietary high-speed bus. A multibus is used for I/O, and data acquisition is handled through an enhanced STD bus

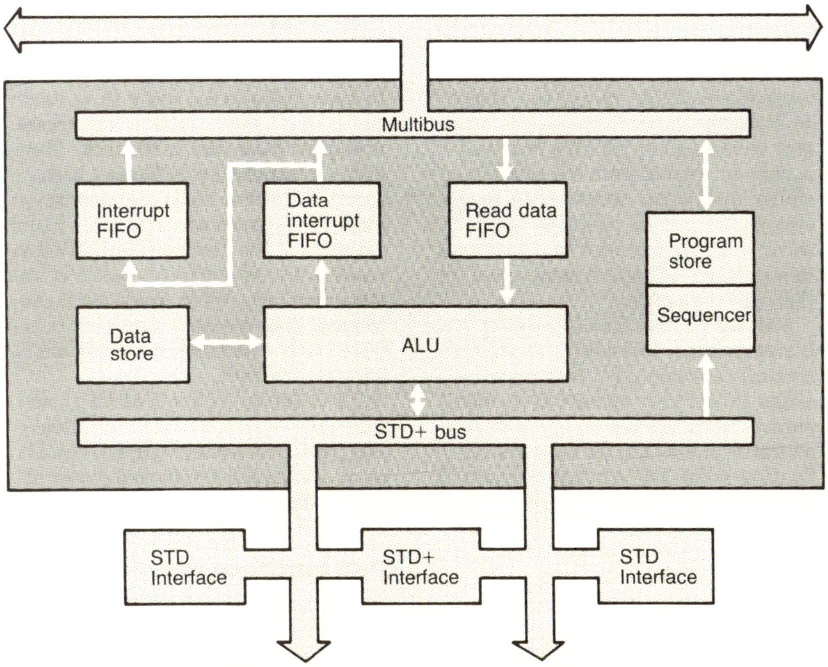

Figure 5. Data acquisition control processor

The DA/CP bit slice processor provides a connection between the Multibus and the STD+ bus. Passing time-critical data to the Multibus through FIFO (first-in, first-out) buffers isolates the processor from DMA latencies. The arithmetic and logic unit (ALU) can perform simple data reduction and make its own decisions in real time. Such design considerations are essential in high-speed environments. The STD bus accepts a variety of vendor-supplied and third-party interface cards

and data-dependent real-time decisions.

Software. Data acquisition and control must be easily implemented from user-written, high-level (Fortran and C) language application programs. A user library is provided to aid in start-up. Subroutine calls are available to assign buffer space, set device modes and clock parameters, manage transfers, identify buffer and task completion routines, and wait for event completion.

Handlers are provided for all Masscomp-supplied STD devices, together with the associated high-level language subroutines. For custom applications or devices, DA/CP handlers must be written. To make user programming as easy as possible, the DA/CP instruction set is designed to look like that of a minicomputer. An assembler with op-codes and syntax familiar to assembly language programmers is available.

Interfaces

A full line of analog and digital interfaces is available for the STD+ bus. The analog interfaces include a 40-kHz, 12-bit analog-to-digital converter (A/D) and a 1-MHz, 12-bit A/D. Both 64-channel and 256-channel multiplexers are available. The 1-MHz A/D system features programmable gain and input range and list-directed channel selection. Both 12- and 16-bit high-speed digital-to-analog converters are offered.

Digital interfaces include a dual-port RS-232C serial-line interface operating at up to 19,200 bits/s, a 16-bit dual-port parallel digital interface, and an IEEE-488 interface.

Accelerated processing

High performance requires a floating-point processor (FPP) and an array processor (AP). Both reside on the high-speed CPU bus.

Floating-point processor. The floating-point computation performance of a system can be increased by up to an order of magnitude through the addition of a FPP. The FPP features multiple sets of 64-bit registers, allowing up to 24 processes to simultaneously maintain intermediate results. All computation is carried out to the full accuracy specified by the IEEE floating-point standards.

Array processor. Vector computation performance of a system can be greatly increased through addition of an AP. Maximum arithmetic processing capability is achieved through parallel operation of the adder and multiplier. Vector operations are performed on data loaded into a local vector storage area that consists of 16 K 32-bit elements. For applications involving larger quantities of data it is possible to load and store data in parallel with vector instruction execution. Single-precision real and complex data types are supported. The software support includes a subroutine library accessible from C, Fortran, or Pascal. A 1024-point complex FFT can be performed in less than 7 ms.

Real-time Unix

The system supports an enhanced Unix operating system (*1*). Enhancements include:

Real-time scheduling. The most important resource that an operating system allocates is CPU time. The conventional Unix scheduler features a mechanism that gives each process a time slice of a specified duration (a quantum). Normal, time-shared processes are automatically promoted or demoted in priority according to a scheduling algorithm that optimizes system throughput while ensuring that all processes are scheduled. In addition there is a fixed-priority real-time scheduler. A real-time process is endowed with a priority. The system selects the highest priority real-time process ready to run and gives unlimited CPU time to that process. A real-time process continues until it terminates, pauses, starts I/O, or is interrupted by a real-time process of higher priority. If no real-time processes can be run, normal time-sharing resumes.

Memory locking. To ensure real-time response, system directives have been added to allow a real-time process to lock itself into physical memory.

Contiguous disk files. To avoid inefficient disk use because of fragmentation, a normal Unix file manager allocates files as discontiguous linked structures. Real-time applications, however, require that data be transferred at maximum speed to and from disk files. Unix must be modified to allow a user to specify contiguous disk storage for a data file.

Interprocess communication. A practical, real-time system requires some form of asynchronous interprocess communication. This usually takes the form of "signals," such as the semaphores used in train operation. The standard Unix signal facility, however, suffers from a major weakness. In certain circumstances signals can be lost. Therefore, an asynchronous system trap (AST) mechanism has been implemented in Unix. ASTs have a delivery priority. Regardless of the order of delivery, the system chooses the AST with the highest delivery priority to deliver to a process and automatically raises the priority of the process. ASTs also have the capability to transfer 32 bits of information between processes and can be used with Unix "pipes."

Reference

(1) Further details and definitions concerning real-time OS requirements may be found in Dessy, R. E. *Anal. Chem.* **1983,** *55*, 650 A and Dessy, R. E. *Anal. Chem.* **1983,** *55*, 883 A.

Originally published in
ANALYTICAL CHEMISTRY **1985** 57(6) 692A
and **1985** 57(7) 805A.

Disks for the Laboratory
Part I

In the past five years there have been tremendous changes in the practice of analytical chemistry as micro and personal computers have invaded the laboratory. The next five years will bring even more profound changes as magnetic- and optical-disk storage technology provides a proper support base for these computers by offering increased capacity (Figure 1) and new methods of transferring data. Inexpensive magnetic disks with capacities of 100 Mbyte will allow the use of more sophisticated operating systems, on-line storage of larger data files that can be manipulated by effective data base management programs, extraction of information from the organized data by powerful statistics and modeling–simulation packages normally found only on large computer systems, and presentation of the results by high resolution two- and three-dimensional graphics software. Optical disks with 1000-Mbyte capacities will enable us to use our instruments more effectively, allow us to have in-house copies of large scientific and bibliographic data bases, permit us to archive internal documents and data, and allow us to inexpensively share these data with other laboratories.

The purpose of this tutorial is to examine the vocabulary and chemistry of magnetic- and optical-disk storage technology and present some of the new tools this technology brings to the analytical chemist. This will allow the chemist to choose the proper disk systems for his or her environment and to evaluate vendors' claims. Next month's capsule reports will examine some of the operating, fiscal, and legal imperatives affected by these new technologies.

Basic disk architecture

As the name implies, disks are flat, circular objects that are mounted on a rotating spindle. As they turn, a write, read, and possibly an erase head are

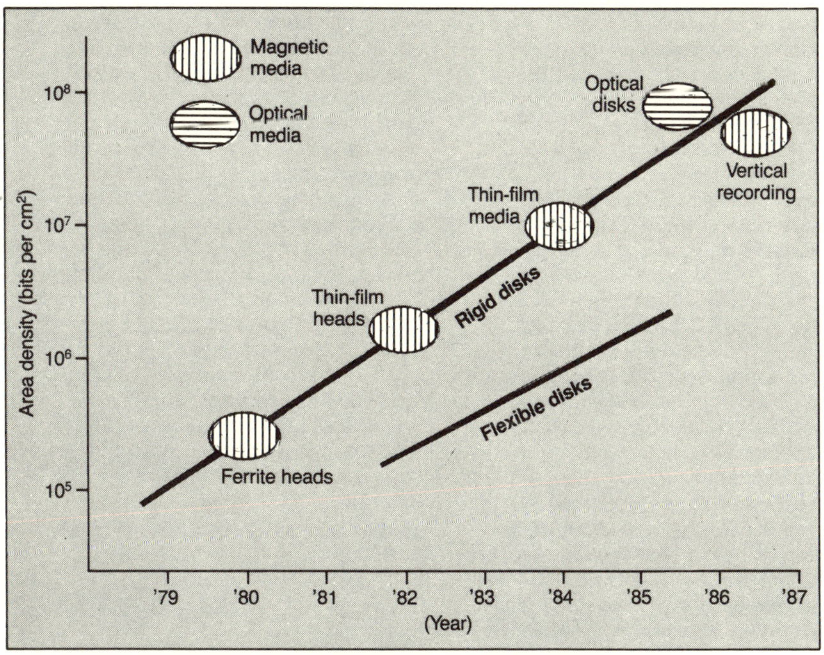

Figure 1. Progress in disk recording densities for the laboratory
Thin-film media, thin-film heads, and perpendicular recording can all improve area recording densities on magnetic media by threefold or more. Magnetic and optical media complement one another

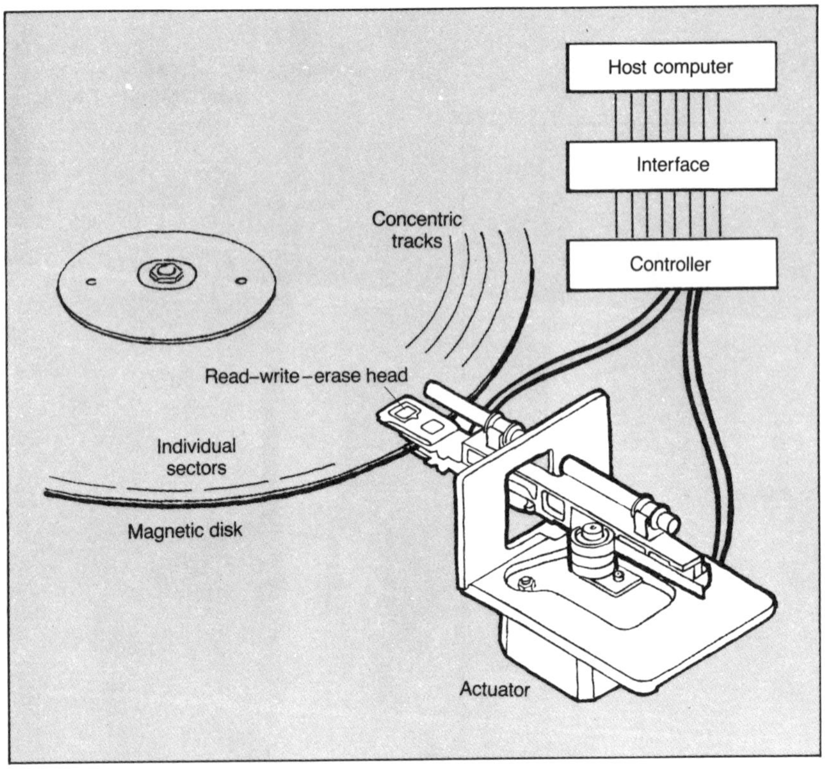

Figure 2. The working parts and vocabulary of a magnetic-disk system

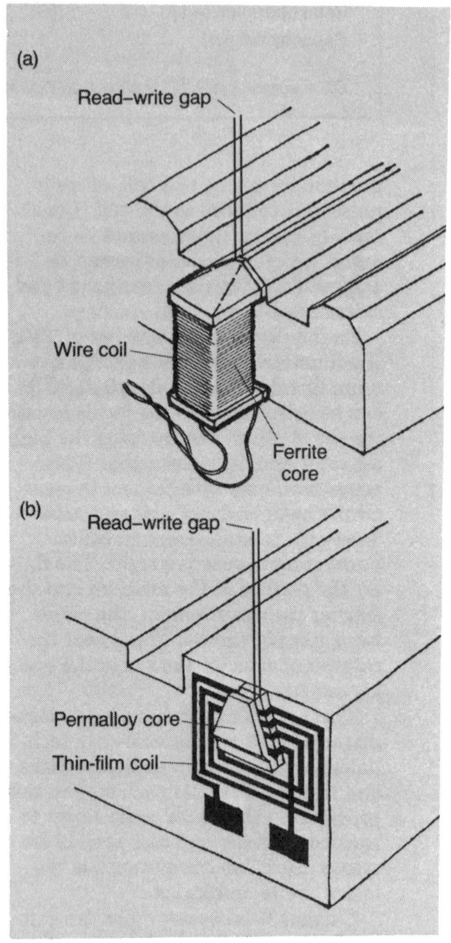

Figure 3. Conventional (a) vs. thin-film (b) head technology

The drawings are not to scale; head, gap, and flexure support are much smaller in (b)

positioned over the disk medium (Figure 2). This head assembly is free to move in and out along a radial line. It is driven by a variety of devices, including magnetic voice coils and stepping motors, that comprise an actuator. As the actuator moves the head to allowed positions, the data are written in tracks that are concentric (magnetic technology) or spiral (optical technology). The density of tracks, measured in tracks per inch (TPI), is one factor determining the data capacity of a disk. The other factor is the density of information in a track, measured in bits per inch (BPI) on optical disks and flux changes per inch (FCI) on magnetic media.

Each track is divided into sections that can store a fixed number of bytes of data (e.g., 256, 512, 1024); these are called sectors. Each sector is uniquely identified by an address written at its beginning that indicates the track and sector number. This is usually preceded by a preamble that is used by the disk's controller for synchronization purposes. Every sector closes with several bytes of data that are used for error detection and correction, then a postamble. Each sector is separated from the next by an intersector gap. The unformatted capacity of a disk is therefore much higher than the data capacity of a formatted disk. In some cases formatting can be done by the user; in others it must be done by the disk manufacturer.

The rate at which data can be read from or written to disk is determined by several factors. First, the proper track must be found. This is affected by the current position of the head and the track it must move to. This gives rise to a head latency time period that is determined by the speed of the head actuator. The speed of the revolving disk determines how long it will take for the beginning of the desired sector to revolve under the head. This is the head latency. Finally, the disk speed determines how long it will take for the last portion of the desired sector to pass by the head as it is being read. It is this figure that is usually quoted as the transfer rate. However, the overall transfer rate is also affected by how fast the disk controller can pass the information back to the host interface. It is this bandwidth that currently limits many small systems.

Magnetic storage technology

The present. A conventional magnetic recording head consists of a C-shaped high-permeability magnetic material (a ferrite) around which a few coils of wire have been wound (Figure 3a). When current flows through the wire the magnetic field emanating from the gap intercepts the magnetic medium on the disk and magnetizes it in the direction of the gap's field. The direction of current flow in the head can be reversed, and information can be stored by magnetizing different regions in the medium in opposite directions. The recorded information is read by another coil. As the magnetic fields from transition regions in the

Table I. Typical, nominal characteristics of current disks

	Flexible	Winchester	Thin film	CD	LD
TPI	100	500	1000	15,000	15,000
BPI				15,000	15,000
FCI	5,000	10,000	50,000		
rpm	300	2,400	2,400	CLV	1,200
Transfer rate (MBPS)	0.075	1.0	5.0	0.15	0.50
Rotational latency (ms)	200	25	25	200	50
Head latency (ms)	100	50	50	>1000	250
Read operation (ms)	1000	200	200	>3000	700
Capacity (Mbyte)	1	50	500	500	1000

CD = compact disk, LD = laser disk, CLV = constant linear velocity

medium intercept this coil, current pulses are induced in the coil (Lenz's law). Magnetic fields cannot be focused, and higher densities can be achieved only by narrower heads and closer head-to-medium spacings.

Radial density, the number of TPI, is a function of head design. Maximum linear recording density (FCI) can be achieved with the heads touching the medium and by using the highest rotational speed possible. This poses problems with respect to wear on the head and medium and problems with tolerance and stability. Some compromise is sought. The flatter the surface of the medium and the cleaner the environment, the closer head spacing can be. The slower the rotational velocity the closer the spacing can be.

These simple facts have led to three distinct types of magnetic-disk technology: rigid disks, Winchester disks, and flexible disks. At each step in this progression the heads come closer to the disk surface, and disk speeds decrease. Most lab computers use the latter two technologies.

Current Winchester disks use gamma ferric oxide particulate recording material (rust) and durable ceramic particles suspended in an epoxy binder (glue). This mixture is spin-coated onto an aluminum substrate, passed under a magnetic field to orient the acicular particles of iron oxide in a circular fashion, cured, and burnished. Area densities are close to 1 Mbit/cm^2.

Media with 1000 TPI can be found, but this represents the limit using traditional heads. Media with 10,000 FCI are obtainable, but this represents the limit of acicular media. Table I is a summary of the characteristics of various disks.

Most flexible disks are made by linear web coating. The iron oxide suspension is applied to continuous rolls of polyester base, and individual disks are stamped out. Because the disks normally rotate inside a lined paper sheath, rotational speeds are much lower (300 rpm), affecting both rotational latency and transfer rate (Table I).

As with magnetic recording tape, chromium- and cobalt-enhanced iron oxide particles have boosted the standard 300 oersteds coercivity to 500–1000. Coercivity, which is measured in oersteds, is a measure of how strong a magnetic field is required to orient the particles in the medium. A high coercivity increases the resolution of the medium, and the writing density is increased. Self-demagnetization is reduced. Coating thicknesses can be decreased, from 100 to 50 microinches.

The future. Particulate gamma ferric oxide normally has an acicular shape, with a long axis of 25 microinches. Magnetization is easiest along this axis. This sets a lower limit on the transition region size (a north–south or south–north transition) and makes any recording occur parallel to the long axis, which itself is parallel to the substrate aluminum or polyester. The inherently rough nature of the surface discourages small spacing between head and medium, lowering FCI values.

A number of alternative media have been developed to reduce these problems, including isotropic iron oxide particulate, plated films (electroplate or electroless), and vacuum-deposited films. All of these "smoother" media increase the FCI value by permitting a smaller separation between head and medium.

Small thin-film heads, made by processes similar to integrated circuit fabrication, allow the head to come closer to the medium. They increase both TPI and FCI values.

Finally, methods for forcing the magnetization perpendicular to the substrate have been developed. This allows a higher FCI value.

Thin-film media and heads

Thin-film media. There are two basic types of thin-film media—plated technology and sputtered technology. The most widely available at present involves plating techniques, either electroplating (electrical reduction) or electroless (chemical reduction).

Electroless media are typified by Alar (Ampex). This involves a 0.075-in. aluminum substrate coated on both sides with 600 microinches of nickel, with a 3-microinch top coat of Co–P. Often a thin sputter coat of graphite is applied to plated material to overcome corrosion and friction/stiction problems. Chemical batch processing is inexpensive, but when the "chemistry" goes awry entire production lots are lost.

Electroplated media can be produced with more precision, and the choice of materials is wider. On the other hand, the process is slower and less conducive to batch runs.

Sputtered media allow even more choice regarding the active media, but the capital investment is so high that widespread use of high-density media is necessary to make it economically successful. Typical disks are made by bombarding a target cathode with ar-

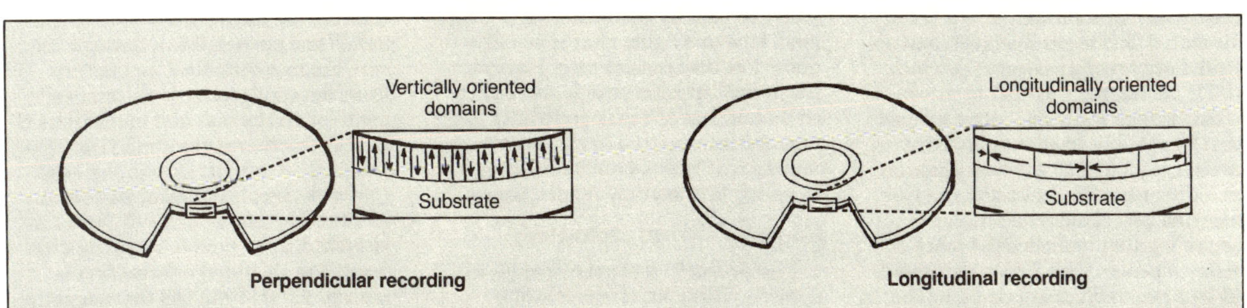

Figure 4. Conventional longitudinal recording vs. perpendicular recording

gon. The 0.075-in. aluminum substrate is first covered with a 20–100-microinch underlayer; then a 2–3-microinch recording layer of Co–Ni, Co–Cr, Co–W, or Co–Pt is deposited. Finally a protective overcoat is applied.

Isotropic iron oxide thin-film media can be made by sputtering a film of iron in an argon–oxygen atmosphere to produce alpha ferric oxide. This is reduced and then reoxidized to gamma ferric oxide.

Ellipsoidal gamma ferric oxide particulates, 2–8 microinches in diameter, have also been prepared (Isomax, Eastman Kodak). They have three easy axes of magnetization. Smaller heads, with closer spacing between head and medium, have been used in flexible-disk environments, increasing density and transfer rates considerably.

Thin-film heads. Whitney technology improves on Winchester technology through the use of small, low-mass heads with a thin-film read–write head and a smaller flexure support. Both allow the head to fly closer to the medium, leading to higher FCI values. The heads are made by depositing and etching procedures, creating a spiral, high-permeability Ni–Fe head (Figure 3b). In multiple platter systems the disks can be mounted closer together, and more tracks can be created at the inner and outer periphery.

Perpendicular recording. The gap on a conventional recording head is wide, compared to the thickness of the medium, so that the magnetic lines of force created by the head are parallel to the medium (Figure 4). Recording densities in such longitudinal systems are limited, in part, by a relatively wide transition zone between adjacent bits recorded in the medium. The lines of force created by the head are opposed by the lines of force associated with the previous bit. A practical limit is approximately 10,000 FCI.

Vertical recording offers relief from these wide zones. The write head is not a ring with an air gap, but a bar. The flux lines are perpendicular to the medium (Figure 4). The medium is usually a Co–Cr thin film with a hexagonal crystal structure and the c-axis (easy magnetization) normal to the substrate. This recording layer is deposited on top of a "keeper" layer of Ni–Fe (recall that a keeper is placed across the poles of a horseshoe magnet for storage). The keeper provides a low-reluctance flux path that helps keep the magnet magnetized. Another approach uses a recording layer of barium ferrite platelets in a binder, with no keeper layer. These particles have a preferred direction of magnetization normal to the platelet plane. In both approaches the recorded "bit"

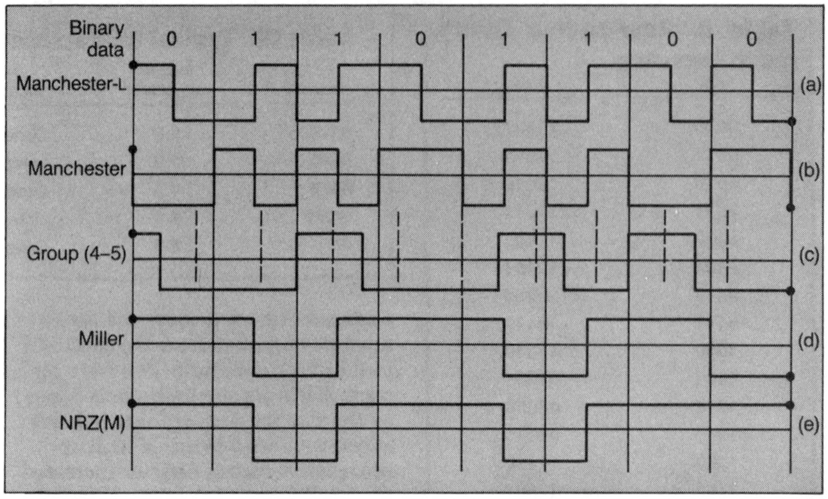

Figure 5. Some of the coding methods used to record binary bit strings onto disk media

(a) Manchester-L coding has a transition in the middle of every bit cell that clearly can be related to a 1 or 0. (b) Manchester coding, in which a transition occurs at the beginning of every data cell and a transition in the middle of the bit cell signifies a logic 1. (c) An example of 4–5 group coding. The eight bits are sliced into groups of four and encoded as five bits (compare Table II). The result is written in nonreturn to zero, mark (NRZ (M)). (d) Miller (sometimes called MFM, modified frequency modulation) uses NRZ (M) but with transitions in the middle of the bit cell, disregard of isolated zeros, and extra transitions at the beginning of a bit cell between two zeros. (e) NRZ (M) records ones in the data stream by transitions in the middle of the bit interval; zeros are ignored

is almost as small as the bar in the head. The transition zone is small because adjacent domains have opposite polarity and tend to pull each other together.

Current competition is between oxide particulates and electroless plated thin film. Over the next two years the sputtered and plated thin-film approaches will compete. Then vertical technologies will appear strongly, simultaneously with the rise of optical recording techniques.

Embedded servos. Disk capacity can be increased either by increasing the FCI value, as described, or increasing the TPI value. The former may be precluded by the bandwidth of the controller—the device that must extract data from the flux changes seen by the read head and then pass these on to the host computer. This will be discussed in a following section. Although an increase in TPI does not affect controller bandwidth, it dramatically increases the demands made on track-following technology.

Inexpensive actuators for low-density drives operate in an open-loop manner. The disk system is usually aligned with a test disk so that an outer, inner, and middle-distance track are read with an optimum read current. All other tracks are found by interpolation. As track density increases it is necessary to add embedded servo capabilities. Positioning information is provided on the disk and/or actuator assembly. Typically it involves bursts of ones written periodically around a track in intersector gaps. These are used in conjunction with closed-loop hardware to fine-tune the read–write assembly over the exact center of the appropriate track. This embedded servo information is often written onto the disk by the manufacturer using special equipment. If this is inadvertently overwritten by the user the disk becomes useless. Such servo information can reduce the disk capacity significantly.

Alternate approaches use a Mylar scalar on the actuator that allows an intelligent controller to calibrate track position periodically, i.e., every minute. The only disk-resident information is a fiduciary track written on a wide, easily found band at the outer periphery, well away from tracks that the user accesses.

Recording techniques

Magnetic recording suffers drastic losses in S/N ratios as the frequencies recorded drop toward dc levels. In digital recording long strings of ones or zeros can lead to no flux changes and to inability of the controller to extract correct timing information. Any temporal drift would make it impossible to recover the data correctly. Providing a clock pulse for every data bit leads to an excessive FCI requirement. There are a large number of coding techniques used today, such as NRZ, MFM, Miller, and Manchester. Although the caption of Figure 5 describes some of them, its purpose is to show that there are too many flux changes in the top recording technique and too few in the bottom. In the other cases some compromise is struck. The goal is a signal that will al-

Table II. Four-to-five (4–5) group encoding

Data	Code
0000	11001
0001	11011
0010	10010
0011	10011
0100	11101
0101	10101
0110	10110
0111	10111
1000	11010
1001	01001
1010	01010
1011	01011
1100	11110
1101	01101
1110	01110
1111	01111

Table III. Typical characteristics of disk protocol standards

Protocol	Transfer rate (MBPS)	Data separator	Data bus	Intelligence
ST-506	0.6	Controller	Serial	Low
SMD	1.2	Drive	Serial	Low
ESDI	1.5	Drive	Serial	Medium
SCSI	1.5	Drive	8-bit parallel	High
IPI	3.0	Drive	16-bit parallel	High.

low a data separator, located in the drive or the controller, to sample the data stream correctly. Incidentally, this is the reason that the FCI value, rather than BPI, is used in discussing magnetic media. The BPI rate is dependent on the coding scheme used.

A group encoding approach is often taken. The incoming data stream is sliced into groups and then transformed into a longer code word that optimizes the FCI rate and avoids long runs of zeros (Table II; Figure 5, Group 4–5). Such run-length-limited (RLL) code can be implemented by algebraic rules or look-up tables. Hardware for 4–5, 8–10, and 8–14 modulation (called EFM for the 8–14 case) is available. The original binary bit string is converted and written to the recording channel, where the new data are termed channel bits. The channel data word is often written in standard NRZ (Figure 5). EFM modulation will become important when we examine optical-disk technology. However, first we must look at the controller, the last element in the chain that can affect data transfer rates.

Standard disk interfaces

The day of host–interface–controller–drive interconnect problems is rapidly disappearing because of standard protocols.

The current interface standard for 5¼-in. Winchester disks is ST506 (Seagate), which limits recording density by supporting data transfer rates less than 0.6 Mbytes per second (MBPS). To implement more advanced RLL recording code and take advantage of higher FCI drive systems, a higher performance interface standard has been needed. The enhanced small device interface (ESDI) started as an upgrade to ST506, but was soon drafted to move the data separator function from the controller to the disk drive itself. This frees the controller from any limitations based on the way the data are encoded and allows vendors a choice of RLL approaches. A second decision increased the intelligence of the recording device, relieving the host or controller of such chores as calculating the number of tracks that must be traversed to move from one location to another.

Similar transparency is provided by two other proposed standards, the small computer systems interface (SCSI) and the intelligent peripheral interface (IPI). Unlike ESDI, however, the objective has not been to create transparent *bit-serial* access to a small handful of devices, but to create a *byte-serial* peripheral subsystem bus that can include a wide variety of devices and to place almost all of the intelligence at the device level (Table III).

SCSI envisions a network of peers, not unlike a local area network. The host is just another node. SCSI uses a standard protocol to move both data and instructions at 1.5 MBPS (5 MBPS with differential drivers). The bus can support up to eight nodes, each with up to eight devices attached. Many of the new magnetic- and optical-disk drives are being built to connect to the SCSI bus, and host–SCSI couplers are available for most small computer systems.

IPI establishes a hierarchical structure with up to 8 slave controllers attached, each with up to 16 devices. Intended for very large systems, it is an outgrowth of dissatisfaction with the storage module drive (SMD) protocol.

Optical-disk technology

Optical-disk (OD) technology offers tremendous storage capacity—10 Gbytes per disk—and area densities of 20 Mbits/cm^2. Early predictions suggested that the storage cost per bit would be much less than magnetic tape and that shelf life would be almost indefinite. This was contrasted to magnetic media, in which gradual degradation limits shelf life to one to three years. Although these predictions have not quite been realized it has been recognized that the nonerasability of some optical media could address problems in the regulated industries, where Good Laboratory Practices and Good Manufacturing Practices, with their implied audit trails, would make welcome a low-cost, high-capacity, write-once medium that would be legally acceptable for storage of laboratory data.

Optical storage technology is being driven by the compact disk (CD), used for music storage and reproduction and the educational aspects of video disks (VD). With VDs, one can interlace video images, video sequences, and stereo–audio with digitally stored information such as text. Mixed audiovisual encyclopedias could present the scientist with interactive text, highly complex graphics, engineering drawings, and photographs elucidating the working of an instrument or procedures for a complex experiment. Other applications involve large scientific data bases and electronic publishing of journals. College-level chemistry and physics courses have been produced (Videodisk Group), and Plato CAI material is forthcoming (IBM/CDC). Bibliographic sources and catalogs of books and theses are the target of a number of existing systems (Library Corp., Carrollton Press, Reference Technology). Archiving systems are available (File Net, 3M, Hitachi) that scan lab documents optically and then can store up to 40,000 pages per optical disk.

CDs are the basis for CD–ROMs (large read-only memories on the same media used for digital audio storage) that can bring text material such as manuals, reference data bases, encyclopedias, dictionaries (with spoken words), catalogs, and journals into the lab at very low cost.

Related technology has brought low-cost laser disk (LD) data storage into the lab, allowing secure archiving of sample information. The entire data archive from a year's work might be kept on one disk.

Inexpensive VD and CD players (Pioneer) and CD–ROM players (Denon, Hitachi, Philips, Panasonic, and Sony) are available for under $2000. These use media that have been produced by mastering techniques akin to record stamping. LD disk systems that allow you to write and store a Gbyte of your own chemi-

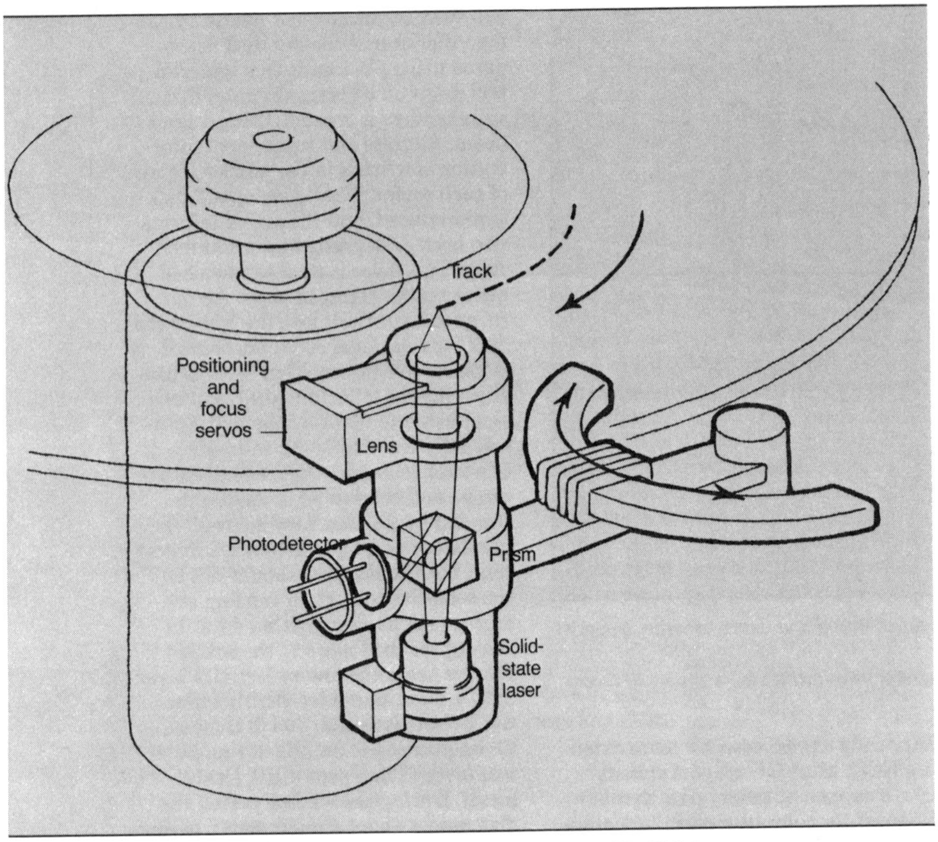

Figure 6. Laser read head used in compact disks and CD–ROMs

A similar head, mounted on a linear motor, is used with VDs and LDs. The LD laser has two powers, a write intensity (~20 mW) and a read intensity (~1 mW). Servo systems keep the laser beam focused on the spiral track and provide vertical and radial control. Depressions in the disk surface modulate the intensity of the light seen by the photodetector as data are read. VDs can be constant angular velocity (CAV) or constant linear velocity (CLV). CDs are CLV; LDs are typically CAV

cal data cost about $10,000 (Thompson-CSF and Optimem). A VD recorder–player system has just been announced (Panasonic/Hitachi, ~$30,000).

OD disk technology can be divided into three basic categories: read-only media (ROM), produced by mastering techniques; write once, read many times (WORM), with which the user can write data on the medium but not erase it; and erasable media (EM), which perform like magnetic disks and can provide read, write, and erase operations. Each of these areas will be explored.

Video disks. The 12-in. VD disks contain a spiral track of tiny pits. They are derived from a master disk that is a photoresist coated glass substrate exposed to a high-intensity writing laser beam that is modulated with information taken from a videotape. A premastering process adds error codes and data code signals to the videotape data and then places the results in a high-speed memory. The buffered data are written to the master via digital-to-analog converters (D/As) using a parallel processor architecture that can sustain a 1.8-MBPS data stream for 0.5 h. The data to be placed on the VD consist of the composite video signal represented as an FM radio frequency carrier. The audio is added, modulated on a lower frequency subcarrier. After limiting, this produces a pulse width modulated signal (analog recording). The master is developed, and in a series of metal-plating and pressing steps, produces a stamping disk that is used to make replicates.

The final VD is a highly reflective surface embossed with pitlike depressions describing a spiral track with a pitch of 1.6 μm. The complete video signal—chrominance, luminance, and synchronization data—along with two channels of audio, is stored on the track by varying the length and spacing of the pits. Each pit is 0.6 μm wide and some 1.2 μm long. The spiral contains 54,000 revolutions, covering a 90-mm space from inner to outer radius. The pits can cause a modulation in the intensity of a scanning laser beam head that is moved by a linear motor actuator (compare to Figure 6). Two servo systems keep this "locked" on the track. The first keeps the laser focused on the reflective surface. A second servo moves the laser radially to correct for eccentricity. The disk can be rotated at constant angular velocity (CAV) or constant linear velocity (CLV). At 1800 rpm, CAV disks can store 54,000 frames, and each track spiral stores one frame of video. The second servo mirror can be used to jump from track to track synchronous with the disk motion to give still-frame, forward–reverse operation. With data code signals written on the disk look-up features, special playback configurations and interactive computer-based player control become possible. VD player interfaces are available for many personal computers. The CLV disk rotates at a variable speed (1800 rpm inside, 600 rpm outside) and can store twice as much data, but it cannot provide still-frame capabilities. The track density is 15,240 TPI, the linear data density is 15,000 BPI, and the transfer rate is over 1 MBPS.

Compact disks. CDs are made using the same basic technology as VDs, but the data are digitally encoded on the 5¼-in. disk. The desired audio signal is sampled at 44 kHz per channel, allowing capture of all audio frequencies below 22 kHz. (Nyquist sampling theory indicates that a time variant signal must be digitized at a rate twice the rate of the highest frequency signal component.) A 16-bit analog-to-digital converter (A/D) is usually used. This binary data word is split into two 8-bit bytes, and each is encoded by 8–14 modulation (EFM) to run-length limit the final code recorded as channel bits (compare to RLL; see above). The two 14-bit channel code segments associated with the two 8-bit bytes from the A/D are interleaved on the disk, so that logically contiguous segments are not physically contiguous. These processes reduce the data capacity of a CD by 50%.

Extended parity information is developed by cross-addition processes. A form of cyclic redundancy check (CRC) is also calculated using what is called Reed–Solomon code. This combination of cross-interleaved Reed–Solomon code (CIRC) is used to provide an error correction and detection code (ECC) that can handle the potentially high error rates encountered on such media. Bit and byte burst errors may reach one in 10^6. CIRC improves the soft error rate to 10^{-13}. Detected errors that are not correctable are handled by interpolation. Undetected errors are about 10^{-16}. Lest this concern the scientist, Beethoven's Ninth Symphony on a Denon recording showed an error rate of 11.2/s, all corrected. Are the disks robust? Baking the CD at 60 °C for 3 h or immersing it in 80 °C water for 3 min did not affect the error rate.

CDs are important, not because they can bring high-quality music into

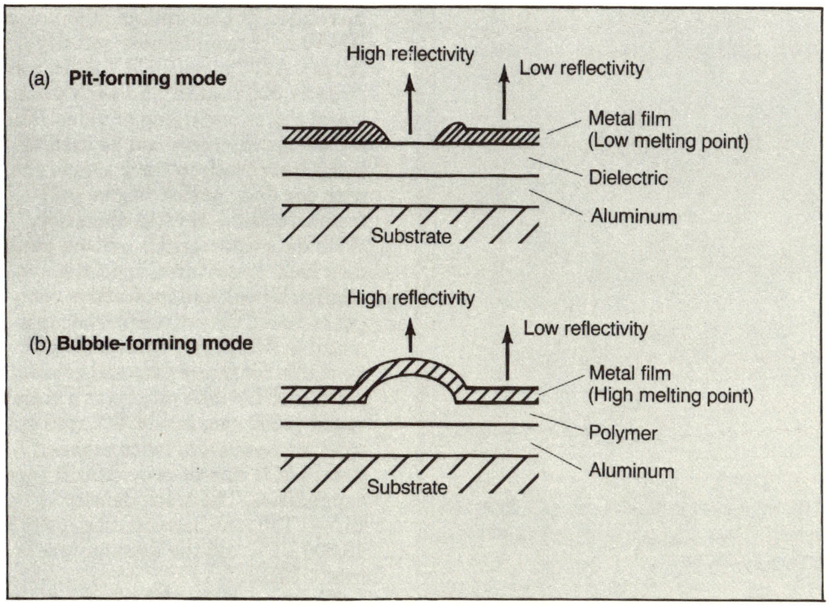

Figure 7. Two techniques are used in WORM disks that allow users to write, but not erase, their own data

One creates ovoidal pits (a) by ablation in a metal film; the other raises bubbles (b), or blisters, by decomposition of an organic polymer beneath the metal layer

the lab but because they are the vehicle for bringing data into the lab.

CD-ROMs are built on CD technology and use the same type of laser read head (Figure 6). They can store 550 Mbyte of useful data. The prerecorded CD disk is divided into 333,000 2-Kbyte blocks. Each block contains, in addition to the 2 Kbyte of data, 12 bytes of synchronizing code, 4 bytes of header, and 288 bytes of auxiliary code. The fourth byte in the header field defines the mode of the data, and if equal to one tells the player that it is computer code. The 288 bytes of auxiliary code can be used for more extensive ECC. Most CD players already have a capped auxiliary port that is designed for computer communication via an RS-232 port.

Laser disks. LD technology to store binary computer oriented data can be divided into two groups—nonerasable, write-once (WORM) disks and erasable media (EM).

WORM systems. WORM media are multilaminate with the laser-sensitive recording material sandwiched between protective layers. The header, servo, and clock information must be provided on unrecorded media by the manufacturer. A master disk is prepared using photosensitive material laid down on a glass substrate. First, a spiral groove is created, using a laser beam. Address and track servo information is written in the header region of each sector. The "developed" disk is reproduced, and the active information layer is deposited on a stamped copy. This layer is then sandwiched with another stamped copy. An optical encoder is built into the hub of the disk for rotational servo purposes.

The active information layer is usually a film of tellurium alloy that can be affected in spots by the writing laser, a 20-mW Al–Ga–As solid-state diode laser (compare Figure 6). Tellurium is used because of its favorable thermal properties. Low heats of vaporization and liquefaction are important, but the medium should not be too sensitive to heat, or reading and storage might alter written data. In one approach (Figure 7) the writing process heats the low-reflectivity layer until a 1-μm-diameter, highly reflective bubble is raised, due to thermal decomposition of an underlying polymer layer (Thompson-CSF, Optimem). During the writing period the disk moves about a micrometer, giving the spot an ellipsoidal shape.

Another technique (Figure 7) ablatively removes material, so that a pit is formed, exposing a reflective under layer (RCA). The writing laser pulse must last long enough to open a small hole by vaporization and keep the metal molten until the rim curls back. The rim velocity is about 1000 cm/s, so a 1-μm-diameter hole must open in 1 billionth of a second. Tellurium is

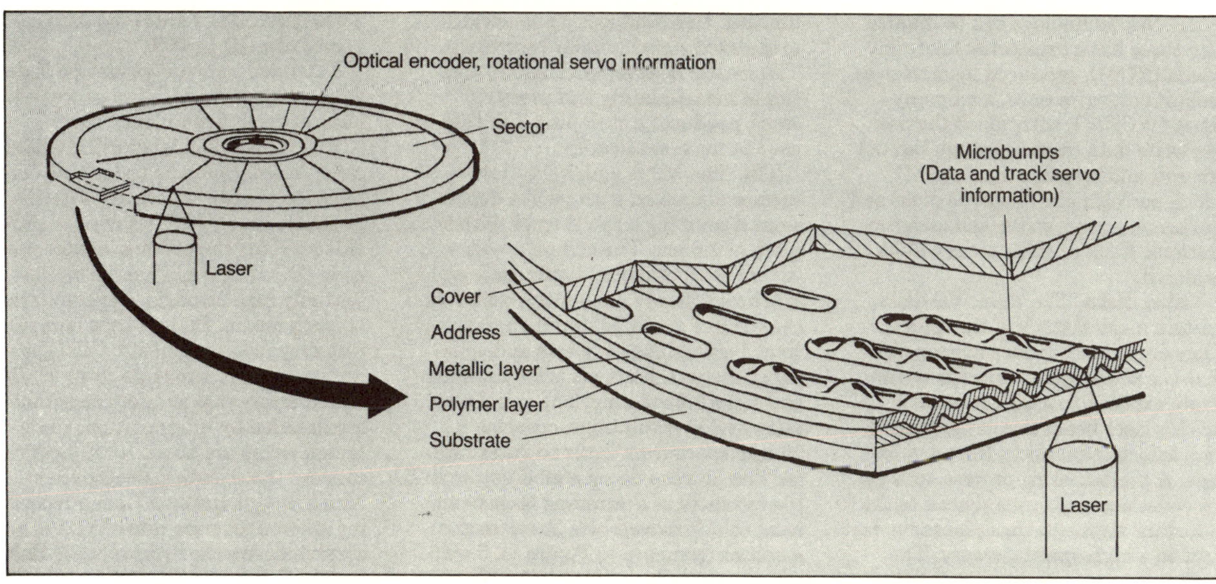

Figure 8. Laser disks for storage of computer data have address information written into the pre-enscribed spiral track by the disk manufacturer

Track servo information is also written onto the track, allowing the controller to keep the laser focused on the midpoint of the spiral trace. An optical encoder at the hub provides rotational servo information for the CAV disk motor

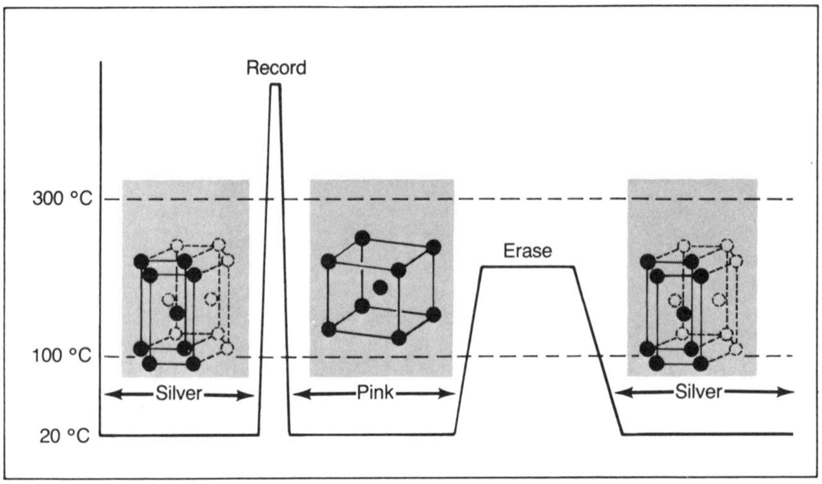

Figure 9. Phase change erasable media (EM) make use of lattice changes in metallic films, such as this Ag–Zn photochromic system. Other EM systems use crystalline to amorphous phase changes or bilayer phase changes

only weakly absorbing at the wavelengths obtainable from most solid-state diode lasers. Instead of using brute force laser power, one approach uses a tellurium coating ¼ wavelength thick on a highly reflective aluminum surface, so that destructive interference makes the film look "black" to the radiation source.

The rotation speed of commercial 12-in. disks (Thompson-CSF and Optimem) is 1122 rpm, with 14,500 TPI and 15,300 BPI, and a transfer rate of about 0.5 MBPS. Figure 8 shows the disk's construction. Using a laser head and actuator, such as shown in Figure 6, rotational latencies are 54 ms and head latencies are 250 ms. LDs are therefore much slower than Winchester technology. However, within a 40-track limit, the mirror servo shown can access a 1-Mbyte band in 5 ms. The 1-Gbyte disks are divided into 40,000 tracks, each with 25 1-Kbyte sectors. CIRC ECC is used with an overhead of 80 bytes/sector. Written data are organized in 100-bit groups, which include 15 bits of servo information for every group (a 25% total servo–ECC overhead). The package includes an SCSI interface.

A 5¼-in. drive targeted to PC users has been announced with a 300-Mbyte capacity at a cost of less than $1000 if purchased in large quantities (Information Storage).

Erasable media. Current efforts in EM are directed toward magneto-optic (Curie and compensation point materials), amorphous to crystalline phase change, and photochromic processes.

Laser-induced thermal crystalline-to-amorphous phase changes in Ge and As doped tellurium oxide films or bilayer phase transitions in Sb_2Se_3/Bi_2Te_3 (Matsushita, Sony, HP, Energy Conversion) lead to changes in the reflective character of the film. These changes can be detected by a low-powered scanning laser. Erasing is performed by heating with a defocused high-power laser and by slow cooling.

A sputtered Ag–Zn or Ag–Al–Cu film can also be used for data storage using a photochromic effect (Figure 9, Hitachi). The Ag–Zn medium has a hexagonal structure when originally laid down and is silvery white. Laser heating converts it to a cubic structure with a pink color. Reading and erasing are accomplished as described above.

For Curie point materials, an active material is selected whose Curie temperature is 100 to 200 °C above room temperature. As the Curie temperature is reached in a small area by heating with a writing laser beam, ferromagnetic order is lost, coercivity decreases, and the material can be magnetized by a weak external magnetic field before it rapidly cools (Figure 10). A low-powered laser is used to read the information back, using the Faraday or Kerr magneto-optic effects (Figure 10). When a polarized light beam passes through, or is reflected from, a magneto-optic material its plane of polarization is rotated. The direction of rotation depends on the direction of magnetization in the material. By using a polarizer–analyzer one can transfer magnetization direction to polarization direction and then to an optical intensity difference. Ga–Tb–Fe, Co–Tb–Fe, and Mn–Cu–Bi thin films are the most popular. Recorders based on this technology have already been introduced (Sanyo, Hitachi, and Sony).

In compensation point materials the rare-earth atoms in the thin-film layer are antiferromagnetically coupled with a transition metal. The net magnetization (Figure 11) is the vector

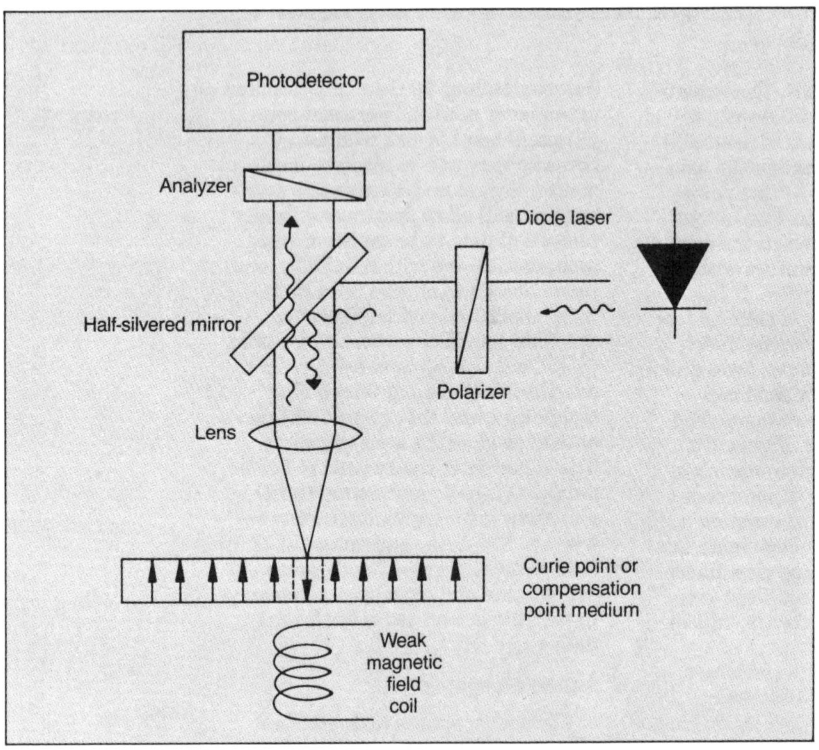

Figure 10. Curie-temperature-based EM disks rely on a writing laser to raise the temperature of the medium to the Curie point in a small area

The coercivity drops, and the area can be magnetically oriented in a preferred direction by a weak electromagnet. The data are read back by a Kerr or Faraday effect subsystem. These use a polarized light beam from a diode laser. As this beam is reflected from the magnetic domains in the media, the plane of polarization is altered. This can be detected by a polarizer–analyzer

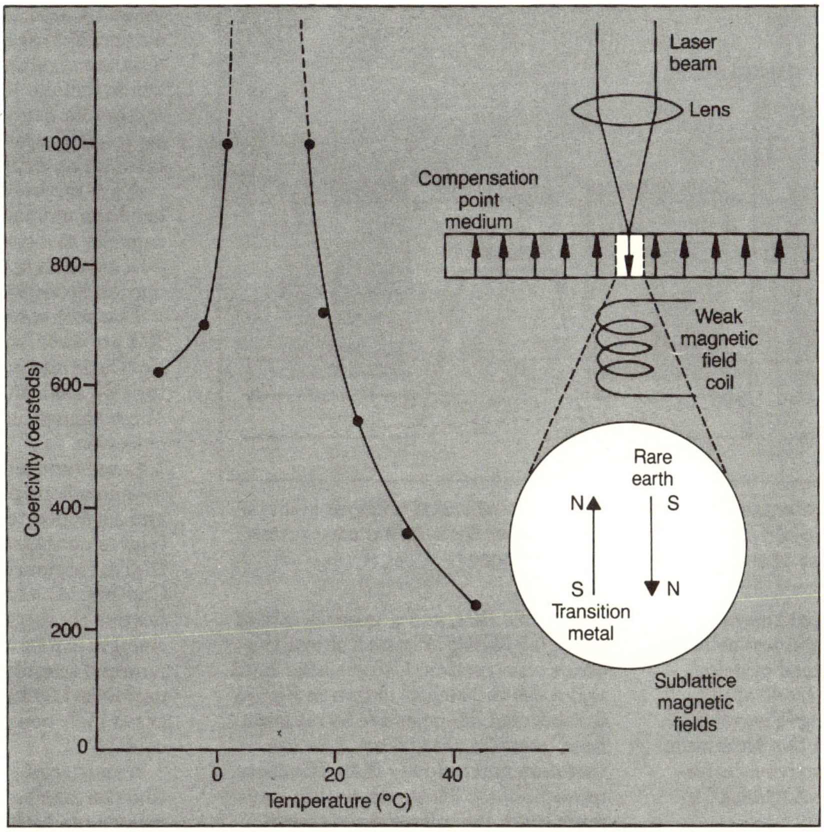

Figure 11. In compensation point EM disks the vector sum of the magnetism exhibited by antiferromagnetically coupled rare-earth and transition metal sublattices is temperature dependent

The coercivity of a spot drops dramatically when its temperature is raised, and the small area can be magnetically reoriented by a weak electromagnet. The data are read back by the subsystems shown in Figure 10

sum of the two materials. The sublattices involved behave differently as the temperature is changed, so that a temperature exists at which the net magnetization is zero. At this point the coercivity is infinite. For LD use an alloy is chosen for which the compensation point temperature is slightly below room temperature. If the temperature in an area is raised a few hundred degrees the coercive force drops to a few hundred oersteds, and a weak external magnetic field can "switch" the film. The data are read out as described above (Figure 10).

Degradation of magneto-optic media occurs when water vapor comes into contact with the information-bearing layer. Phase change materials are subject to fatigue and slow phase change conversions at ambient temperatures. Ablative disks are subject to corrosion.

Many laboratories are concerned about the 5–10-year (?) lifetimes quoted for WORM optical disks. However, they do recognize that LD technology will develop a solution to the record-keeping problem, and many laboratories are therefore rapidly consolidating the data base functions that have often been scattered in cottage-industry fashion by the uncoordinated activities of mini and personal computers. When LD disk technology firms up they will be ready to implement a storage and distribution strategy that will allow fiscally and legally important data to be captured with assurance, stored with reliability, and disseminated to all who need it.

In laboratory environments the standard interface protocols discussed (SCSI and IPI) allow current users to mix flexible disks and Winchester technology, and they permit add-on optical disks as the need warrants. The Winchester can be used to buffer data that is to be written to the LD and cache it for highly interactive retrievals. Magnetic- and optical-disk storage technologies thus complement one another and open up new horizons in the storage and use of analytical data.

Acknowledgment

The author wishes to thank Ken Hinson, a member of the Virginia Tech Chemistry Department's Laboratory Automation and Instrument Design Group, and David Hooley, Sohio, for discussions that helped mature this tutorial.

Part 2

Last month's tutorial presented the chemistry, physics, and engineering technology associated with magnetic and optical disks. These rapidly changing devices are permitting us to store and access more programs and data in the electronic laboratory. With that capability come new problems in administration and rapidly changing fiscal, legal, and regulatory imperatives. The following capsules are intended to raise questions of which users and vendors should be aware. Some of these issues are fraught with emotion and legal uncertainties, but they cannot be ignored for they will not go away. The subjects of archiving, security, validation and certification, and protection will be explored.

The Standard Oil Company
Warrensville Laboratory
Cleveland, Ohio 44128

Contributor: David Hooley

Archiving

Nearly all major analytical instruments incorporate substantial computer systems for control, data acquisition, and subsequent data workup. In the past three years we have seen the size of these computer systems in our laboratory grow from less than 64 Kbytes of memory and 10 Mbytes of mass storage to systems with 10 times that capacity. Raw data acquired per day rose from less than 100 Kbytes to several Mbytes. Future instrument acquisitions will increase the data generated by an order of magnitude in 1985.

Although all of the instruments operate successfully as independent, stand-alone systems, our analysts are universally experiencing the following problems:

- Data acquisition times are often short in relation to data analysis. Commonly a $500,000 instrument is only used for 5 min every hour because its $20,000 computer system must be used to analyze the data. A great deal of human interaction with the computer is necessary for efficient data workup, making heavy demands

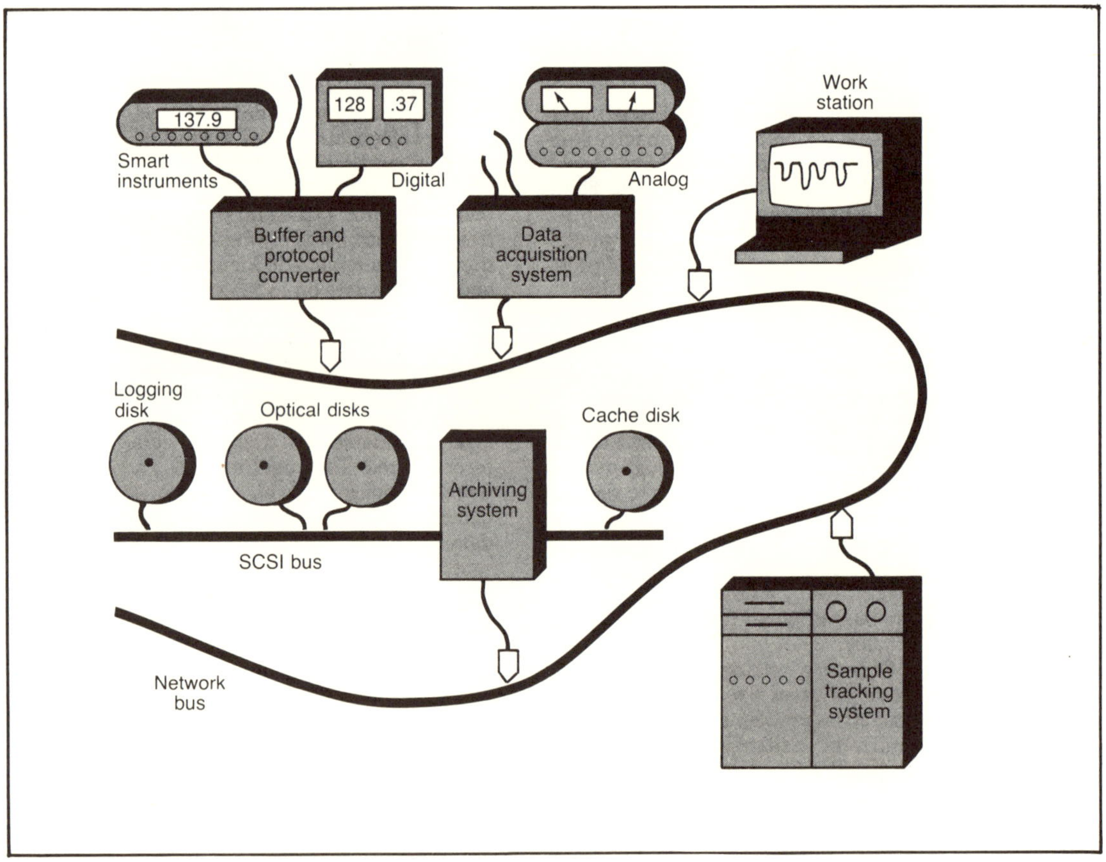

Figure 1. An optical-disk-based archiving system
The network bus interconnects data acquisition, manipulation, and sample-tracking facilities. The archiving system consists of a disk server that provides a logging facility for temporary recording of update material, the archiving optical disks, and a high-speed cache disk for interactive retrieval. All of these can be threaded on a small computer standard interface (SCSI) bus. This standard protocol allows a mixture of various vendors' products and relieves the attached computer of transfer overhead

on the instrument's computer during this period. Real-time archiving and "batch" postprocessing could alleviate some of these problems.

- Equipping each instrument with its own archiving device would cost as much as implementing network links but would add the costs of mechanical maintenance and operation.

- Data backup and archiving to existing central computer facilities take an increasing amount of the skilled analyst's time and instrument computer time. For this reason it is often avoided. Archived data are often not readily accessible because the only person who can quickly find these data in the present, inadequately indexed archive is busy with more pressing tasks. Each instrument has its own archiving format and procedures, making the collection of all the data in the entire laboratory for a particular sample very difficult.

- Raw data archiving is a universal need of the laboratory. This has become increasingly obvious as the environmental- and health-related sample load of the laboratory increases. Our Molecular Spectroscopy Group feels that progress beyond mere archiving is essential.

The technology to accomplish data storage on this scale, and larger, has existed for many years, albeit at a considerable price. Valuable manpower must be expended in creating and maintaining the large-scale computer systems for which large-capacity storage systems traditionally have been targeted.

The development of 1-Gbyte optical-disk technology provides a good, low-cost answer to this series of problems. The write-once nature of this medium is ideal. One of the greatest concerns of both the analyst and the legal community is the ease of changing or destroying data in a computer system. Once data are acquired and stored on such media, the data do not change. With such technology available, there is no reason not to record and keep all raw data for future reference, no matter how rarely that occurs. A computer system to manage several 1-Gbyte optical-disk drives can be configured for less than $50,000.

Of course, getting the data to the archiving system is essential and may pose some problems. Invariably, a multivendor situation, with data communications protocols ranging from none to completely unknown, is encountered in the laboratory. A great deal of continuing effort in the area of networking will be required to incorporate current instrumentation and integrate future acquisitions into the system.

The archiving system under construction at Sohio is dedicated to archiving alone. Because of this, it is being accomplished in a reasonable time at low cost. It is designed to accept and replay data with very little consideration of the content or format of that data, except for identification of the samples to allow efficient indexing. Even this will be limited in scale; other systems on the network will be responsible for more extensive indexing.

Figure 1 shows the components of the system. The archiving computer system provides the interface to the optical disk and is the only centralized feature of the system. Processing power and instrument control are decentralized for greater reliability and easier implementation. Distributed data bases are not easily manageable; therefore copies of system software and reference libraries are kept in a central location where updating, file locking, and access security can be more easily and efficiently implemented and controlled.

The network provides data transmission and contention services for instruments attempting to archive data. A separate magnetic-disk system is used to log temporarily all data exactly as they are received. This allows reconstruction of the data base, without information loss, should a system crash or corruption of the index occur. Instrument interfaces to the network vary, depending on the data rate and protocol characteristics of the instrument. In most cases, an intelligent buffer, typically a microcomputer system, is required to pick up the data from the instrument in whatever form they are generated, add identification, and pass the data on to the archiving system through the network.

The high-performance read–write magnetic disk maintains an index to the data written on the optical disk. Pointers in the binary tree index used must be updated with every raw data record stored. This requires read–write technology. The high-performance disk can also cache optical-disk data. This is essential for several reasons. Inexpensive optical disks are medium-performance systems; therefore, data that are used repeatedly, particularly indexes, must be kept on devices with shorter latency periods and higher transfer rates to maintain system performance as the amount of data grows. It is also convenient to periodically write data to the optical disk in large batches, with subsequent read comparisons to check that the data were recorded correctly. The buffering, or cache, operation ensures that such operations need not occur at peak demand times, as this could degrade system performance.

Part of the system described is in operation on a Q-bus LSI-11 with a 70-Mbyte Winchester disk. The optical disk (Optimem) is interfaced via a small computer system interface (SCSI) bus host adapter.

Burroughs Wellcome Company
Wellcome Development Laboratories
3030 Cornwallis Rd.
Research Triangle Park, N.C. 27709

Contributor: Steven A. Benezra

Validation and certification

Laboratory microcomputers at Burroughs Wellcome are supported by a ring local area network. The coordinator is a DEC PDP 11/70. When an analyst wishes a file to be archived it is placed on the logical section of a magnetic disk allocated to the analyst or the analyst's laboratory. A shadow disk is being installed to ensure that no data are lost by disk crashes. Archived data are off-loaded weekly onto magnetic tape, and this is stored off site. This provides protection for corporate data. However, of great concern is the validation and certification of those data.

From Jan. 15 to Jan. 18, 1984, in Crystal City, Va., the Pharmaceutical Manufacturers Association (PMA) held a conference titled "Concepts and Principles for the Validation of Computer Systems in the Pharmaceutical Industry." More than 600 individuals from industry and government attended the conference. The PMA realized that validation of computer systems was a "hot and sensitive" issue, but was surprised by the large attendance. All aspects of computer-controlled applications from process control to laboratory testing were covered at this meeting. This capsule report is limited to the impact of that meeting on laboratory applications of computers and the steps being taken at Burroughs Wellcome to address the issues involved.

The term *validation*, as agreed to by the Food and Drug Administration (FDA) and the pharmaceutical industry, is defined as "establishing documented evidence that a system does what it purports to do." This deceptively simple definition, when applied to computer systems, has caused much debate among computer experts within the pharmaceutical industry and among knowledgeable individuals in the FDA and those in the industry it regulates. Fortunately the relationship between the FDA and the pharmaceutical industry on this topic is not an adversarial one but a cooperative effort attempting to establish realistic guidelines for the pharmaceutical industry.

The problem of validation of computer systems can be broken down into two parts—hardware and software. The validation of software, the more difficult of the two parts, need be done only once. If a software package does what it purports to do for an application, it will do so time after time as long as the code or the operating system is not changed and the hardware does not fail or is not changed. Hardware, on the other hand, must be periodically calibrated, maintained, and validated because components of the hardware system will deteriorate and fail.

A significant amount of time has been spent at Burroughs Wellcome validating a software package called CHROM. The software was written in-house and is used extensively in the research, development, and medical laboratories for chromatography data acquisition and reduction. The approach to validation of CHROM was simple. A single chromatogram of four well-separated components was generated by a high-performance liquid chromatograph; analog data were captured on strip chart recorders, and the corresponding digital data were recorded on disk. The use of magnetic media allows the chromatograms to be replicated without reinjecting the mixture and introducing errors caused by the components of the chromatographic system. The data can be archived, and the "standard sample mix" is always available for revalidation should a change in CHROM be made.

CHROM-generated areas were compared to traditional manual methods of determining peak areas, e.g., cut and weigh, peak height × width at half height, and triangulation. Thirteen individuals were given the raw data in the form of replicated strip chart recording traces and a floppy disk containing computer-readable data. Each analyst reduced the raw data to peak information manually and by using CHROM. The collected results were submitted to the Statistics Department for analysis. There was no statistical difference between the areas determined by the traditional methods and those determined by computer. There was virtually no variance in the computer-determined areas, but there was variance between analysts using the traditional manual techniques. The variance that did occur in the computer-generated areas resulted because individuals were allowed to set parameters that determined the beginning and end of a peak.

The validation of CHROM using this procedure provided confidence that the software performed well within the overall error of the chromatographic analysis. More work is in progress using computer-generated Gaussian peaks with tailing and incomplete separation. Even though the computer can obtain reproducible results on less-than-ideal chromatograms, we believe that no sophisticated software is a replacement for good chromatographic practice. One should always strive for well-separated symmetrical peaks; if these are not obtained, the errors caused by data reduction of the nonideal case should be recognized. The computer is an aid to the analyst, not a replacement.

The version of CHROM that will be used in the future for the analysis of samples that come under FDA Good Laboratory Practices (GLP) regulations will be a compiled version that cannot be altered by the analyst. GLP regulations also require the archiving of raw data. The term *raw data* has caused debate between the FDA and the pharmaceutical industry. Some individuals consider raw data as strip chart recordings; others consider raw data to be data stored on magnetic media. The growing consensus is that raw data can be the information stored on magnetic or optical media provided that such data have been shown to be an accurate representation of the data generated by the analyst or instrument. This is the area where hardware validation enters the picture. Standard operating procedures are now being written for the routine calibration of A/D and D/A converters, and other components of the computers, which acquire data and control instruments.

Although we feel confident that we are making a reasonable effort in the validation of computer systems and preserving computer-generated data, we cannot say this about commercial laboratory software systems. When software or instrument vendors are asked how the software for their integrator or computer system has been validated the response is often a blank stare or a long silence. It would be preferable that the vendor do an acceptable job of validating the software and make the validation available to regulatory agencies. In lieu of that, the responsibility devolves to the user. This is not a simple task. For those vendors willing to share source code and algorithms, it is often impossible to understand the logic that went into the creation of the finished product. Some vendors will not provide code to the purchaser of the system because it is considered proprietary.

There is a potential solution to this problem. The FDA requires detailed synthetic procedures for the manufacture of a drug in all states of its development. If an intermediate is purchased from an outside source for use in the synthesis and the vendor does not want to disclose the synthetic procedure to the purchaser, a Drug Master File (DMF) exists to which only the FDA has full access. Vendors can deposit synthetic procedures in this file, and users of the synthetic intermediate are given only DMF reference numbers to complete their records for the regulatory agency. A similar process could be set up for third-party software. The vendor of the software could establish a Software Master File containing source code and validation procedures that could be accessed only by the FDA. The software would be validated by the vendor, and details of the software need not be divulged to the user. This is merely the extension of the procedures now demanded by many companies that require vendors to place source software in escrow, to avoid difficulties if the supplier ceases business operations in the area.

The concept of validation of computer systems is complex and will not be fully resolved in the near future. As the paperless lab evolves there will be many questions that will need to be answered such as, Are the data tamper proof? At what stage can analysts review their data to ensure that they are accurate? If data are considered "original" on a floppy disk, are they still original data when transferred to a large data base and, if so, how does one validate the transfer process?

The only way to address these questions is through an open dialogue among the pharmaceutical industry, computer and instrument vendors, and the FDA. Meetings such as the PMA Crystal City Conference are the start of this dialogue, and it must continue.

Merck Sharp and Dohme Research Laboratories
Division of Computer Resources
West Point, Pa. 19486

Contributor: Dennis M. Gross

Security

The pharmaceutical industry, like many other industries that deal with large volumes of data, quickly embraced computers as tools for storing and retrieving information as well as creating new informational relationships. The early systems were batch oriented and depended on keypunched cards for their input. The person who submitted the batch job was also the individual who retrieved the output. This limited access to the data and served as a form of security.

However, as input devices and mass storage technology matured, the operating systems (OSs) and our attitudes toward data security did not keep pace. Laboratory managers and vendors alike often have not coped adequately with the technology, psychology, and politics of data protection. Therein lies the problem of security and its relationship to the issue of rapidly evolving mass storage techniques.

Data security involves protection of data from unauthorized destruction, disclosure, and modification. These areas of data exposure can be affected accidentally or intentionally.

Magnetic tape is our oldest convenient mass storage medium. It is primarily used today for archival storage of off-line information. Retrieval of records from tape requires a message to the system operator or tape librarian—a message that can and should be authenticated by the system manager. However, once that tape is made available to an authorized user, risk exposure exists for destruction or modification of the data unless some control mechanisms are put into place. These mechanisms are not easy to implement for they entail a corporation performing a total risk analysis of its computer resources and assessing the sensitivity and critical nature of the stored data. And, once these mechanisms are in place, they can provide only the first level of protection—physical access control. They cannot stop determined users from reading the information and rearchiving the data in new accounts where they can read, write, delete, or reroute the records. Additional measures obviously need to be instituted at the system level to prevent this.

Most users do not like to archive data on tape because of the time it takes to reload—hence the proliferation of large mass storage systems using rigid disks. The advantage of having data available on-line is the rapidity with which the data can be accessed. A DEC RP07 disk has 516 Mbytes of formatted storage and an average access time of 32 ms with a peak transfer rate of 1.3 Mbytes/s. These media present their own problems. On the one hand, we install them because of the speed with which data can be accessed, but we now must tell end users that multilayered access controls must be installed. To the users this apparently self-defeating policy is often further exacerbated by the cumbersome methodology used to provide the protection.

Access controls of late have centered on add-on software. Many of our present day OSs for mini- and midicomputers are evolved from, or in some instances are patched versions of, batch-oriented systems. In an effort to provide security many new firms have come into existence selling products, both software and hardware, that the suppliers of the computers and their OSs should have been providing years ago. Products such as Lock-11 (Ontrack), Top Secret (CGA), and RACF (IBM) are system-level products that work in conjunction with the computer's OS to provide access control that can be tailored to the end user's needs. Many data base packages such as ADABAS (Software AG) even provide their own security module. Some of these solutions are not totally integrated. In many instances, their continuing functionality beyond password-level protection is totally dependent on the OS, and the software products they protect, not changing appreciably in the next release. However, these add-ons can determine which individuals have permission to access a data base and who has read–write or simply read authorization. They can even delimit the time of day and the particular terminals authorized for the tasks the user wishes to perform.

Hardware security systems often require insertion of a plastic "key" card into a slot on the terminal. Some of these cards are "smart" and contain their own microprocessor chips (e.g., coder cards, Colby and Codercard) that work in conjunction with the host to clear access for the prospective user. One card (ACE, Security Dynamics) generates a time variant number computed by a program specific to each smart card; the user must key in the number as well as his card number and a password. A new challenge–response system uses a small optical reader wand that can interpret challenge characters displayed on the user's terminal and display, for manual entry by the user, a response number (LazerLock, United Software Security). The challenge and response alphanumerics change for each access.

When modem communications are involved, several systems insist on "calling back" the user at a predetermined telephone number (Multisentry, TACT; SAM, LeeMAH; Defender, Digital Pathways). Voice print identification is still a year from maturation (Telesignal Processing). Going one step further, some systems implement encryption–decryption schemes; these require complementary scrambling and unscrambling code at both ends. Of course, the security of any system is inversely proportional to the skill and determination of the user who wishes to penetrate it.

Unfortunately, it is far easier to secure an entire data base than to protect only a part of it. Most software security packages are capable of working down to the file level, some reach to the record level, and a few protect at the field level. This granularity often leads to overprotection, or underprotection, of many pieces of information.

Rapid improvement in these tools is desperately needed as optical disks become more prevalent. These disks must mature with respect to shelf-life and protocol standards. But as they are accepted, they will present an even greater risk exposure than that offered by any existing magnetic storage. Because of their enormous storage capacity, data that might have been archived and placed under strict access controls will stay on-line for even longer periods. Whereas magnetically recorded data have a shelf life of 3 years, optically encoded data can be maintained reliably for 5–10 years—a very long risk exposure period.

The write-once, read-many-times attribute of today's computer-oriented optical disks fortunately can reduce this risk exposure because records cannot be erased or modified and must be read from within the established security system.

However, it is obvious that the carefully developed strategies for protecting data have, of late, run into the personal computer revolution. The move away from centralized computing to a rapidly expanding population of either networked or stand-alone computing resources has strained the already tenuous fabric of corporate data security policies. Many of the advances in storage technology that are appearing for minicomputers and mainframes are entering the marketplace just as rapidly for personal computers. Witness the introduction of the new IBM PC/AT with a 20-Mbyte Winchester and a 1.2-Mbyte floppy disk or the host of streaming tape systems for personal computers with storage capacities of 20–50 Mbytes in a removable format. The table-top personal computer can easily compete with many smaller minicomputers with respect to on-line storage capacity.

Therein lies a new and most troubling problem. Even if an end user is allowed only read access to critical data bases on the central corporate computer, the use of a personal computer as a terminal gives the end user the ability to down-load, and store on his or her disk, everything that flashes by on the PC's monitor. The best-planned software-based access control policies on the central computing resource will have trouble preventing this. Although the need to down-load critical data may not be malicious, e.g., local graphical analysis or incorporation into a document being prepared on the PC, the risk exposure is obvious. Physical access to most PCs with Winchester disks involves merely depressing a switch. If flexible disks are used, they are easily transported to other PCs. The end result can be extremely large volumes of critical and highly sensitive data stored on a PC disk volume with no access control at all. The appearance of read–write optical disks for PCs will only compound the problem further.

Data security packages for personal computers are beginning to appear. Typical security systems available for PCs include those using passwords (Dataguard, Village Information; Systemate, Systemate Inc.), physical keys (Secureware, Remote Systems), hardware encryption (Micro-Guard, Omni Inc.), RS-232 line encryption (Datacryptor, Racal-Milgo), or software encryption and audit trails (Protec, Sophco; Watchdog, Fischer-Innis), and dial-back techniques (GTX-100, Lockheed-Getex; and SAM, Leemah).

Solutions to these security problems involve many facets. Societal pressures and security awareness training are a beginning. Misunderstanding of computer operations is prevalent. Because data are intangible, and often have no assigned monetary value, they are often treated accordingly. Articulate pressure on vendors is essential. In most instances the solutions require extensive consideration by top management. Traditional authority, control, and information flow centered around human contact have been lost. The problems stretch across traditional division lines and require entirely new conventions and regulations. Users and administrators must realize that adequate security should not and cannot become an impediment to getting the job done, but without it neither the integrity nor validity of the data can be ensured.

Laboratory Technologies Corporation
255 Ballardvale St.
Wilmington, Mass. 01887

Contributor: Frederick A. Putnam

Protection

The problem. Vendors want their software protected. Users want access to that software to make changes specific to their needs or to meet validation requirements.

There are apparent conflicts between vendors' and users' needs for protection. Research users involved in methods development express a desire for software that can be copied at will and for total access to software source codes. On the other hand, commercial software developers that use software tools provided by primary vendors need to have their software protected against uncontrolled modification to ensure reliability, because software can only be validated or certified by subjecting a stable version to an exhaustive series of test cases. Even these users have diverse needs that can be met only by providing some degree of programmability to their software. But how can they take advantage of programmability if they don't have access to source code from the primary vendor or if they have protected their software against modification by the end user?

Software vendors need different protection for their software—protection from unauthorized duplication, because their only income is derived from sales of software. One way this can be achieved is by restricting access to source codes. The best examples of this come from the personal computer field, where source code is virtually never made available.

For object (machine readable only) software, the most common way to prevent unauthorized duplication is to use disk media that have embedded "key" codes that can't be copied. The software looks for these key codes and terminates unless it finds them. The key code disks are now commercially available from companies that specialize in software protection. Less sophisticated copy protection schemes write sectors out of sequence, place magnetically written keys in intersector or intertrack gaps, or "garble" a sector so that the error checking algorithms reject it. Advanced schemes may use a laser to place small holes in the magnetic medium (e.g., Prolok, Filelok; Vault Corp.). Many users object to these approaches because they cannot "back up" their original disks.

Some software will run only on a specific computer, preventing users from working at home on their own compatible personal computers. To solve this problem a hardware key is available that attaches transparently to the RS-232 port. The system interrogates for the presence of this device periodically, and execution is terminated if it is absent. Now any number of disk copies can be made, but only one can be running at a time. These devices even function in network environments, where computers may not have disks.

Regardless of vendors' attitudes or actions, users have real needs for software customization. If the software vendors don't fulfill these needs, users will buy elsewhere, write their own software, or continue using manual methods. Surprisingly, there is a simple solution to the entire problem.

The solution: a layered open-system design. The protection and access required to meet all the above needs can be achieved through a layered approach. Each layer is independently specified in a rigorous fashion, so that its implementation is independent from the rest of the layers. Each layer can have a different level of protection. Some layers remain open to change.

One example of a layered open system design is the structure of the PC-DOS operating system for the IBM personal computer. Most of PC-DOS is proprietary object code. This code has been validated through extremely rigorous testing and cannot be modified or examined by users. It is very well protected. Part of the operating system, however, is not protected at all. The basic input–output system (BIOS) and device drivers are supplied in source code. IBM also supplies thoroughly documented methods by which users may replace the standard PC-DOS device drivers with their own.

By dividing the software control program for the computer into layers and by keeping certain layers open, IBM was able to keep its software protected and provide users with the access to source code that they needed to adapt the computer to a wide variety of needs.

Another example of the layered open-system approach is the Labtech Notebook (LTN) software package designed by a team at Laboratory Technologies Corporation. This software package is primarily for real-time analog data acquisition and control. It also does some data analysis (FFTs and nonlinear curve fitting). The software was designed to provide an easily learned package for use on a personal computer. The level and type of protection in each layer can be adjusted independently for the application at hand.

Each layer in LTN (Figure 2) communicates with the other layers through disk data files. These data files have a number of formats, one of which is human-readable ASCII format. ASCII files have the great advantage that they can be created and modified with conventional text processors and can be read and written easily by all programming languages and most applications packages. The formats of the other communicating data files are also rigidly specified. Each of the layers can receive its data files from either standard LTN modules or customized modules that the users might create. The various layers and their protection levels are described in the following paragraphs.

LTN programming language is an interactive, incrementally compiled language (similar to FORTH in structure and to Pascal in syntax). At this level users can write programs at will, and there is complete access to their completely unprotected source code files. The nucleus of the language is

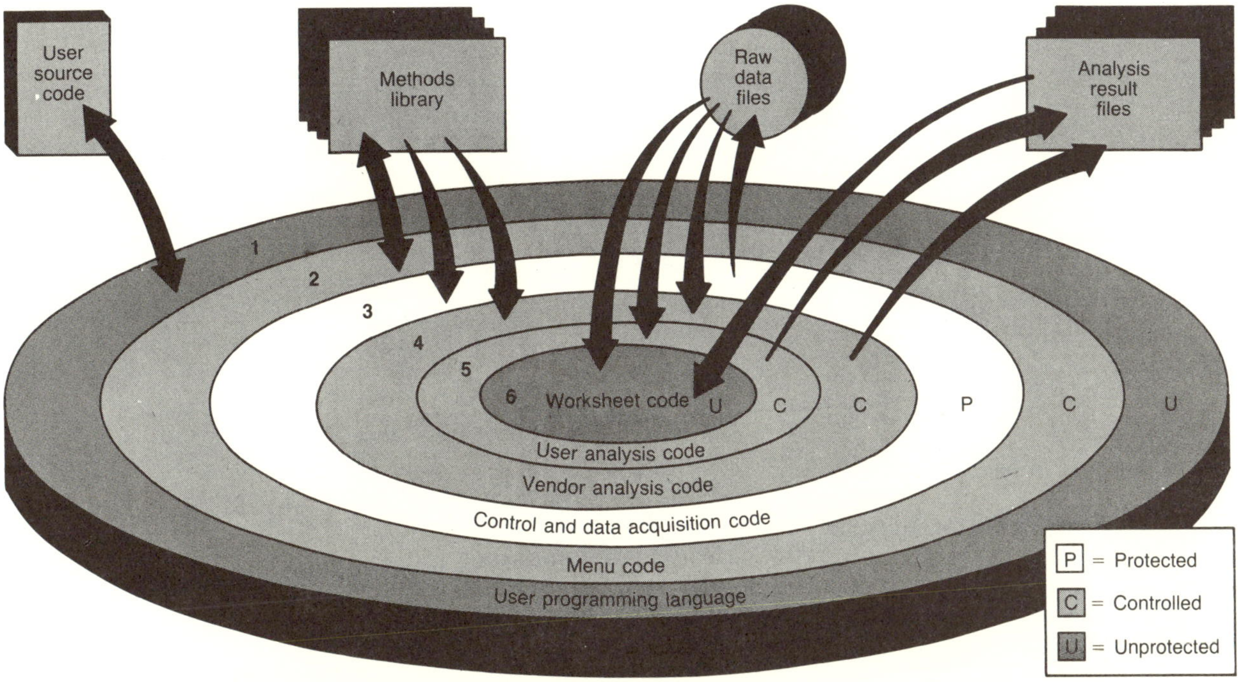

Figure 2. A layered approach to software access and protection problems

Each layer has a protection level. Protected layers are distributed by the software vendor in executable format only and are protected from illegal copying by various methods. Unprotected layers are distributed in source code format and can be freely altered by users to fit their specific requirements. Controlled layers can be set to execute only, read, or read–write. The appropriate level and access mechanism are chosen by the vendor or user, depending on software origin. This approach gives programmers the protection they require and users the access they need

protected by a key lock system as described above.

LTN menu code provides menus that are protected from modification by naive end users but that can be modified extensively by expert users, vertical application developers, and instrument manufacturers. The menu code creates and manipulates the method library, which interfaces to the next two layers.

The LTN data acquisition and control code responds to setup information in the method library and gathers data in real time from laboratory instruments and sensors. It also allows the computer to be used for any other program while the data are being gathered. Because this code is extremely critical and advanced (the integrity of the user's data depends on it), and also because it is proprietary, it is completely protected from user modification and supplied entirely in machine-readable form.

LTN data analysis codes operate on data files created from real-time experiments, on data that have been brought in from other systems or even data typed in at the keyboard. Each of the data analysis codes is a separate layer and has a different level of protection.

User-supplied analysis codes may consist of off-the-shelf software packages like Bolt, Beranek and Newman's RS/1 or Lotus Development's 1-2-3 (protected). Alternatively, they may be analysis programs developed by users in their favorite languages (protected or not, depending on the user's needs).

LTN 1–2–3 worksheets are supplied in unprotected source code as part of LTN. These perform spreadsheet data analysis and graphic display and are provided in source form so that end users can modify them at will to evolve their own custom data workup spreadsheets and graphs. Once these worksheets have been developed, 1–2–3 allows them to be protected on a cell-by-cell basis.

In summary, protection of software is an interesting area with many different and seemingly conflicting requirements. Software technology has evolved, however, to accommodate all these varied requirements without compromising the effectiveness of the systems.

Index

Index

D

E

F

Production by Meg Marshall
Indexing by Karen McCeney
Jacket design by Pamela Lewis

Elements typeset by Hot Type Ltd., Washington, D.C.
Printed and bound by Maple Press Co., York, Pa.